Pastoral systems in marginal environments

Pastoral systems in marginal environments

Proceedings of a satellite workshop of the XXth International Grassland Congress, July 2005, Glasgow, Scotland

edited by:
J.A. Milne

Subject headings:
Temperate grasslands
Semi-arid environments
Grazing management

ISBN 9076998744

First published, 2005

Wageningen Academic Publishers
The Netherlands, 2005

Organising Committee

Professor John Milne, Macaulay Institute, Aberdeen,
Professor Cled Thomas, DS International, Edinburgh,
Dr Tony Waterhouse, SAC, Stirling, and
Ms Leslie Gechie, SAC, Ayr

Acknowledgements

The Organising Committee wish to acknowledge the generous sponsorship of the Workshop and the book from The Scottish Executive Environment and Rural Affairs Department, The Stapleton Trust, Scottish Natural Heritage, The Macaulay Development Trust, SAC and the South West Scotland Grassland Society.

Stapledon Memorial Trust

Supporting the
land-based industries
for over a century

Foreword

Pastoral systems are some of the most fragile human ecosystems that exist and are under threat from the expansion of cultivation, changes in social patterns and climate change. These ecosystems are of major importance since they contain a rich biological and cultural diversity.

The aim of this book is to take a holistic view of pastoral systems by bringing together papers written by specialists in plant and animal ecology, who have an interest in the application of their research, with papers taking an economic and social perspective. The papers in this book were presented at the Satellite Workshop of the XXth International Grassland Congress held in Glasgow from 3-6 July 2005. The workshop was entitled "Pastoral systems in marginal environments" and contained nine plenary invited papers, which are given in full in this book, and 29 offered papers and 74 posters which are presented in Abstract form.

The focus is on marginal environments where the issues are in greatest relief with the papers tackling key issues in semi-arid and disadvantaged temperate areas. The key issues relate to identifying the biological constraints of these pastoral systems, understanding soil/plant/animal relationships, exploring biodiversity, landscape and social issues in multi-functional systems and providing solutions to constraints through a number of case studies. By comparing and contrasting these two environments, the book will be taking a completely new approach to understanding how pastoral systems function and how they will evolve in the future.

The programme of the Workshop was organised by Professor Cled Thomas, DS International, Edinburgh, Dr Tony Waterhouse, Scottish Agricultural College (SAC), Stirling and myself. We believe that the book will be of value to all those with an interest in pastoral systems by providing an up-to-date account of current understanding of these multi-functional systems and new insights into how they function and how they will develop in the future.

Professor John Milne
Macaulay Institute,
Aberdeen.

Table of contents

Keynote presentations

Constraints to pastoral systems in marginal environments

A.J. Ash and J.G. McIvor
CSIRO Sustainable Ecosystems, 306 Carmody Rd, St. Lucia, Qld 4067, Australia
Email: Andrew.Ash@csiro.au

Abstract

Variability in climate, landscape productivity and markets and the large spatial scale of most pastoral operations in marginal environments provide challenges and constraints for management that are quite distinct from those in intensively managed grasslands. Dealing with these constraints requires an ecological rather than an agronomic approach. Another feature of pastoralism in marginal environments is the tight coupling between biophysical and socio-economic drivers. As a consequence, constraints to livelihoods and sustainability in marginal environments are more driven by the complexity of interactions between management decisions, climate, environmental response and external drivers than by the direct biological constraints. In this paper we examine some of the main environmental and socio-economic constraints to pastoralism in marginal environments and put forward the view that these constraints are most appropriately managed by considering these pastoral environments as linked socio-ecological systems.

Keywords: socio-economic systems, pastoralism, policy, environment

Introduction

Marginal environments in semi-arid and arid regions of the world are commonly characterised as rangelands. These marginal environments include the semi-arid tropical and temperate savannas of Africa, Australia, South and North America as well as the low rainfall temperate and tropical deserts, steppes and prairies in Africa, Central Asia and North America. A distinguishing feature of these marginal or rangeland environments is that rainfall is usually too low and/or too variable for regular cropping and as a result they are largely used for livestock production. The vegetation in these semi-arid and arid environments is grasses, shrubs and trees that occur in mixtures that range from open grasslands with little tree or shrub cover, to shrub communities with little herbaceous material, and to savanna woodlands where trees or shrubs form a variable layer over a grassy understorey.

In addition to large climatic variability, which includes extremes of rainfall and temperature, rangelands are on the whole nutrient poor with low and patchy productivity. This combination makes them generally unsuitable for improvement and rangelands have remained as relatively intact ecosystems. This contrasts with more mesic livestock environments, which are intensively developed and have high inputs of seed and fertiliser. The contrast between marginal and more endowed environments is therefore quite distinct with constraints in mesic, benign environments being overcome through *agronomic* approaches and widespread use of technology. However, in marginal environments, constraints are acknowledged and a much stronger *ecological* approach to management has been adopted. This does not mean that serious efforts to overcome the environmental constraints of semi-arid environments have not been attempted. In parts of Africa, Australia and North and South America considerable resources have been expended in finding and breeding more productive pasture species, applying fertiliser or using mineral supplements, removing trees to increase livestock production, or improving fencing and water infrastructure to reduce spatial variability. However, these efforts at lifting pasture and livestock productivity have met with mixed success with the "improvements" sometimes leading to unintended consequences such as overgrazing and land degradation (Landsberg *et al.*, 1998).

A feature of most rangeland environments is a long history of herbivore use by humans either with domestic livestock or through hunting of native herbivores. As a result, rangelands and their management involve both natural and social forces. Because of the tight coupling of these linked socio-ecological systems it is inappropriate to discuss biological constraints to production in isolation from socio-economic factors so we also address some of the social and economic constraints to pastoral systems in marginal environments.

Governments and land management policies strongly influence pastoral systems (Walker & Janssen, 2002). This is particularly so in the area of tenure because pastoral tenure throughout the world is on the whole not as secure as that in more developed environments where freehold ownership is strong. We will also address some of these government and policy issues and how they interact with biological constraints.

A theme of this paper is that marginal environments used for pastoralism are complex adaptive systems (Walker & Janssen, 2002; Stafford Smith, 2003), and that this complexity is evident both in the biophysical and socio-economic system components and becomes amplified once the interactions between system components are considered.

Environmental constraints

There are a large number of environmental constraints that limit production in marginal environments. These include poor soil fertility, large climatic variability that drives limited and variable forage production, highly variable forage quality, spatial variability in resources including water distribution, vegetation that is susceptible to disturbance, and susceptibility to woody weeds. Many of the constraints such as poor soil fertility and limited production of low quality forage have been adequately covered in previous reviews and they will not be addressed in this paper. We believe that the key environmental constraints to production in marginal environments are associated with understanding and managing variability and in this section we focus on these issues.

Temporal variability

The amount and variability of rainfall is the major constraint to net primary productivity and livestock production in semi-arid environments. Figure 1 highlights how the variability of rainfall is significantly greater in semi-arid and arid environments than in more mesic environments. There is a fairly good relationship between increasing rainfall and decreasing rainfall variability using the coefficient of variation (CV) as the measure of inter-annual rainfall variability. Ellis (1994) suggested that once CVs of inter-annual rainfall reach 0.3 to 0.33 (as commonly occurs in rangelands), rainfall is so variable and droughts frequent enough that such systems are better characterised by their rainfall variability than by total rainfall. It is also clear from Figure 1 that rainfall variability in Australia is greater than for other continents. In Africa and Australia, climate variability is strongly influenced by the El Niño Southern Oscillation (ENSO) with severe droughts occurring once every 3 to 5 years.

Low rainfall combined with extreme temperatures limits productivity in cold rangelands such as occur extensively through central Asia. In a similar way to how droughts exert particular constraints on productivity in warmer semi-arid environments every few years, extreme snow and ice events, known in central Asia as *dzhut*, occur about every ten years and lead to significant livestock losses (Kerven, 2004).

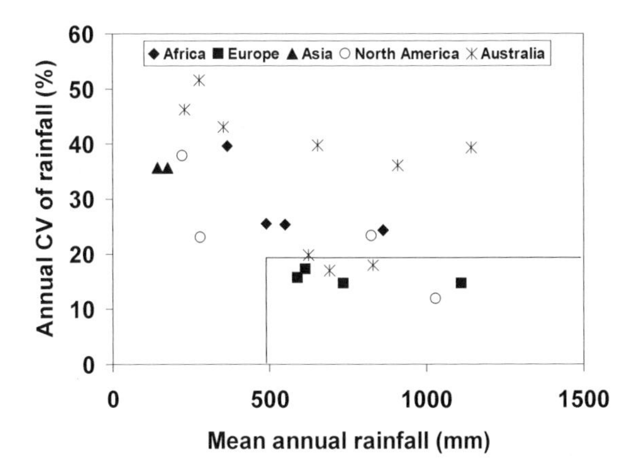

Figure 1 Relationship between annual rainfall and inter-annual rainfall variability for locations in semi-arid and arid environments. The locations in the small rectangle in the bottom right are environments with intensively managed and improved grasslands.

While climate data provides one measure of variability in marginal environments, ultimately it is the variability in forage supply that is more closely tied to livestock production. There are few long-term datasets of forage production for semi-arid and arid environments. Table 1 shows that for the limited amount of data that is available, forage production is considerably more variable than rainfall in semi-arid and arid grasslands.

Table 1 Comparison of variability in rainfall and above-ground net primary production (ANPP) of four semi-arid or arid grassland environments. Colorado, Toowoomba and Bakhyz sites are measured ANPP; the Queensland values are modelled ANPP, using AussieGrass (Carter *et al.*, 2000).

	Colorado, USA	Toowoomba, South Africa	Bakhyz, Turkmenistan	Queensland, Australia
Vegetation	Shortgrass prairie	Savanna grassland	Desert steppe	Savanna woodland
Years	1939-1990	1950-1980	1948-1982	1890-2002
Mean annual rainfall (mm)	321	628	310	633
Coefficient of variation (CV) of interannual rainfall (%)	31.0	21.0	30.0	37.0
Mean annual above-ground net primary production (kg/ha)	660	1890	611	2042
CV(%) of annual above-ground net primary production	44.0	41.4	59.6	45.9

Large temporal variability in forage production poses significant challenges to livestock managers because of the difficulty in balancing animal numbers with forage supply. Regular adjustment of animal numbers to track forage supply may be advantageous (Illius *et al.*, 1998) but it is difficult in many commercial livestock operations to vary animal numbers by a large amount and optimisation studies indicate, for individual enterprises, it is only economic to increase or decrease herd numbers by 20 to 30% on an annual basis (Stafford Smith *et al.*, 2000). Varying animal numbers can be achieved by sale or agistment (i.e. pasture is rented by another pastoralist usually in another region not affected by dry conditions). While agistment provides pastoralists with the ability to retain ownership of stock during a drought, there is an ecological advantage of sale over agistment in that animal numbers tend to build up more slowly after drought, which can aid in post drought pasture recovery. There is evidence from a study in Namibia that a slow build up in animal numbers on communal grazing lands following drought is intentional to provide rest to the rangeland in the post-drought recovery period (Burke, 2004). In northern Australia, where large companies own a number of pastoral properties there is greater opportunity to vary stock numbers from year to year through modern transhumance approaches made possible by advanced road transport.

Another option in commercial operations is to stock conservatively on a consistent basis so that in good years the pasture resource is under-utilised. This strategy ensures that going into drought years pasture is in good condition and is capable of tolerating one or two years of relatively high utilisation. Whether this grazing strategy is more economic than a variable stocking strategy that tracks forage supply is dependent on market factors such as relative sale and purchase prices of livestock and transport costs (Stafford Smith *et al.*, 2000).

A third option for commercial pastoralists in eastern Australia and parts of Africa is to be proactive and use seasonal climate forecasts to manage stock numbers before droughts develop (e.g. Boone *et al.*, 2004). Modelling studies where seasonal climate forecasts are used to adjust stock numbers demonstrate economic and environmental benefits but these are usually modest because of limited ability to react to a forecast (Stafford Smith *et al.*, 2000).

In traditional transhumance or nomadic systems, pastoralists managed the constraint of temporal variability in climate and forage supply by exploiting spatial heterogeneity and moving their animals to neighbouring regions experiencing a different climate cycle (Ellis & Galvin, 1994). Coping with temporal variability by exploiting spatial variability is also practised in commercial situations. For example, it is not uncommon for cattle in temperate areas of North America to graze elevated mountain pastures or meadows in summer and lower lying arid lands in winter, or in northern Australia for cattle to be shifted from one region to another through agistment arrangements. For those unable to shift livestock because of sedentarisation or with access only to communal grazing lands, livestock mortalities are high during droughts. This is followed by a period of post-drought recovery in animal numbers creating a cycle of fluctuating livestock populations (Ellis & Swift, 1988). Alternatively, animal numbers are maintained on pasture during drought and provided with supplementary feeds (e.g. urea-molasses or mineral/protein blocks, conserved forages). However, overcoming the constraints of a variable forage supply in this way can create other problems in semi-arid environments in the form of overgrazing (Landsberg *et al.*, 1998).

Spatial variability in use of the pasture resource

A key distinguishing feature of most marginal grazing environments is the greater spatial scale of management units (paddocks/pasture) compared with more intensively managed pastures. This large spatial scale of management units poses a number of constraints for pastoral management. In many extensively grazed situations, water for domestic livestock is not evenly or adequately distributed and this tends to focus grazing activity around watering points, whether they be artificial or natural sources of water (Thrash, 2000; Pringle & Landsberg, 2004). Uneven grazing results producing piospheres, with areas within walking distance from water points being prone to overgrazing and degradation.

As a result of inadequate water distribution there can be areas of paddocks that are ungrazed or underutilised and this can have negative consequences for both individual animal performance and carrying capacity (Hart *et al.*, 1993). However, areas that are too far from water for domestic livestock to reach do provide refuges for grazing sensitive plant and animal species and this can be positive for biodiversity conservation (James *et al.*, 1999).

Large management units also mean there is often a diversity of vegetation communities within a paddock. Grazing animals exhibit strong preferences for different plant communities and can exert considerable grazing pressure on the more fertile parts of the landscape. This preference for particular vegetation communities can lead to their overuse and eventual degradation. As these preferred parts of the landscape lose their productivity in response to overuse (Ash & McIvor, 1998), animals shift their attention to the next preferred parts of the landscape and a cycle of sequential degradation is established (Ash *et al.*, 2004).

Diet selection also occurs at finer spatial scales in the form of patch grazing and species selection. Patch grazing becomes a problem where areas are repeatedly grazed without time for recovery between grazing events and can cause localised degradation. Overgrazing can lead to loss of plant productivity, particularly if the grazing occurs when plants are susceptible to defoliation (Ash & McIvor 1998). It can be difficult to manage this patch grazing because of its relatively fine scale compared with the scale of paddocks but fire used in rotation in paddocks can overcome some of the problems of patch grazing in tropical grasslands (Andrew 1987).

Clearly, foraging behaviour in complex plant communities in rangelands provides significant challenges to the sustainable management of the vegetation and soil resources, but how does it affect animal production? We have already mentioned that overgrazing and loss of perennial pasture species can reduce net primary productivity and hence carrying capacity. However, spatial variability and how it interacts with foraging behaviour can also provide opportunities for animals in addition to constraints. For example, patch grazing results in short leafy swards of high quality, and in situations where plant systems have co-evolved with large herbivores, productivity from grazed patches can even be enhanced (McNaughton, 1984). In addition, a mixture of plant communities within a paddock can provide diet diversity and 'key resources' that can help buffer animal production during drought (Scoones, 1995). There is now increasing evidence for this spatial buffering with Stokes *et al.* (2004) demonstrating that diet quality, as measured by Near-Infrared Spectroscopy (NIRS), declines to a greater extent in the dry season in smaller homogenous paddocks than it does in large heterogenous paddocks.

Rangelands as complex ecological systems

A striking feature of rangeland environments is that they often do not obey equilibrium concepts of vegetation dynamics that for so long governed rangeland management (Dyksterhuis, 1949). Equilibrial vegetation dynamics dictates that changes in vegetation are reversible such that if mistakes are made with grazing management and undesirable vegetation change occurs, a relaxation of the disturbance can allow the vegetation to return to a "desirable" condition. However, in marginal environments subject to large climatic variability, disturbances such as grazing or fire can lead to irreversible vegetation change, at least in the temporal context of a management generation (20-50 years). Westoby *et al.* (1989) used state and transition models to explain this non-equilibrial behaviour in vegetation dynamics and its importance to management. Transitions between vegetation states can occur quite rapidly, especially where thresholds are crossed. Alternatively, the changes can be gradual over periods of years to decades (Watson *et al.*, 1996). Continuous, gradual change gives pastoralists time to adapt management while event driven change poses more constraints because changes can be set in train long before management is aware of a problem. For example, in chenopod shrublands, change in vegetation in response to management is gradual (Watson *et al.*, 1996); in contrast, establishment of a woody weed can occur as a result of a single event such as a very wet year.

In either scenario, monitoring is a key to managing this complexity in vegetation dynamics. For environments where event-driven change is important it is especially critical to be monitoring at the right time and in the right place. This provides a significant challenge for pastoral management because of the extensive nature of most pastoral enterprises. There are usually few resources made available to monitor vegetation dynamics and even where a monitoring program is established it is usually in a few locations of small spatial scale. The value of a few sites in providing feedback for management on which major grazing management decisions are made is questionable.

The equilibrium/non-equilibrium dynamics in rangelands also affects the coupling of animals to forage resources in a temporally and spatially variable environment. In intensive grazing systems there is quite tight coupling between animal numbers and animal performance This is best described by the linear model of Jones & Sandland (1974) in which animal production per head declines linearly with increasing stocking rate. This model holds true in rangeland environments at the small spatial scale of most grazing experiments but the rate of decline in animal performance per unit increase in stocking rate is much lower than for improved pastures (Ash & Stafford Smith, 1996). The little available evidence at larger spatial scales suggests that this stocking rate – animal performance relationship becomes less coupled at low to moderate stocking rates (Ash *et al.* 2004) and the coupling or density dependence only becomes important at high stocking rates. At lower stocking rates diversity buffers diet quality and allows animal production to be maintained over a fairly wide stocking rate range; at higher stocking rates, low forage availability limits intake and animal performance (Illius & O'Connor, 1999).

Taking this further, in arid pastoral regions of east Africa it has been argued that climate variability is such that there is no linkage between animal populations and vegetation dynamics because animal numbers build up in wet years and crash during droughts before forage availability and vegetation condition is affected (Ellis & Swift, 1988). Illius & O'Connor (1999) argue that while there might not be a tight coupling between animal numbers and forage resources for much of the time, density dependence is exhibited through the amount and quality of "key resources', which are critical during drought i.e. there is a limit to the spatial buffering that these key resources can provide.

While these concepts have been the source of much discussion amongst social scientists and ecologists in recent years it is the implications for pastoral management that are of interest in this paper. In intensively managed, improved pasture systems an optimum stocking rate is relatively easily determined because of the sensitivity of animal production to increasing utilisation and declining diet quality. However, with much weaker coupling in rangeland systems it is more difficult to determine the optimum number of animals and how to sensibly vary this number in the face of large climatic variability. Also, because the feedback effects of overstocking can take a few years to manifest themselves in terms of changed plant composition and productivity, grazing management in rangelands is complex and challenging. Some pastoral managers have responded to these challenges by grazing conservatively (e.g. Landsberg *et al.*, 1998) to avoid or minimise mistakes, especially during droughts i.e. stock numbers are set to cope with the bad years.

Social, economic and policy constraints

So far we have covered the complexities and constraints that are environmental. However, pastoralism in marginal environments is as much shaped and constrained by social, economic and policy issues as it is by biophysical factors. Ultimately it is the interaction of all these factors that determines the success or otherwise of pastoral livelihoods. In this section we will highlight a few of the important social, economic and policy factors that interact with the biophysical environment to influence constraints on pastoral systems.

Land tenure arrangements

Throughout arid and semi-arid rangelands, land use intensification and fragmentation is occurring either through population growth, through policy initiatives to provide more livelihood opportunities for pastoralists, through market forces to increase economic growth and net regional welfare, or through conservation of landscapes that are of high environmental value. For example, in Africa, where pastoralism is of great importance, fragmentation is expected to increase in the coming decades and this will put pastoral livelihoods under even more pressure (Reid *et al.*, 2003). We will use two case studies to highlight how policy-driven land tenure arrangements can constrain pastoralism.

Case study 1: Flock mobility in Central Asia (Kerven *et al.*, 2003)
Prior to the 1700s, Kazaks were mobile pastoralists who migrated hundreds to thousands of kilometres to take advantage of different growth seasons of pastures. As interaction with the Russians increased during the period from 1700-1900, there was a gradual decline in long distance nomadism, which was compounded by the excision of the most productive pastures along water courses for cultivation. During the period of Stalinism from 1918-1938, most pastoralists were sedentarised and animal numbers declined dramatically. However, in the 1940s migratory pastoralism was again encouraged by the government and was supported by the application of science and technology to improve livestock productivity. In the 1960s more policy pressure was exerted to increase livestock production. This involved irrigation of pastures in semi-desert rangelands and required significant investment in infrastructure (irrigation, planting and harvesting equipment) and services. This intensification again disrupted traditional migratory pastoralism. With the collapse of the Soviet Union in the 1990s, the infrastructure to maintain a sheep population of 35 million rapidly disintegrated and sheep numbers in Kazakstan crashed to 9 million (see Kerven 2004), with a similar scale decline in pastoralist numbers. This example, albeit quite extreme, serves to highlight how changes in policy and land tenure can significantly constrain and disrupt pastoral livelihoods.

Case study 2: Closer settlement in south-west Queensland, Australia
Most pastoral development in Australia's rangelands occurred between the mid-1800s and the early 1900s. During this initial pastoral expansion individual holdings were very large and relatively undeveloped in terms of water and fencing infrastructure. During the early 1900s, governments viewed pastoral lands, which were mostly leasehold, as underutilised and began

introducing closer settlement schemes. These schemes involved the resumption and sub-division of large properties for subsequent allocation by ballot to aspiring new pastoralists. The mulga lands of south-west Queensland was one such region subject to closer settlement, which gained impetus after both World Wars. The fragmentation of pastoral holdings continued until the 1960s. However, the nominal stocking rates for many of these new properties were insufficient to support a family unit. Because of the environmental constraints of a variable rainfall and low soil fertility there were few opportunities to increase productivity through technological innovation, so owners raised stock numbers in an attempt to increase their incomes. The numbers exceeded carrying capacity leading to overgrazing and degradation, particularly during periods of drought,with the result that both land condition and livelihoods suffered. A recent response has been for the government to introduce new policies to encourage property amalgamation and diversification to achieve "living areas".

Historically, covenants on much leasehold land were about taking measures to increase livestock production (levels of fencing, tree clearing, minimum stocking rates etc) but legislation governing leasehold land is now much more focussed on resource management and property planning. This imposes new challenges, constraints and opportunities for pastoral management (Queensland Department of Natural Resources and Mines, 2003). This example highlights how land tenure policy is often closely linked to biophysical outcomes and how we can learn from past mistakes to bring about more appropriate land tenure arrangements.

Policies

As indicated earlier, pastoral enterprises are complex adaptive systems. As such, these systems change in response to external forces but are also undergoing co-evolution and change as a result of their own, dynamic interactions. As a result, policies that aim to achieve a particular outcome can have unintended consequences when applied in rangelands because of the interactions and feedbacks between social, cultural and environmental issues. A good example of this in Australian rangelands is drought policy. In 1992 a National Drought Policy was announced which had three main planks: (a) adoption of self-reliant approaches to managing drought and climate variability; (b) maintenance of the agricultural and environmental resource base; and (c) facilitation of early recovery of rural industries following major climatic stress. To support this policy a range of interest rate subsidies and taxation instruments were introduced. However, an analysis of the tax instruments (Stafford Smith, 2003) illustrated that in trying to improve the financial situation for pastoralists the instruments increased the risk of degradation because they encouraged stock to be retained in a dry period. This brief example serves to highlight how policy can interact with the management of the biophysical resource to cause constraints in the sustainable management of rangelands.

Markets

Over decades and even centuries all forms of agriculture have suffered from a continual decline in commodity prices and rising costs. Viability has been maintained through improvements in productivity, through acquiring additional land to achieve economies of scale or has occurred in many developed countries through the introduction of farm subsidies. In marginal environments, this cost-price squeeze is exacerbated because the production of animals often occurs a considerable distance from key markets, which increases costs associated with transport. Also, the quality of meat is often lower than from more intensively managed livestock systems which means prices received are generally lower. There are exceptions to this where niche markets can be established for rangeland meat based on it being a "clean, green" product (e.g. see http://www.obebeef.com.au/).

A particular constraint for rangeland enterprises is the cyclical nature of market prices and climate and how they interact. This is illustrated in Figure 2 which highlights how there have been very few periods in the past 30 years in northern Australia where good prices for rangeland beef have coincided with good rainfall and pasture growth seasons. The challenge for pastoral managers when faced with the variable nature of climate and markets, which are the strongest external drivers of enterprise profitability, is to take advantage when good prices and good seasons coincide and to minimise losses at other times.

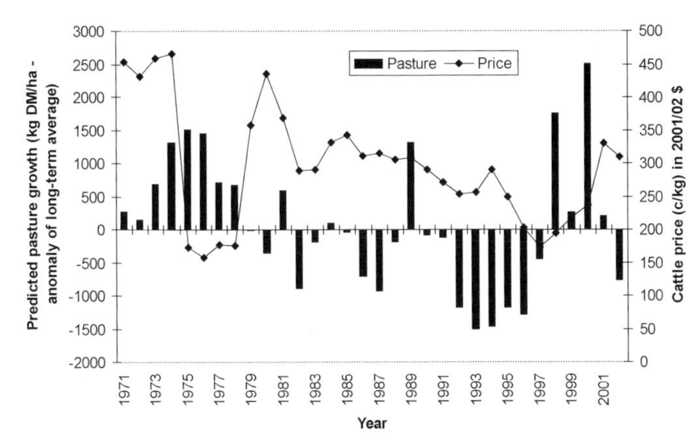

Figure 2 Historical record of cattle prices and pasture growth for north-east Queensland (from Ash & Stafford Smith, 2003)

A framework for integrating constraints in linked socio-ecological systems

In this paper we have tried to demonstrate that biophysical and socio-economic constraints are both important in the sustainable management of pastoral enterprises in marginal environments and that these constraints strongly interact. In these linked socio-ecological systems the large number of constraints can usually be represented by a few key "slow" variables. Examples of slow biophysical variables are soil condition or perennial grass cover while debt-equity ratio is an example of a slow socio-economic variable. In contrast, examples of fast variables are annual forage production or cash flow. Fernandez *et al.* (2002) presented a simple framework for describing how biophysical and socio-economic slow variables and thresholds can be linked and used to explain desertification in semi-arid and arid landscapes. Figure 3 represents this framework using an enterprise in a semi-arid savanna as an example. The "slow" variables are perennial grass cover and farm equity. We envisage three hypothetical pastoral managers. Pastoral Manager A has a high level of equity in his property and grazes conservatively. When faced with constraints such as poor seasons or poor prices the manager may move towards either the biophysical or socio-economic threshold but the seasons or prices recover before a threshold is crossed. Pastoral Manager B also has high equity but is a greater risk taker and maintains stocking rates closer to the threshold. However, this manager's ecological understanding and grazing management skills are not well advanced and when faced with a run of dry years, pasture condition declines and the biophysical threshold is crossed. In a more degraded state pasture productivity is reduced and large sums of money are borrowed to supplement the herd. Equity is reduced and over time the socio-economic threshold is also crossed and the enterprise ends up in a biophysically and socio-economically degraded state. Pastoral Manager C has recently bought a property and has a large level of debt with high interest rate payments and the operation is very close to the socio-economic threshold. Interest rates rise and prices fall so the socio-economic threshold is crossed. Management responds by pushing already high stocking rates even higher to increase returns. Pasture condition declines and the biophysical threshold is crossed into a degraded state.

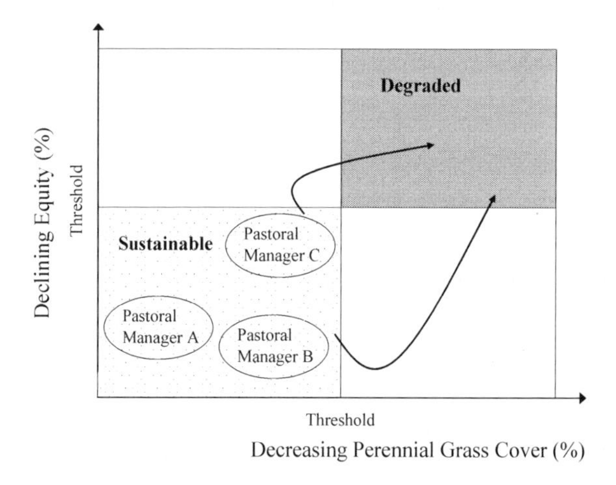

Figure 3 Conceptual representation of socio-ecological thresholds in a semi-arid commercial pastoral enterprise (adapted from Fernandez *et al.*, 2002)

This socio-ecological degradation framework highlights that management responses to both biophysical and socio-economic constraints in marginal environments must be considered in the context of system thresholds. Where a pastoral management system is operating a long way from either threshold, responses to constraints are not required immediately whereas pastoral systems operating close to the thresholds need immediate and well considered action as constraints emerge.

Conclusion

Livestock management in marginal environments is particularly challenging because of the complexity brought about by a low productivity and heterogenous landscape, a highly variable climate and a high degree of uncertainty associated with markets, tenure and policy and institutional arrangements. There is often a limited ability to exert much management control over many of these complexities and uncertainties. The challenge for management is to understand the complexities in the context of relatively simple response strategies e.g. opportunistic stocking versus more conservative stocking. This requires a systems approach that links socio-ecological drivers but our science to support this systems approach is still in its infancy and there still as many questions as there are answers.

References

Andrew, M.H. (1987). Selection of plant species by cattle grazing native monsoon tallgrass pasture at Katherine, N.T. *Tropical Grasslands*, 20, 120-127.

Ash, A.J. & J.G. McIvor (1998). How season of grazing and herbivore selectivity influence monsoon tallgrass communities of northern Australia. *Journal of Vegetation Science*, 9, 123-132.

Ash, A.J. & M. Stafford Smith (1996). Evaluating stocking rate impacts on rangelands: animals don't practice what we preach. *The Rangeland Journal*, 18, 216-243.

Ash, A.J. & D.M. Stafford Smith (2003). Pastoralism in tropical rangelands: seizing the opportunity to change. *The Rangeland Journal*, 25, 113-127.

Ash, A.J., D.M. Stafford Smith & J.E. Gross (2004). Scale, heterogeneity and secondary production in tropical rangelands. *African Journal of Range and Forage Science* (in press).

Boone, R.B., K.A. Galvin, M.B. Coughenour, J.W. Hudson, P.J. Weisberg, C.H. Vogel & J.E. Ellis (2004). Ecosystem modelling adds value to a South African climate forecast. *Climatic Change*, 64, 317-340.

Burke, A. (2004). Range management systems in arid Namibia – what can livestock numbers tell us? *Journal of Arid Environments*, 59, 387-408.

Carter, J.O., W.B. Hall, K.E. Brook, G.M. McKeon, K.A. Day, & C.J. Paull (2000). Aussie GRASS: Australian Grassland and Rangeland Assessment by Spatial Simulation. In: G. Hammer, N. Nicholls & C. Mitchell (eds.) Applications of Seasonal Climate Forecasting in Agricultural and Natural Ecosystems – the Australian Experience. Kluwer Academic Press, Netherlands, 329-349.

Dyksterhuis, E.J. (1949). Condition and management of rangeland based on quantitative ecology. *Journal of Range Management*, 2, 104-115.

Ellis, J.E. (1994). Climate variability and complex ecosystem dynamics: implications for pastoral development. In: I. Scoones (ed.) Living with Uncertainty. Intermediate Technology Publications, London, UK, 37-46.

Ellis, J.E. & D.A. Swift (1988). Stability of African pastoral ecosystems: Alternate paradigms and implications for development. *Journal of Range Management*, 41, 450-459.

Ellis, J. & K. Galvin (1994). Climate patterns and land use practices in the dry zones of Africa. *BioScience*, 44, 340-349.

Fernandez, R.J., E.R.M. Archer, A.J. Ash, H. Dowlatabadi, P.H.T. Hiernaux, J.F. Reynolds, C.H. Vogel, B.H. Walker, & T. Wiegand (2002). Degradation and recovery in socio-ecological systems. In: M. Stafford Smith & J. Reynolds (eds.) An Integrated Assessment of the Ecological, Meteorological and Human Dimensions of Global Desertification, Dahlem Press, Berlin, 297-323.

Hart, R.H., S. Clapp & P.S. Test (1993). Grazing Strategies, Stocking Rates, and Frequency and Intensity of Grazing on Western Wheatgrass and Blue Grama. *Journal of Range Management*, 46, 122-126.

Illius, A.W., J.F. Derry & I.J. Gordon (1998). Evaluation of strategies for tracking climatic variation in semi-arid grazing systems. *Agricultural Systems*, 57, 381-398.

Illius A.W. & T.G. O'Connor (1999). On the relevance of nonequilibrium concepts to arid and semiarid grazing systems. *Ecological Applications*, 9, 798-813.

James, C. D., J. Landsberg & S.R. Morton (1999). Provision of watering points in the Australian arid zone: a review of effects on biota. *Journal of Arid Environments*, 41, 87-21.

Jones, R.J. & R.L. Sandland (1974). The relation between animal gain and stocking rate: Derivation of the relation from the results of grazing trials. *Journal of Agricultural Science, Cambridge*, 83, 335-342.

Kerven, C. (2004). The influence of cold temperatures and snowstorms on rangelands and livestock in northern Asia. In: S. Vetter (ed.) Rangelands at Equilibrium and Non-Equilibrium – Recent Developments in the Debate Around Rangeland Ecology and Management. Programme for land and Agrarian Studies, University of the Western Cape, Capetown, South Africa, 41-55.

Kerven, C., I. Ilych Alimaev, R. Behnke, G. Davidson, L. Franchois, L. Malmakov, E. Mathijs, A. Smailov, S. Temirbekov & I. Wright (2003). Retraction and expansion of flock mobility in Central Asia: costs and consequences. In: N. Allsopp, A.R. Palmer, S.J. Milton, K.P. Kirkman, G.I.H. Kerley, C.R. Hurt, & C.J. Brown (eds.) Proceedings of the VIIth International Rangelands Congress. Document Transformation Technologies, Durban, South Africa, 543-556.

Landsberg, R.G., A.J. Ash, R.K. Shepherd & G.M. Mckeon (1998). Learning from history to survive in the future: management evolution on Trafalgar Station, north-east Queensland. *The Rangeland Journal*, 20, 104-118.

McNaughton, S. J. (1984). Grazing lawns: animals in herds, plant form and coevolution. *American Naturalist*, 124, 863-886.

Pringle, H.J.R. & J. Landsberg (2004). Predicting the distribution of livestock grazing pressure in rangelands. *Austral Ecology*, 29, 31-39.

Queensland Department of Natural Resources and Mines (2003). Draft State Rural Leasehold Land Strategy. Queensland Government, Brisbane, 43pp.

Reid, R.S., P.K. Thornton & R.L. Kruska (2003). Loss and fragmentation of habitat for pastoral people and wildlife in East Africa: concepts and issues. In: N. Allsopp, A.R. Palmer, S.J. Milton, K.P. Kirkman, G.I.H. Kerley, C.R. Hurt, & C.J. Brown (eds.) Proceedings of the VIIth International Rangelands Congress. Document Transformation Technologies, Durban, South Africa, 557-568.

Scoones, I. (1995). Exploiting heterogeneity - habitat use in cattle in dryland Zimbabwe. *Journal of Arid Environments*, 29, 221-237.

Stafford Smith, M. (2003). Linking environments, decision-making and policy in handling climatic variability. In: L. Courtenay Botterill & M. Fisher (eds.) Beyond Drought – People, Policy and Perspectives. CSIRO Publishing, Australia, 131-152.

Stafford Smith, M., R. Buxton, G. Mckeon & A. Ash (2000). Seasonal climate forecasting and the management of rangelands: Do production benefits translate into enterprise profits? In: G.L. Hammer, N. Nicholls & C. Mitchell (eds.) Applications of Seasonal Climate Forecasting in Agricultural and Natural Ecosystems. Kluwer Academic Publishers, The Netherlands, 272-290.

Stokes, C.J., A.J. Ash & R.J. Mcallister (2004). Fragmentation of Australia's rangelands: risks and trade-offs for land management. In: G.N. Bastin, D. Walsh, & S. Nicolson (eds.) Living in the Outback: Proceedings of the 13th Biennial Australian Rangeland Society Conference. Australian Rangeland Society, Alice Springs, 39-48.

Thrash, I. (2000). Determinants of the extent of indigenous large herbivore impact on herbaceous vegetation at watering points in the north-eastern lowveld, South Africa. *Journal of Arid Environments*, 44, 61-72.

Walker, B.H. & M.A. Janssen (2002). Rangelands, pastoralists and governments: interlinked systems of people and management. *Philosophical Transactions of the Royal Society of London*, 357, 719-725.

Watson, I.W., D.G. Burnside& A. McR. Holm (1996). Event-driven or continuous; which is the better model for managers? *The Rangeland Journal*, 18, 351-369.

Westoby, M., B.H. Walker & I. Noy-Meir (1989). Opportunistic management for rangelands not at equilibrium. *Journal of Range Management*, 42, 266-274.

Soil/plant interactions

P. Millard and B.K. Singh
Macaulay Institute, Craigiebuckler, Aberdeen AB15 8QJ, Scotland,
Email: p.millard@macaulay.ac.uk

Abstract

The interactions between grassland vegetation and soil microbial communities are reviewed. Recent methodological developments for measuring soil microbial community structure are discussed and their application to the study of interactions between grassland vegetation and soils considered at three different scales. First, the evidence that different grassland communities condition soil microbial diversity is reviewed. Secondly, evidence for interactions between individual grass species and soil microbes is discussed at the level of the rhizosphere, by considering results from vegetation substitution experiments. Finally, interactions occurring in the rhizoplane are considered and research discussed showing that, while the impact of plant species on the mycorrhizal community is comparatively strong, co-selection between plant species and the bacterial community structure is weak. It is concluded that, while individual plant species can affect the activity of the soil microbial community, its structure is determined more by a range of environmental factors such as soil fertility, pH and possibly soil organic matter quality.

Keywords: grassland vegetation, soil microbial community structure, rhizosphere, rhizoplane

Introduction

In addition to providing a substrate for plant growth, soils provide many ecosystem services, such as purifying waters, sequestering carbon and cycling nutrients, the majority of which are mediated through the activity of the soil microbial community. The role of microbial communities is particularly important in soils that do not receive inputs of nutrients from fertilizers, such as extensively grazed, upland grasslands, where the availability of nutrients such as N and P to plants is dependent largely upon microbial activity. One of the main drivers of soil microbial activity is the availability of carbon from vegetation. Studies of soil respiration have clearly demonstrated the importance of recently assimilated carbon from plants in driving soil respiration (Bhupinderpal-Singh *et al.,* 2003). Interactions between grassland vegetation and soils as mediated via soil microbes can be considered as a simple model (Figure 1). Plants transfer carbon to soil through litter returns, root turnover and rhizodeposition (a collective term encompassing the secretion of exudates, mucilage and dead cells). In return the microbes provide mineral nutrients, through atmospheric N_2 fixation and mineralisation of soil organic matter (SOM), which can in turn be utilized by the plants. Because plants are seldom carbon-limited, whereas the activity of soil microbes often is, this model represents a simple trade-off between carbon inputs to the soil from plants, which in return gain nutrients released from SOM turnover.

While the model in Figure 1 is undoubtedly too simple, the importance of plant community structure in determining soil microbial diversity and functioning is largely unknown. This is due in part to the limitations imposed by the methods available to study soil microbial diversity and partly due to the other complex environmental factors which interact to influence soil microbial diversity and activity.

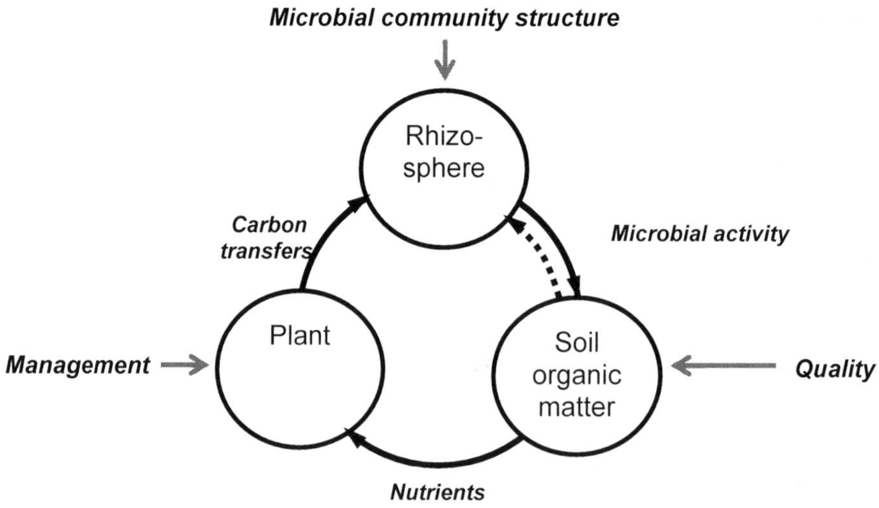

Microbial community structure

Rhizo-
sphere

Carbon
transfers

Microbial activity

Plant

Management

Soil
organic
matter

Quality

Nutrients

Figure 1 Conceptual model of plant–soil interactions, as mediated by microbes in the rhizosphere

This chapter will review some recent research on the interactions between plants and the soil microbial community in grazed upland grasslands. Recent methodological developments for measuring soil microbial community structure will be discussed and then their application to the study of interactions between grassland vegetation and soils considered. These interactions will be considered at three different scales. First, evidence that different plant communities condition the structure of the soil microbial community will be considered. Secondly, evidence for interactions between individual grass species and soil microbes will be discussed at the level of the rhizosphere (the volume of soil surrounding plant roots that is influenced by their activity). Finally, interactions occurring in the rhizoplane (the surface of the root itself) will be considered, as it is here that the plant-specific influence on soil microbial communities is likely to be greatest. The impact of grassland vegetation on soil microbial community structure and activity will be discussed by considering results from the MICRONET project, a ten-year study of the interactions between upland grassland vegetation and soil microbial communities.

Methods for measuring soil microbial community structure

Traditional methods for assessing soil microbial community structure have concentrated on measuring the size of the microbial biomass in soil and culturing "representative" bacteria and fungi, to assess how distinct different communities are. As a consequence, the whole research area of plant–soil microbial interactions has, until recently, been limited by the availability of suitable techniques. However, in the last few years the approaches for studying soil microbes have moved from biochemical and microbiological determinations, such as enzyme activities, microbial biomass, and respiration coefficients, towards the investigation of soil microbial diversity and community structure (Hill *et al*., 2000; Crecchio *et al*., 2004). These approaches have used either phenotypic measurements (based upon various profiling methods) or genotypic methods based upon molecular approaches.

Considering phenotypic approaches, BIOLOG (a metabolic assay) has been adapted to investigate the functional diversity of the soil microbial communities (Garland & Mills, 1991). Community level physiological profiles (CLPP) are obtained by determining the use of a broad spectrum of single C-sources by microbial communities extracted from soil and cultured. This is conveniently achieved in 96 well microtitre plates containing the substrates and a redox indicator to monitor the utilisation of carbon. Despite recent improvements (O'Connel & Garland, 2002; Campbell *et al.*, 1997), the BIOLOG method still suffers from bias problems similar to those of culture-plating methods, as according to most liberal estimates less than 5% of soil microorganisms are cultivable (Prosser, 2002). One recent development to solve this problem has been to adapt the method to measure the effect of adding single C-sources on respiration from a column of whole soil (Campbell *et al.*, 2003).

Another phenotypic profiling method is phospholipid fatty acid (PLFA) analysis, which provides a cultivation-independent broad-scale analysis of diversity and shift in community structure (Frostegård *et al.*, 1993). For this approach, several signature lipids have been identified which represent the presence of broad microbial groups (e.g. Vestal & White, 1989). The PLFA profile of a soil is derived from the whole viable microbial community and each species contributes to the profile in proportion to its biomass (Hedrick *et al.*, 2000). Although this method is very useful at its intrinsic level of resolution, it does not provide a detailed or fine resolution of soil microbial community structure (Bossio *et al.*, 1998). For example, there are a limited number of PLFA biomarkers that can be used for fungi and so only total fungal biomass can be obtained.

These limitations in the phenotypic methods have been overcome to some extent by using rRNA gene analysis for microbial diversity studies. PCR amplification of rRNA gene from soil DNA samples, combined with fingerprinting techniques such as denaturing gradient gel electrophoresis (DGGE), terminal restriction fragment length polymorphism (TRFLP), amplified rDNA restriction analysis (ARDRA), cloning and sequencing provide detailed information about the species composition of whole communities (Torsvik & Ovreas, 2002). These techniques, especially DGGE and TRFLP, are the molecular methods most extensively used for studying changes in microbial community structure and diversity (Anderson & Cairney, 2004). The DGGE technique separates PCR products of the same size but different sequences by chemical denaturation. Following staining of the gel, banding patterns may be used to compare different communities or the same community following a perturbation (Prosser, 2002). The TRFLP technique is an automated and sensitive method which can be used to compare microbial communities and monitor changes in community structure. A fluorescently-labelled primer is used for the PCR and after restriction digestion fragments of different length are generated. The sequencer recognises only the fluorescently labelled terminal fragment and, therefore, in principle each fragment represents a unique genome in the sample (Blackwood *et al.*, 2003). Although DGGE and TRFLP represent rapid and suitable techniques for resolving PCR-amplified products from complex microbial communities, the major limitation of the techniques is that in soil ecosystems, the number of distinct genomes is so great that the complexity of rDNA of different fragments can exceed the resolving power of the existing techniques (Torsvik & Ovreas, 2002). However, a clear advantage of the rRNA-based techniques is that very small soil samples are needed for the analysis, meaning that it is possible to sample from both the rhizosphere and the rhizoplane of plant roots. This is important because it is in these areas that the main interactions between plant roots and soil microbes are mediated.

No single method, at present, can give a complete and accurate picture of the microbial community structure. Therefore, a combined approach provides a better assessment of microbial community structure and minimises the drawbacks from different methods. Combinations of both these phenotypic and genotypic approaches have recently given a clearer insight into the extreme complexity of soil microbial communities and their interaction with vegetation. These interactions will now be discussed by considering research at the plant community scale, the scale of the rhizosphere and within the root rhizoplane.

Assessing the impact of plant communities on soil microbial community structure

In grasslands, plant community structure has been found to affect the size and composition of associated microbial communities (Grayston *et al.*, 2001; 2004; McCulley and Burke, 2004), with increases in microbial diversity being associated with more diverse plant communities (Kowalchuk, 2002). A common finding in studies along soil fertility gradients of upland grasslands is that the biomass of the soil microbial community is higher under low fertility conditions than under high fertility conditions that are maintained by regular nitrogen additions (Bardgett *et al.*, 1996; 1997; Grayston *et al.*, 2004). Associated with these changes in total biomass are also shifts in microbial community structure, with high soil fertility and nutrient availability favoring the bacterial community and low soil fertility favouring the fungi (Bardgett *et al.*, 1996; 1998; Grayston *et al.*, 2001; 2004). Such variations in soil microbial communities have been attributed to quantitative and qualitative differences in substrate supply between upland grasslands (Bardgett *et al.*, 1998). In particular, it has been suggested that differences in plant species composition and species-dominance between grasslands are likely to exert strong selective pressures on the soil microbial community through plant-specific changes in the quantity and variety of compounds lost through rhizodeposition and litter and root senescence (Grayston *et al.*, 2001). In contrast, little or no relationship was found between microbial community structure and floristic groups in chalk grassland (Chabrerie *et al.*, 2003). From these studies it has not been possible to determine the direct effects of vegetation on soil microbial communities, as opposed to indirect effects mediated by other environmental variables (such as soil fertility or pH). However, from a study of microbial community DNA it is clear that there is great spatial variation in community structure within grassland soils (Clegg *et al.*, 2000; Ritz *et al.*, 2004). For example, Clegg *et al.*, (2000) found the DNA similarity (as assessed by cross-hybridization experiments) was no greater between replicate plots within unimproved grassland at one site as between unimproved grasslands separated by several 100 km. However, two other recent studies considering different land use systems have found contradictory results to those of Clegg *et al.* (2000). Green *et al.* (2004) found that fungal (Ascomycete) diversity exhibited spatially predictable aggregation patterns over scales ranging from 10 m^2 to 10^{10} m^2. Horner-Devine *et al.* (2004) reported similar bacterial diversity–area relationships. This is an interesting observation because it suggests that microbial communities follow the rule of community composition decay with geographical distance that is found for plant and animal communities.

A number of studies have attempted to draw links between plant communities and below-ground microbial diversity, but most have focused on plants growing in monocultures, mainly in agricultural soils. These studies have provided varying evidence about whether the major factor influencing the composition of the microbial community is plant species (Smalla *et al.*, 2001; Wieland *et al.*, 2001; Miethling *et al.*, 2003) or soil characteristics (Brodie, 2002; Buckley & Schmidt, 2003; Girvan *et al.*, 2003), while others indicate that the importance of the plant community depends on soil type (Marschner *et al.*, 2001). In any study of soil microbial community structure in relation to grassland plant diversity, it is difficult to separate out the influence of the vegetation *per se* from other factors such as soil pH or fertiliser inputs. One reason is that the effect of soil nutrient status on microbial community structure within a single upland grassland site can be more important than the composition of proximal vegetation (Ritz *et al.*, 2004). Therefore, management inputs, such as lime and fertilizers, which are associated with pasture improvement, would be expected to directly alter the composition of the soil microbial community. Another main driver of soil nutrient status in grasslands at the sward scale is large herbivores. Grazing animals can affect soil microbes either directly, through dung and urine returns or soil compaction (Jarvis, 2000), or indirectly by altering the carbon inputs from vegetation (Bardgett *et al.*, 1998). Urine excretion is possibly the greatest cause of the high variability in microbiological species composition found at small scales in upland grasslands. In extensively grazed grasslands urine returns can account for 5-25 kg N /ha annually, mainly as urea (Whitehead, 1995). However, a urine patch represent a locally high input of N, typically equivalent to 300-500 kg N /ha (Haynes & Williams 1993), which leads to an increase in soil pH caused by urease enzymes in the soil. Soil P availability can also be greatly enhanced, due to desorption of organically-bound P from the mineral fractions in the soil (Shand *et al.*, 2002). Williams *et al.* (2000) found that carbon utilisation patterns of soil microbial communities in upland pastures were altered by treatment with urine, generally leading to an increase in substrate utilisation 2 to 5 weeks after urine addition. Urine deposition in intensively grazed pasture also leads to increased heterogeneity in the vegetation, reflecting the increased patchiness of nutrient supply (Marriott *et al.*, 1997). However, the impact of urine on the soil microbes has been found to be confined largely to phenotypic measures of the community, while the background genetic structure of the community did not appear to be affected (Ritz *et al.*, 2004).

Soil microbial community structure at the rhizosphere scale

Given the changes in soil pH and fertility associated with pasture improvement is there any evidence that individual plant species in upland grasslands condition soil microbial diversity around their roots? Pot experiments using sieved soils under controlled conditions have often shown that plant roots can condition the diversity or activity of soil microbes (Grayston *et al.*, 1998, Marilley *et al.*, 1998, Bardgett *et al.*, 1999). However, field experiments have not always shown such conclusive results. In one such field experiment vegetation was removed from replicated plots in an area of unimproved grassland. The upland site (at the Sourhope Research Station, Scotland) had been managed by occasional grazing by sheep for at least 120 years and the vegetation was a permanent *Festuca ovina–Agrostis capillaris–Galium saxatile* grassland (National Vegetation Classification U4a). The site was described by Grayston *et al.* (2004). The soil from each plot was mixed thoroughly to break up any historical urine patches and monocultures of either *A. capillaris* (one of the dominant species found at the site) or *Lolium perenne* (not previously present at the site) grown. Other plots were left fallow. During the next two years cores were taken to extract rhizosphere soil. A range of phenotypic and genotypic measures of soil microbial diversity were then made, in order to determine the impact of the different plant species on microbial community structure. The data showed a great temporal variation in microbial community size and structure and large differences between the plots growing grass and fallow soil. However, no statistically significant differences were found between the microbial communities in the *A. Capillaris* or *L. perenne* plots at any of the dates, while the plots with vegetation were different from the fallow soil (Figure 2).

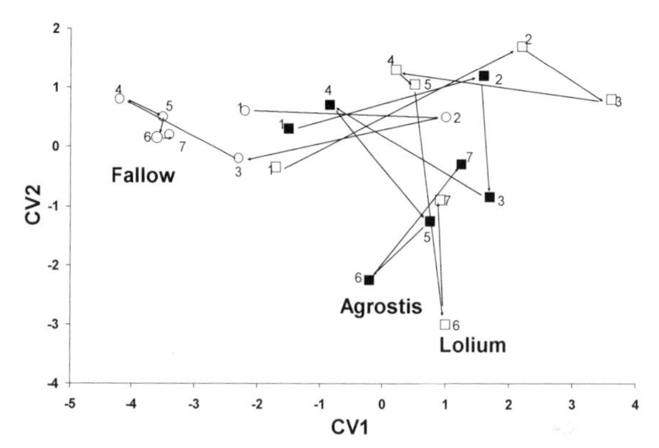

Figure 2 The impact of growing two different grass species in monoculture on soil microbial diversity. The figure is a plot of ordination of canonical variates (CV) generated by canonical variate analysis of PLFA data from rhizosphere soil that was fallow (○) as a control, or had a monoculture of *Agrostis capillaries* (■) or *Lolium perenne* (□). The numbers refer to different sampling dates (1-November 1998, 2-January 1999, 3-April 1999, 4- May 1999, 5-July 1999, 6- October 1999 and 7-May 2000). Each point is the mean of six replicate plots. The graph has been redrawn from Grayston *et al.* (submitted).

There are several possible interpretations of these results. First, while the presence of plants changes the soil microbial community, the impact of different plant species is small when considered against the background of the heterogeneity of the soil physiochemical environment. Secondly, it might be possible that while soil microbial *activity* is regulated by rhizodeposition from plant roots, *community structure* might be more affected by the quality of the recalcitrant carbon derived from root and leaf litter. The SOM is a much larger pool of carbon than that available from rhizodeposition. If plant communities take decades to evolve, soil microbial communities underneath them might take a similar length of time, with the main drivers being root turnover and litter decomposition, rather than rhizodeposition. A third possibility is that by sampling rhizosphere soil it was not possible to measure the plant-specific influence on the soil microbial community. If such interactions were mediated via exudates, the greatest effect would be expected to be in the rhizoplane, where specific nutritional selection would occur, prior to diffusion of less plant-specific breakdown products into the rhizosphere soil.

The concept of 'rhizosphere' implies a spatial relationship between plants and microbes, focusing on the interface between root and soil. There have been few studies which have attempted to quantify the spatial structure of rhizosphere microbial communities in grasslands. Ritz *et al.* (2004) studied the spatial properties of genetic, phenotypic and functional aspects of microbial community structure in an area of unimproved, upland grassland. They found geostatistical ranges extending from approximately 0.6-6 m, dispersed through both chemical and biological properties. The CLPP data tended to be associated with ranges greater than 4.5 m, while there was no relationship between physical distance in the field and genetic similarity based upon DDGE profiling. However, analysis of samples taken as closely as 1 cm apart suggested some spatial dependency in community DNA-DGGE parameters below an 8-cm scale (Ritz *et al.*, 2004). These results were consistent with studies of the spatial dependency of soil microbial properties carried out in other systems, including a forest soil (Saetre & Bååth, 2000) and a chaparral system (Klironomos *et al.*, 1999). Taken together, these studies suggest a high level of spatial complexity in the microbial communities in soil and suggest that a complex set of interactions impact on them.

Soil microbial community structure at the rhizoplane scale

One of the problems with the research described above was that soil was sampled by taking cores. As a consequence the rhizosphere soil was mixed with bulk soil. However, there have been few studies that have considered microbial diversity at a smaller spatial scale than a few centimetres. One such field experiment was undertaken recently, to assess the interaction between plant species and the community composition of bacteria and fungi colonising the rhizoplane of grass roots in an unimproved upland grassland soil, previously shown to be dominated by seven grass species. Soil cores were taken, utilising a spatially explicit sampling design. The DNA was extracted from individual root fragments isolated from the cores and used for plant identification and measurements of bacterial and fungal community composition (Ridgeway *et al.*, 2003). In this way the specific associations between individual plant roots and the soil microbial community could be studied, while avoiding the difficulties associated with identifying root fragments based upon their morphology.

Principal co-ordinate analysis was used to identify underlying patterns in the similarity between plant species and the microbial community in their root rhizoplane, based upon both the species of the individual plant roots sampled and the roots of other plant species found within the same core. This allowed an analysis of the variation in bacterial and fungal communities due to individual roots or surrounding roots. For fungi the total distance information in community structure accounted for was 23% and 16% for the individual and surrounding roots, respectively. For bacteria the values were lower at only 2% and 18% for the individual and surrounding roots, respectively. The results show direct evidence that the structure of specific microbial groups (those capable of nitrite reduction and arbuscular mycorrhizal fungi) were strongly affected by the identity of the plant they were associated with, while the general bacterial community was only weakly correlated. This suggests that there will be a functional difference in the microbial community associated with different plant species. These differences in the microbial communities caused by plants were greater and clearer for the mycorrhizal fungi than for specific groups of bacteria. This brings into question what are the main drivers of soil microbial diversity in grassland soils? One explanation is that in such unimproved, upland grassland soils microbial *community structure* is determined largely by the quality or composition of the soil organic matter (which in turn reflects the previous vegetation and land management history over decades to centuries), while the *activity* of the microbial biomass is driven by the current vegetation, possibly through grass growth and carbon transfers below-ground.

Conclusion

Research has shown that the soil microbial community structure in upland grasslands is very diverse. At the plant community scale, there is a relationship between vegetation and soil microbial community structure. This has been seen as differences in both the size and composition of the microbial community when comparing soils under different vegetation types (for example unimproved versus improved grassland). However, within a vegetation type, there is also evidence of great spatial heterogeneity in soil microbial diversity, with samples collected a few metres apart being as dissimilar as those collected from within the same vegetation type but several hundred kilometres apart. From these studies at the plant community scale, it is not clear if the differences in microbial community structure and activity under different vegetation types are due to direct effects of the plants on the microbes, or indirect effects mediated via changes in (for example soil fertility and pH) as a consequence of pasture improvement.

Studies at the scale of the rhizosphere have shown differing results. Pot experiments have often demonstrated an impact of individual plant species on soil microbial diversity. However, a field experiment with vegetation substitution treatments showed significant temporal variation in microbial community size and structure, and large differences between plots growing grass and those with fallow soil. However, plant species were found to have no significant effect on either the size or the composition of the biomass at any date of sampling over a 21-month period. The impact of individual plant species on the diversity of microbes in their root rhizoplane has also been assessed. The results showed direct evidence that while the structure of specific microbial groups (such as mycorrhizal fungi or nitrite-reducing bacteria) were significantly affected by the identity of the plant they were associated with, the general bacterial community was not. These results overall suggest that, while individual plant species have an effect upon the diversity of soil microbial communities, a wide range of other factors also influence their composition. Both soil pH and fertility have been shown to be important in this respect. It also appears that there is a functional redundancy within the microbial community, with several or many groups of bacteria able to perform similar functions.

Acknowledgements

The Scottish Executive Environment and Rural Affairs Department (SEERAD) fund research on Plant-Soil Interactions at the Macaulay Institute through their grant-in-aid. Some of the research described in this chapter was undertaken as part of the SEERAD MICRONET programme.

References

Anderson, I.C. & J.W.G. Cairney (2004). Diversity and ecology of soil fungal communities: increased understanding through the application of molecular techniques. *Environmental Microbiology*, 6, 769-779.

Bardgett, R.D., P.J. Hobbs & A. Frostegård (1996). Changes in soil fungal:bacterial biomass ratios following reductions in the intensity of management of an upland grassland. *Biology and Fertility of Soils*, 22, 261-264.

Bardgett, R.D., D.K. Leemans, R. Cook & P.J. Hobbs (1997). Seasonality of soil biota of grazed and ungrazed hill grasslands. *Soil Biology and Biochemistry*, 29, 1285-1294.

Bardgett R.D., J.L. Mawsdsley, S. Edwards, P.J. Hobbs, J.S. Rodwell & W. J. Davies (1999). Plant species and nitrogen effects on soil biological properties of temperate upland grasslands. *Functional Ecology*, 13, 650-660.

Bardgett R.D., DA.Wardle & G.W. Yeates (1998). Linking above-ground and below-ground food webs: how plant reponses to foliar herbivory influence soil organisms. *Soil Biology and Biochemistry*, 30, 1067-1078.

Bhupinderpal-Singh, A. Nordgren, M.O. Löfvenius, M.N. Högberg, P.-E. Mellander & P. Högberg (2003). Tree root and soil heterotrophic respiration as revealed by girdling of boreal Scots pine forest: extending observations beyond the first year. *Plant, Cell and Environment*, 26, 1287 – 1296.

Blackwood, C.B., T. Marsh, S.-H. Kim & E.A. Paul (2003). Terminal restriction fragment polymorphism data analysis for quantitative comparison of microbial communities. *Applied Environmental Microbiology*, 69, 926-932.

Bossio, D.A., K.M. Scow, N. Gunapala & K.J. Graham (1998). Determinants of soil microbial communities: effects of agricultural management, season, and soil type on phospholipid fatty acid profiles. *Microbial Ecology*, 36, 1-12.

Brodie, E., S. Edwards & N. Clipson (2002). Bacterial community dynamics across a floristic gradient in a temperate upland grassland ecosystem. *Microbial Ecology*, 44, 260-270.

Buckley, D. H & T.M. Schmidt (2003). Diversity and dynamics of microbial communities in soils from agro-ecosystems. *Environmental Microbiology*, 5, 441-452.

Campbell C.D., S.J. Chapman, C.M. Cameron, M.S. Davidson & J.M. Potts (2003). A rapid microtiter plate method to measure carbon dioxide evolved from carbon substrate amendments so as to determine the physiological profiles of soil microbial communities by using whole soil. *Applied and Environmental Microbiology*, 69, 3593-3599.

Campbell, C.D., S.J. Grayston & D.J. Hirst (1997). Use of rhizosphere carbon sources in sole carbon utilisation tests to discriminate soil microbial communities. *Journal of Microbiological Methods*, 30, 33-41.

Chabrerie, O., K. Laval, P. Puget, S. Desaire & D. Alard (2003). Relationship between plant and soil microbial communities along a successional gradient in a chalk grassland in north-western France. *Applied Soil Ecology*, 24, 43-56.

Clegg C. D., K. Ritz & B.S. Griffiths (2000). %G+C profiling and cross hybridization of microbial DNA reveals great variation in below-ground community structure in UK upland grasslands. *Applied Soil Ecology*, 14, 125-134.

Crecchio, C., A. Gelsomino, R. Ambrosoli, J.S. Minati & P. Ruggiero (2004). Functional and molecular responses of soil microbial communities under differing soil management practices. *Soil Biology and Biochemistry*, 36, 1873-1883.

Frostegård, A., E. Bååth & A. Tunlid (1993). Shifts in the structure of soil microbial communities in limed forests as revealed by phospholipid fatty-acid analysis. *Soil Biology & Biochemistry,* 25, 723-730.

Garland, J.L. & A. L. Mills (1991). Classification and characterization of heterotrophic microbial communities on the basis of patterns of community-level sole-carbon-source utilization. *Applied and Environmental Microbiology,* 57, 2351-2359.

Girvan, M. S., J. Bullimore, J.N. Pretty, A.M. Osborn & A.S. Ball (2003). Soil type is the primary determinant of the composition of the total and active bacterial communities in arable soils. *Applied and Environmental Microbiology*, 69, 1800-1809.

Grayston, S.J., C.D. Campbell, R.D. Bardgett, J.L. Mawdsley, C.D. Clegg, K. Ritz, B.S. Griffiths, J.S. Rodwell, S.J. Edwards, W.J. Davies, D.J. Elston & P. Millard (2004). Assessing shifts in microbial community structure across a range of grasslands of differing management intensity using CLPP, PLFA and community DNA techniques. *Applied Soil Ecology,* 25, 63-84.

Grayston S.J., C.D. Clegg, K. Ritz, B.S. Griffiths, A.E. McCaig, J.I. Prosser, D.J. Elston & P. Millard (2005) Temporal changes in soil microbial communities under grass monocultures in the field. *Applied Soil Ecology* (submitted).

Grayston, S.J., G.S. Griffith, J.L. Mawdsley, C.D. Campbell & R.D. Bardgett (2001). Accounting for variability in soil microbial communities of temperate upland grassland ecosystems. *Soil Biology and Biochemistry*, 33, 533-551.

Grayston, S.J., S. Wang, C.D. Campbell & A.C. Edwards (1998). Selective influence of plant species on microbial diversity in the rhizosphere. *Soil Biology and Biochemistry*, 30, 369-378

Green, J.L., A.J. Holmes, M. Westby, I. Oliver, D. Briscoe, M. Dangerfield, M. Gilling & A.J. Beattie (2004). Spatial scaling of microbilal eukaryotic diversity. *Nature*, 432, 747-756.

Haynes, R. J. & P.H. Williams (1993). Nurient cycling and soil fertility in the grazed pasture ecosystem. *Advances in Agronomy*, 49, 119-199.

Hedrick, D.B., A. Peacock, J.R. Stephen, S.J. Macnaughton, J. Bruggemann & D,C. White (2000). Measuring soil microbial community diversity using polar lipid fatty acid and denaturing gradient gel electrophoresis data. *Journal of Microbiological Methods*, 41, 235-248.

Hill, G.T., N.A. Mitkowski, L. Aldrich-Wolfe, L.R. Emele, D.D. Jurkonie, A. Ficke, S. Maldonado-Ramirez, S.T. Lynch, & E.B. Nelson (2000). Methods for assessing the composition and diversity of soil microbial communities. *Applied Soil Ecology*, 15, 25-36.

Horner-Devine M.C., M. Lage, J.B. Hughes & B.J.M. Bohannan (2004). A taxa-area relationship for bacteria. *Nature*, 423, 750-753.

Jarvis S.C. (2000). Soil-plant-animal interactions and impacts on nitrogen and phosphorous cycling and recycling in grazed pastures. In: G. Lemaire, J. Hodgson, A. de Moraes, P.C. de F. Carvalho & C. Nabinger (eds.) Grassland Ecophysiology and Grazing Ecology. CABI Publishing, Wallingford, 317-337.

Klironomos, J.N., M.C. Rillig & M.F. Allen (1999). Designing below ground field experiments with the help of semi-variance and power analysis. *Applied Soil Ecology*, 12, 227-238.

Kowalchuk, G. A., D.S. Buma, W. de Boer, P.G.L. Klinkhamer & J.A. Van Veen (2002). Effects of above ground plant species composition and diversity on the diversity of soil-borne microorganisms. *Antonie Van Leeuwenhoek,* 81, 509-520.

Marilley, L., G. Vogt, M. Blanc & M. Aragno (1998). Bacterial diversity in the bulk soil and rhizosphere fractions of *Lolium perenne* and *Trifolium repens* as revealed by PCR restiction analysis of 16s rDNA. *Plant and Soil* 198, 219-224.

Marriott, C.A., G. Hudson, D. Hamilton, R. Neilson, B. Boag, L. Handley, J. Wishart, C.M. Scrimgeour & D. Robinson (1997). Spatial variability of soil total C and N and their stable isotopes in an upland Scottish grassland. *Plant and Soil*, 196, 151-162.

Marschner, P., C.H. Yang, R. Lieberei & D.E. Crowley (2001). Soil and plant specific effects on bacterial community composition in the rhizosphere. *Soil Biology and Biochemistry*, 33, 1437-1445.

McCulley, R. L. & L.C. Burke (2004). Microbial community composition across the Great Plains: landscape versus regional variability. *Soil Science Society of America Journal*, 68, 106-115.

Miethling, R., K. Ahrends & C. Tebbe (2003). Structural differences in the rhizosphere communities of legumes are not equally reflected in community-level physiological profiles. *Soil Biology and Biochemistry*, 35, 1405-1410.

O'Connel, S. & J.L. Garland (2002). Dissimilar response of microbial communities in Biolog GN and GN2 plates. *Soil Biology and Biochemistry*, 34, 413-416.

Prosser, J.I. (2002). Molecular and functional diversity in soil micro-organisms. *Plant Soil*, 244, 9-17.

Ridgeway, K.P., M.J. Duck & P.W. Young (2003). Identification of roots from grass swards using PCR-RFLP and FFLP of the plastid trnL (UAA) intron. *BMC Ecology*, 3, 8.

Ritz, K., W. McNicol, N. Nunan, S. Grayston, P. Millard, D. Atkinson, A. Gollottee, D. Habeshaw, B. Boag, C.D. Clegg, B.S. Griffiths, R.E. Wheatley, L.A. Glover, A.E. McCaig & J.I. Prosser (2004). Spatial structure in soil chemical and microbiological properties in an upland grassland. *FEMS Microbiology Ecology*, 49, 191-205.

Saetre, P. & E. Bååth (2000). Spatial variation and patterns in soil microbial community structure in a mixed spruce-birch stand. *Soil Biology and Biochemistry*, 32, 909-917.

Shand C. A., B.L. Williams, L.A. Dawson, S. Smith & M.E. Young (2002). Sheep urine affects soil solution nutrient composition and roots: differences between field and sward box soils and the effects of synthetic and natural sheep urine. *Soil Biology and Biochemistry*, 34, 163-171.

Smalla, K., G. Wieland, A. Buchner, A. Zock, J. Parzy, S. Kaiser, N. Roskot, H. Heuer & G. Berg (2001). Bulk and rhizosphere soil bacterial communities studied by denaturing gradient gel electrophoresis: plant-dependent enrichment and seasonal shifts revealed. *Applied and Environmental Microbiology*, 67, 4742-4751.

Torsvik, V. & L. Ovreas (2002). Microbial diversity and function in soil: from genes to ecosystems. *Current Opinion in Microbiology*, 5, 240-245.

Vestal, J. R. & D.C. White (1989). Lipid analysis in microbial ecology. *Bioscience*, 39, 535-541.

Whitehead, D. C. (1995). Grassland Nitrogen. CAB International, Wallingford, UK. 397 pp.

Wieland, G., R. Neumann & H. Backhaus (2001). Variation of microbial communities in soil, rhizosphere, and rhizoplane in response to crop species, soil type and crop development. *Applied and Environmental Microbiology*, 67, 5849-5854.

Williams, B. L., S.J. Grayston & E.J. Reid (2000). Influence of synthetic sheep urine on the microbial biomass, activity and community structure in two pastures in the Scottish uplands. *Plant and Soil*, 225, 175-185.

How herbivores optimise diet quality and intake in heterogeneous pastures, and the consequences for vegetation dynamics

R. Baumont[1], C. Ginane[1], F. Garcia[1,2] and P. Carrère[2]

[1]*Institut National de la Recherche Agronomique, Unité de Recherches sur les Herbivores, 63122 Saint-Genès-Champanelle, France, Email: baumont@clermont.inra.fr*

[2]*Institut National de la Recherche Agronomique, Unité d'Agronomie, 63039 Clermont-Ferrand, France*

Abstract

Understanding the interplay between foraging behaviour and vegetation dynamics in heterogeneous pasture is an essential requirement for evaluating the value of the resource for large herbivores and for managing that resource. The orientation of selective grazing behaviour between intake and diet quality depends on the spatial and temporal scales considered. In the short-term scale of a grazing sequence, there is evidence that large herbivores tend to optimise the intake rate of digestible materials by adaptation of their biting behaviour and by patch choice. On a day-to-day scale, there is evidence that large herbivores tend to prioritise the quality of the diet to minimise digestive constraints within the time that they can spend grazing. On a pasture scale, the search for areas giving the best trade-off between quantity and quality of intake leads to the optimisation of their foraging paths, in particular by modulating their sinuosity in response to heterogeneity. Repeated grazing of preferred patches creates a positive feedback on forage quality and enhances heterogeneity. Long-term consequences on vegetation dynamics, botanic composition and grassland quality are less understood.

Keywords: ruminant, heterogeneous pastures, grazing behaviour, intake, vegetation dynamics

Introduction

Grazing management aims to provide herbage in quantity and of sufficient quality to satisfy animal needs while sustaining the grassland. On grassland of high productivity, extensive management for environmental purposes, such as reducing pollution and enhancing biodiversity, can be achieved by lowering grazing pressure, resulting in the development of pasture heterogeneity. Marginal environments, such as semi-arid areas, wetlands or uplands, are characterized by a low productivity and do not suffer high grazing pressures. When grazing pressure is low, the larger area offered to large herbivores makes the actual grazing pressure vary spatially and temporally, as they can make their own choices on what to eat. The uneven use of the grassland by large herbivores will lead to enhanced heterogeneity in biomass availability and quality due to edaphic factors. Understanding the interplay between foraging behaviour and vegetation dynamics is therefore an essential requirement for evaluating the resource value for the animals and for managing that resource.

The interaction between grazer and vegetation is dynamic and bidirectional. The structure, quality and distribution of plant material affect the quantity and quality of the grazed diet, while grazing affects the structure and composition of the vegetation. Frequently grazed plants and areas will diverge from the less frequently and ungrazed plants and areas, creating spatial patterns at different scales (Marriott & Carrère, 1998). Based on the Optimal Foraging Theory (Stephens & Krebs, 1986), it can be postulated that animals try to maximise the intake of energy and minimise the related costs. To achieve this, foraging behaviour consists of a series of discrete decisions at the successive spatio-temporal scales of bite prehension through to patch choice and plot utilisation. All the decisions represent trade-offs, in particular between diet quantity and quality, since in heterogeneous grasslands, areas of low biomass and high quality coexist with areas of high biomass but poor quality (Wallis de Vries & Daleboudt, 1994). In the present review, we will focus on the trade-offs at the main relevant temporal scales of plant/animal interactions: the short-term scale of bite prehension and patch choice within a grazing sequence, and the longer-term scale of intake over a day and beyond. Large herbivores integrate into their decisions the knowledge they have gained on the nutritional consequences of their diet choices (Provenza, 1995) and on the resource availability and spatial distribution (Dumont & Petit, 1998). We will then examine how foraging behaviour affects how animals use pastures and the consequences on vegetation dynamics.

Optimising biting behaviour and patch choices during grazing sequences

A functional way to represent a heterogeneous pasture is as a mosaic of patches. A patch can be defined as an area over which intake rate is relatively constant (Illius & Hodgson, 1996) which implies a relative homogeneity in the structure and composition of the vegetation. When grazing a patch, what does the animal try to achieve? It is often postulated that grazing behaviour aims to maximise intake rate.

In order to explain how sward characteristics within a patch affect intake rate, many authors have used an analytical breakdown that splits intake rate into bite mass and time per bite, then bite mass into bite volume and bulk density of the sward, and then bite volume into bite area and bite depth (for reviews, see Prache & Peyraud (2001) and Penning & Rutter (2004)). Bite depth tends to be a constant proportion of sward height slightly modulated by sward density. Bite area is dependent on the size of the animal's dental arcade and on sward height and density. Time per bite can be split into the sum of the time required to collect and sever a bite, which is considered independent from bite mass, and the time required to masticate a bite, which is dependent on its mass and its resistance to chewing (Parsons *et al.*, 1994). Finally, intake rate increases with bite mass which in turn increases with both sward height and density.

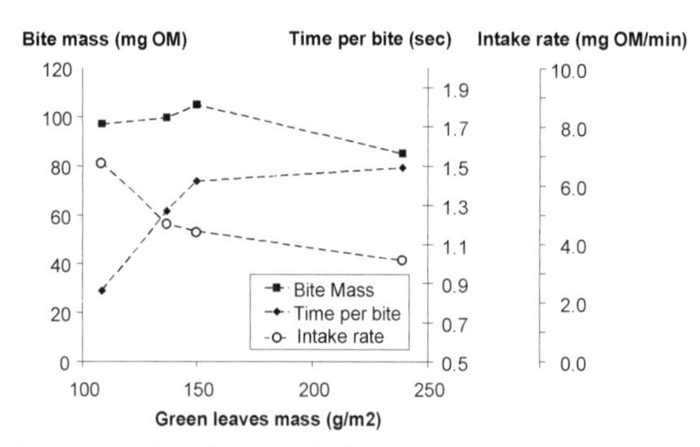

Figure 1 Bite characteristics and intake rate in sheep grazing a maturing and accumulating cocksfoot sward at low stocking rate (5 ewes on 3000 m²) from April to September (from Garcia *et al.*, 2003a)

The geometry of the biting process combined with a representation of the vertical distribution of sward biomass supports the mechanistic modelling of intake rate (Parsons *et al.*, 1994; Baumont *et al.*, 2004) and gives satisfactory predictions on vegetative swards. Based on this approach, it can be predicted that intake rate could be maximised by increasing bite mass. However, bite mass may be limited by the pseudostem, which has been suggested to constitute a physical barrier to bite depth due to the greater resistance to defoliation related to its layered structure and higher fibre content (Illius *et al.*, 1995). Even for large ruminants that have enough strength to sever pseudostems (Griffiths *et al.*, 2003), deeper biting should also decrease the quality of the plant material ingested, as the nutritive value of the grass generally decreases from the top to the bottom of the sward (Delagarde *et al.*, 2000). When the composition of the sward is more complex, as is the case on maturing swards containing reproductive material, sheep have been shown to significantly increase time per bite in relation to selective behaviour for green leaves (Prache *et al.*, 1998; Garcia *et al.*, 2003a). Bite mass remained stable throughout the course of the season, although biomass strongly accumulated in the sward (Garcia *et al.*, 2003a, Figure 1). This behaviour in favour of bite quality decreases intake rate, indicating that sheep did not adopt a strategy of intake rate maximisation only. Rather, they would try at the bite level to optimize both the quality of the plant material ingested and the intake rate.

During a grazing sequence, animals frequently face a choice between patches differing in vegetation structure and/or quality. When patches differ only by their sward height, cattle have been shown to select the feed that provided the highest food intake rate (Distel *et al.*, 1995). Similar results have been reported for sheep (Kenney & Black, 1984) and goats (Illius *et al.*, 1999) presented with a choice of different forages or plant species, when the forages giving higher intake rate also had a higher quality and energy intake rate. In contrast, preferences of sheep between forages providing similar intake rates were in accordance with differences in nutritive value (Baumont *et al.*, 1999). However in these experiments, animals did not really face a trade-off between quality and quantity, unlike when they have a choice between frequently and infrequently grazed patches.

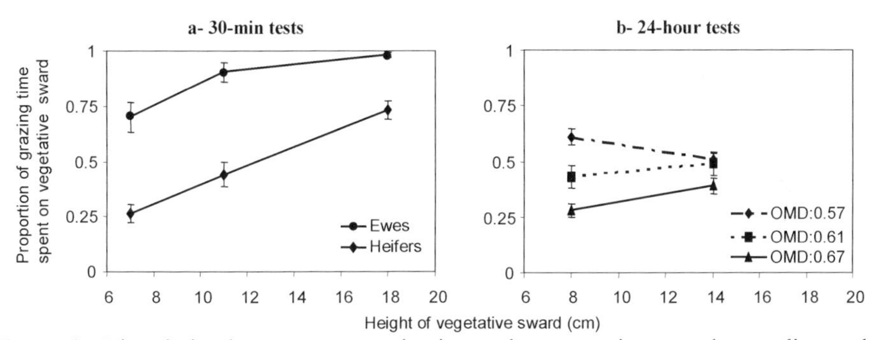

Figure 2 Diet choice between a reproductive and a vegetative sward according to height: a- effects of species (sheep and cattle) during a short-term test (from Dumont *et al.*, 1995 a; b); b- effects of the decreasing quality of the reproductive sward (OMD = organic matter digestibility) on heifer's choices on a day-to-day scale (from Ginane *et al.*, 2003)

Frequently grazed patches remain of high quality (digestibility) but low availability, and provide a low intake rate. Conversely, available biomass accumulates on infrequently grazed patches that can allow a high intake rate of lower quality plant material. To simulate this situation, Dumont *et al.* (1995a; b) offered sheep and cattle a choice between a reproductive sward of high height/low quality and vegetative swards of low height/high quality. These experiments revealed that differences in quality were important, and sheep clearly preferred the vegetative swards except at the lowest height. Heifers, which are disadvantaged on short swards where bite depth is limited (Illius & Gordon, 1987), showed an overall lower preference for vegetative swards than sheep, and their switch to the reproductive sward was more pronounced (Figure 2a). Garcia *et al.* (2003b) investigated in sheep how short-term preferences between more or less intensively grazed swards evolve during the grazing season. During spring and early summer, differences in quality were low or absent and animals preferred the less grazed and tall patches that allowed easier selection of green leaves. In late summer, their preference switched to the more intensively grazed patches that were of higher quality due to vegetative regrowth. Criteria characterizing relative quality, such as relative abundance of green leaves or relative digestibility, were able to explain the observed choices during the grazing season. This suggests that animals integrate both intake rate and quality at the patch choice stage, and should therefore act as energy intake rate maximisers (Tolkamp *et al.*, 2002).

This should particularly apply when the preferred patches are dispersed spatially, implying moving costs for the grazing animal in terms of time and energy. Short-term tests have shown that sheep and cattle are able to integrate these costs and modify their choices accordingly. They decreased their preference for a good-quality hay, either when the amount offered (reward) per distance walked decreased (Dumont *et al.*, 1998) or when the difference in quality between the reward and another lower quality hay available without moving decreased (Ginane *et al.*, 2002a). In both experiments, the ewes and heifers selected the food option that maximised their rate of energy intake, as predicted by the optimal patch choice model. However, the choices were suboptimal and conformed to an overmatching pattern in favour of the good-quality forage (Senft *et al.*, 1987).

Balancing digestive and time constraints to optimise intake and diet choice

At the day-to-day scale grazing animals have to satisfy various nutritional needs in the time that they can spend grazing. Optimal trade-offs between quantity and quality may vary with the time scale, i.e. between short-term rate of food intake and long-term rate of nutrient assimilation (Wallis de Vries & Daleboudt, 1994; Newman *et al.*, 1995; Wilmshurst *et al.*, 1995). The regulation of diet choice and intake integrates digestive and nutritional feedbacks which govern the balance between motivation to eat and satiety, and which modulate feed preferences (Baumont *et al.*, 2000). The longer time scale also integrates behavioural compensatory mechanisms incorporating walk speed between patches (Roguet *et al.*, 1998), biting rate and grazing time (Taweel *et al.*, 2004).

Herbivores faced with a quantity-quality trade-off on a long-term scale were shown to selectively graze high quality patches of low to intermediate height or biomass (Wallis de Vries & Daleboudt, 1994, Wilmshurst *et al.*, 1995; Ginane *et al.*, 2003). This behaviour does not maximise short-term intake rate but would allow the animals to maximise their energy intake on a daily basis (Fryxell, 1991). Indeed, digestible organic matter intake probably has to be considered as the currency the animals want to maximise on a daily basis and beyond. Digestible organic matter intake integrates both the quality and the total quantity of food ingested, and a given level may result from a wide range of theoretically possible strategies from maximising quality to maximising quantity. Maximising quality implies high selective behaviour for parts of plants or patches of high digestibility that are often of low accessibility. This option reduces intake rate and increases the time spent grazing. Maximising quantity implies less selective behaviour and the processing of less digestible material through the digestive tract. The trade-off between quantity and quality has to take into account the link between behavioural and digestive constraints (Baumont *et al.*, 1990). Progress in integrating intake and digestion has been achieved by mechanistic modelling (Illius & Gordon, 1991; Sauvant *et al.*, 1996). The latter proposed a self-regulated intake model in which the decision whether to eat or not is taken every minute by comparing a motivation-to-eat function with a satiation function based on digestion and a metabolic sub-model. Time spent eating is governed by the balance between motivation to eat, which depends primarily on energy demand, and satiation which integrates the energy supply and the fill effect of the ingested forage, based on its digestion kinetics in the rumen (Baumont *et al.*, 1997). This model has recently been extended to grazing integrating the intake rate response to sward characteristics (Baumont *et al.*, 2004). A simulation, using this model of how intake is regulated from short sward of high quality to tall sward of lower quality, is illustrated in Figure 3. If dry matter intake increases with sward height, despite the decrease in sward quality, digestible organic matter intake is maximised for the combination of highest quality and height. When the sward is shorter, the increase in grazing time does not fully compensate for the decrease in intake rate. When the sward is higher and of lower quality, intake rate and dry matter intake increase but digestible organic matter intake decreases. The higher satiation effect of ingesting lower quality plant material limits the time spent grazing. Predictions made using this model are in favour of prioritising quality, in accordance with the model developed by Hutchings & Gordon (2001) stating that the 'digestibility' strategy is the most efficient.

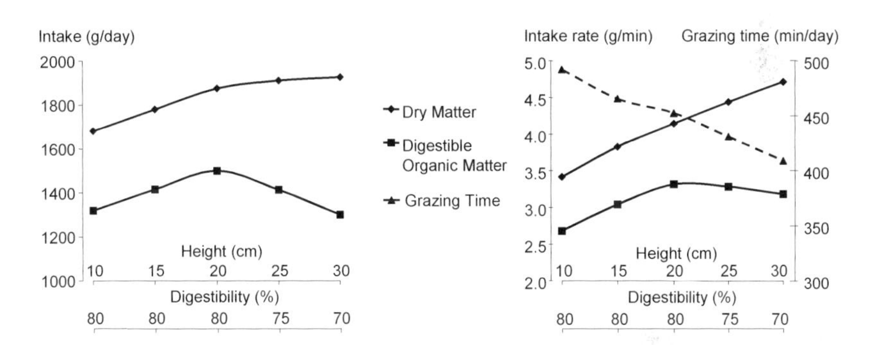

Figure 3 Prediction of intake rate, grazing time and daily intake in sheep with concurrent variations in sward height and digestibility. Data simulated using the model developed by Baumont *et al.* (2004).

In the field, grazing time is widely used by ruminants as a way of adapting to a decrease in availability of the feeding resource (Allden & Whittaker, 1970; Penning *et al.*, 1991; Rook *et al.*, 1994; Gekara *et al.*, 2001). In choice situations too, cattle and sheep have been shown to increase their grazing time on the preferred sward as its accessibility decreased while a lower quality alternative was simultaneously offered (Hester *et al.*, 1999; Rook *et al.*, 2002; Ginane *et al.*, 2003). By experimentally investigating animal choices as the pressure of constraints increases, it may be possible to estimate which factor in the quantity-quality trade-off is first prioritised. An experiment conducted throughout the grazing season with sheep at different stocking rates showed that they constantly maximised the quality of their diet in conditions of either low quantity-high quality (high stocking rate) or high quantity-low quality (low stocking rate) (Garcia *et al*., 2003a). In choice experiments with a vegetative sward height constraint, heifers have been shown to maintain or lengthen the proportion of their grazing time spent on short vegetative swards compared to reproductive swards, thereby revealing their priority for diet quality (Ginane *et al*., 2003, Figure 2b). When the daily time available for grazing was strongly limited, heifers maintained their choice for the vegetative sward at the expense of total intake (Ginane & Petit, 2005).

However, since grazing time is not indefinitely increasable, especially for producing animals with high nutritional requirements that need a long basal grazing time (Gibb *et al*., 1999) and, since digestive regulation limits the large intake of rapidly ingestible material, animals are unlikely to behave in an all-or-nothing way, and the optimal trade-off would be to ingest both alternatives in relative proportions depending on the nature and intensity of the harvesting and food-processing constraints. Furthermore, mixed diets are the general rule in choice situations (Duncan *et al*., 2003) and the nutritional hypotheses put forward in the literature vary greatly according to the choice situation. For example, sheep have been shown to eat straw (Cooper *et al*., 1995) or 10-mm polyethylene fibres (Campion & Leek, 1997) to prevent rumen disorders and restore normal rumination activity when fed a high concentrate diet. The partial preference of heifers for clover *versus* grass may be due to a prevention of sub-clinical bloat status (Rutter *et al*., 2004). Finally, goats at turn out appear to seek herbage species that are relatively low in protein and rich in fibre in order to reduce the variation in ingesta composition as far as possible given the large seasonal variations in vegetation composition (Fedele *et al*., 1993). An underlying mechanism would be the ability of animals to learn the post-ingestive consequences of their previous choices. Faced with trade-offs between food concentrations of energy and protein (Wang & Provenza, 1996) or energy and toxin (Ginane *et al*., 2005), herbivores showed they were able to perceive these characteristics and to adapt their diet choices accordingly. As post-ingestive stimuli need to be periodically reinforced, the animal regularly has to re-evaluate the benefits and costs of the different choices.

Optimising spatial utilisation of a pasture

The search for areas that allow the best trade-off between intake quantity and quality induces repeated grazing on such areas. It can be hypothesised that when animals perceive sward heterogeneity, their foraging walks are no longer random but structured to respond efficiently to the sward structure (Parsons & Dumont, 2003). Three behavioural mechanisms are involved in optimising the spatial utilisation of the resource: the modulation of foraging velocity (Shipley *et al*., 1996), the use of spatial memory and visual cues (Edwards *et al*., 1996; Dumont & Petit, 1998), and the modulation of foraging path sinuosity (Ward & Saltz, 1994). These behavioural mechanisms concur to modulate spatial utilisation through resource abundance or resource heterogeneity and complexity (Dumont *et al*., 2002).

A persistent issue is to identify the spatial scales at which the animals perceive the heterogeneity of the pasture, and to characterise how animals modulate their foraging paths through resource abundance and heterogeneity. Garcia *et al*. (2005) have used fractal analysis to analyse the foraging paths of ewes grazing a continuously-distributed and spatially-limited resource. This method, which investigates the functional heterogeneity of a habitat (Marell *et al*., 2002), can identify the heterogeneity at which the animal responds. It also provides insight into the hierarchical levels of foraging behaviour (Nams, 2005). In this study, the vegetation did not exhibit any spatial distribution before the experiment and ewes adopted a random walk at the beginning of the grazing season. This corresponds to the absence of any optimal searching scale, and remains the most advantageous as it reduces the costs of searching in homogeneous non-patchy environments (Foccardi *et al.*, 1996). The vegetation structure became more complex after a few weeks of grazing, and the sheep modulated their foraging paths through resource abundance and/or sward structure. A breakpoint was identified at 5 metres, for which the fractal dimension is always low, meaning that the animal's path is straighter at that scale (Figure 4). Within a scale of 0-5 metres, the modulation of sinuosity was not linked to sward abundance and structure, and sheep mainly developed behavioural adaptations at bite and feeding station scales (Garcia *et al.*, 2003a). Within a scale of 5-12 metres, the behavioural mechanisms involved the modulation of foraging path sinuosity, which implies an adaptation of spatial utilisation in relation to the perception of the environment. Grazing paths were tortuous on tall swards in summer (higher fractal dimension), and straighter on heterogeneous, well-structured swards showing visual cues in the autumn. The breakpoint for fractal dimension across spatial scales may thus indicate the hierarchical threshold in spatial adaptation of the foraging behaviour of grazing herbivores (Garcia *et al.*, 2005). This experiment suggests that the determinants of sward heterogeneity organisation, described in Adler *et al.* (2001), are rather more complex in grassland systems than in moorlands or forests, where the distribution of the resource is discrete and more easily perceptible by the foraging animal. Fractal dimensioning proved to be a useful synthetic tool for identifying the scales of inter-patch and intra-patch movements.

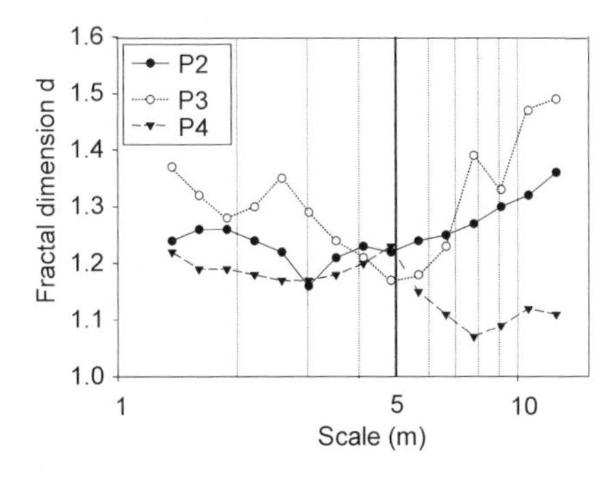

Figure 4 Evolution of the fractal dimension of foraging paths across spatial scales between 1 to 12 m in ewes grazing a cocksfoot sward managed at a low stocking rate (5 ewes on 3000 m^2) in May (P2), July (P2) and September (P4) (from Garcia *et al.*, 2005).

Consequences on vegetation dynamics

Few studies have documented the effects of large herbivores on the spatial heterogeneity of the grazed vegetation (Adler & Lauenroth, 2000) and the consequences on vegetation dynamics at different spatial and temporal scales (Parsons *et al.*, 2000). Repeated grazing of preferred patches and partial rejection of others leads to a bimodal frequency distribution of patch states in the plot (Parsons & Dumont, 2003). When grazing pressure is low, this means that large herbivores focus their grazing activity, only on a part of a pasture. A macro-heterogeneity, characterised by the coexistence of well grazed areas (low quantity, high quality) and partially-rejected areas (high quantity, low quality), will emerge. The spatial organisation of these areas could be influenced by the localisation of several attractive points such as water and sleeping areas.

Foraging behaviour determines the severity and frequency of defoliation on patches and thus the quality and quantity of the biomass resulting from the post-grazing regrowth. When animals regraze previously defoliated areas, they maintain the sward in a more juvenile and more digestible state (Donkor *et al.*, 2003). This, together with other possible mechanisms including a reduction of senescent material and an increase in below-ground available nitrogen, creates a positive feedback between grazing and forage quality (Adler *et al.*, 2001). This positive feedback promotes the continued use of previously grazed patches.

In many cases the reduction of growth is less than expected from the proportion of biomass removed, which means that the vegetation is able to develop a compensatory response to defoliation (Ferraro & Oestersheld, 2002). The different mechanisms involved in this compensatory response may be linked to plant environment (the decrease of self-shading), plant physiology (an increase of photosynthetic rate, the reallocation of growth from other parts of the plants, reduction of leaf senescence and greater light use efficiency) and morphogenetic adaptation (an activation and proliferation of axillary meristems: tillering and clonal development). The compensatory response increases with defoliation intensity, and a longer recuperation time after defoliation favours the occurrence of a compensatory response (Ferraro & Oestersheld, 2002). Garcia *et al.* (2003b) have shown that sheep graze patches at relatively low frequency but high severity, rather than the reverse.

While patch grazing may produce short-term positive feedbacks, changes in composition may cause negative feedbacks. Pastor *et al.* (1997) suggested that when the short-term increases in forage quality caused by grazing are outweighed by the compositional shift towards unpalatable or low nitrogen plant species, patch grazing cannot persist. This is more likely to occur in ecosystems where very distinct functional plant groups compete (i.e. grasses *vs.* shrubs). While there is evidence that grazing may influence plant diversity, it is not clear whether changes in spatial pattern drive this effect. At the patch scale, grazing may affect plant diversity by reducing local competition between species (Collins *et al.* 1998), but also through selective defoliation which creates an asymmetric competition for the preferred species. At a larger scale, these modifications may be caused by the uneven use of the grassland by grazing animals, an uneven distribution of excreta from grazing animals or an uneven dispersal of plant seeds through the faeces across a grassland (Shiyomi *et al.*, 1998).

A more functional approach which describes species from a functional rather than a taxonomic perspective should help to capture the long-term evolution of the grazed ecosystem (Lavorel *et al.*, 1997). The use of quantitative traits (measurable characteristics on individuals), to which continuous numeric values can be assigned, has recently been advocated (Lavorel & Garnier, 2002). In a pasture managed for the long-term with a gradient of grazing intensity, Louault *et al.* (2005) identified three important functional groups based on four significant traits: lamina dry matter ratio, specific leaf area, elongated plant height and the start of flowering. The first group corresponds to competitive species that are tolerant to grazing, the second to small-sized conservative species, which avoid being grazed, and the third to large conservative species. The first two groups coexist in well-grazed pasture, whereas the third is present in tall non-defoliated areas. This leads to the hypothesis that the structural heterogeneity created by grazing could modify the community process, and induce some persistent divergence in pasture diversity.

Conclusion

Over the last two decades, investigations of biting behaviour, diet selection and intake at pasture have led to great advances in the understanding of plant-animal relationships. Selective grazing tends to optimise diet quality at the different levels of feeding behaviour. However, most of the studies were conducted in simple experimental conditions – mono or bi-specific, vegetative or reproductive swards – and often on a short-term basis. In more complex situations like natural grassland of high diversity, predictions of diet selection, intake and the large herbivores' impact on the vegetation remain hazardous. The nutritive value – and thus animal performance – of a diet containing a high number of various plants is difficult to predict, as the digestive effects of forage associations are poorly understood. Forage diversity should stimulate intake (Ginane *et al.*, 2002b), but the respective roles of digestive and behavioural factors have yet to be established. As plant diversity increases, the ability of large herbivores to discriminate and make appropriate associations between plant characteristics and nutritional consequences should decline. Further studies need to be conducted to increase our understanding of the relative importance of pre- and post-ingestive cues in diet selection in complex situations. Integrative modelling linking intake and digestion should be further developed to improve the prediction of animal response to various types of pastures.

The development of a predictive understanding of diet selection in complex situations should allow a more effective use of herbivores as "landscape engineers". This implies extending our current knowledge to wider temporal and spatial scales, and integrating the related complexity. Modern techniques, for example associating GPS localisation and marker techniques to estimate diet composition, as proposed by Milne (2002), should provide deeper analysis of the relationship between plant diversity, vegetation heterogeneity and diet selection. Progress in modelling and computer science should allow the development of long-term and spatially-explicit models that can be usefully applied to simulate the effects of plants, animals and management characteristics (Baumont *et al.*, 2002).

References

Adler P.B. & W.K. Lauenroth (2000). Livestock exclusion increases the spatial heterogeneity of vegetation in Colorado shortgrass steppe. *Applied Vegetation Science*, 3, 213-222.

Adler P.B., D.A. Raff & W.K. Lauenroth (2001). The effect of grazing on the spatial heterogeneity of vegetation. *Oecologica*, 128, 465-479.

Allden, W. G. & I. A. Whittaker (1970). The determinants of herbage intake by grazing sheep: the interrelations of factors influencing herbage intake and availability. *Australian Journal of Agricultural Research,* 21, 755-766.

Baumont R., N. Séguier & J.P. Dulphy (1990). Rumen fill, forage palatability and alimentary behaviour in sheep. *Journal of Agricultural Science, Cambridge*, 115, 277-284.

Baumont, R., J.P. Dulphy & M. Jailler (1997). Dynamic of voluntary intake, feeding behaviour and rumen function in sheep fed three contrasting types of hay. *Annales de Zootechnie*, 46, 231-244.

Baumont R., A. Grasland & A. Détour (1999). Short-term preferences in sheep fed rye-grass as fresh forage, silage or hay. In: D. van der Heide (ed.) Regulation of feed intake, CAB International, Wallingford, 123-128.

Baumont, R., S. Prache, M. Meuret & P. Morand-Fehr (2000). How forage characteristics influence behaviour and intake in small ruminants: a review. *Livestock Production Science*, 64, 15-28.

Baumont, R., B. Dumont, P. Carrere, L. Perochon, C. Mazel, C. Force, S. Prache, F. Louault, J.F. Soussana, D. Hill & M. Petit (2002). Development of a spatial multi-agent model of a herd of ruminants grazing a heterogeneous grassland. *Rencontres Recherches Ruminants*, 9, 69-72.

Baumont R., D. Cohen-Salmon, S. Prache & D. Sauvant (2004). A mechanistic model of intake and grazing behaviour integrating sward architecture and animal decisions. *Animal Feed Science and Technology,* 112, 5-28.

Campion, D.P. & B.F. Leek (1997). Investigation of a fibre appetite in sheep fed a long fibre-free diet. *Applied Animal Behaviour Science*, 52, 79-86.

Collins S.L., A.K. Knapp & W.K. Lauenroth (1998). Modulation of diversity by grazing and mowing in native tallgrass prairies. *Science*, 2180, 745-747.

Cooper, S.D.B., I. Kyriasakis & J.V. Nolan (1995). Diet selection in sheep: the role of the rumen environment in the selection of a diet from two feeds that differ in their energy density. *British Journal of Nutrition*, 74, 39-54.

Delagarde, R., J.L. Peyraud, L. Delaby & P. Faverdin (2000). Vertical distribution of biomass, chemical composition and pepsin-cellulase digestibility in a perennial rye-grass sward: interaction with month of year, regrowth age and time of day. *Animal Feed Science and Technology*, 84, 49-68.

Distel, R.A., E.A. Laca, T.C. Griggs & M.W. Demment (1995). Patch selection by cattle: maximisation of intake rate in horizontally heterogeneous pastures. *Applied Animal Behaviour Science,* 45, 11-21.

Donkor, N.T., E.W. Bork & R.J. Hudson (2003). Defoliation regime effects on accumulated season-long herbage yield and quality in boreal grassland. *Journal of Agronomy and Crop Science*, 189, 39-46.

Dumont, B. & M. Petit (1998). Spatial memory of sheep at pasture. *Applied Animal Behaviour Science*, 60, 43-53.

Dumont, B., P. Carrère & P. D'Hour (2002). Foraging in patchy grasslands: diet selection by sheep and cattle is affected by the abundance and spatial distribution of preferred species. *Animal Research*, 51, 367-381.

Dumont, B., A. Dutronc & M. Petit (1998). How readily will sheep walk for a preferred forage? *Journal of Animal Science,* 76, 965-971.

Dumont, B., M. Petit & P. D'Hour (1995a). Choice of sheep and cattle between vegetative and reproductive cocksfoot patches. *Applied Animal Behaviour Science,* 43, 1-15.

Dumont, B., P. D'Hour & M. Petit (1995b). The usefulness of grazing tests for studying the ability of sheep and cattle to exploit reproductive patches of pastures. *Applied Animal Behaviour Science*, 45, 79-88.

Duncan, A.J., C. Ginane, I.J. Gordon, & E.R. Ørskov (2003). Why do herbivores select mixed diets? VIth International Symposium on the Nutrition of Herbivores (. L.'t Mannetje (ed.), Merida, Mexico, 195-209.

Edwards, G.R., J.A. Newman, A.J. Parsons & J.R. Krebs (1996). The use of spatial memory by grazing animals to locate food patches in spatially heterogeneous environments: an example with sheep. *Applied Animal Behaviour Science*, 50, 147-160.

Fedele, V., M. Pizillo, S. Claps, P. Morand-Fehr & R. Rubino (1993). Grazing behaviour and diet selection of goats on native pasture in Southern Italy. *Small Ruminant Research*, 11, 305-322.

Ferraro D.O. & M. Oesterheld (2002). Effect of defoliation on grass growth. A quantitative review. *Oikos*, 98, 125-133.

Foccardi, S.P., P. Marcellini & P. Montanaro (1996). Do ungulates exhibit a food density threshold? A field study of optimal foraging and movements patterns. *Journal of Animal Ecology*, 65, 606-620.

Fryxell, J.M. (1991). Forage quality and aggregation by large herbivores. *American Naturalist* 138, 477-498.

Garcia, F., P. Carrère, J.F. Soussana & Baumont, R. (2003a). The ability of sheep at different stocking rates to maintain the quality and quantity of their diet during the grazing season. *Journal of Agricultural Science, Cambridge,* 140, 113-124.

Garcia, F., P. Carrère, J.F. Soussana & Baumont, R. (2003b). How do severity and frequency of grazing affect sward characteristics and the choices of sheep during the grazing season? *Grass and Forage Science,* 58, 138-150.

Garcia, F., P. Carrère, J.F. Soussana & Baumont, R. (2005). Characterisation by fractal analysis of foraging paths of ewes grazing heterogeneous swards. *Applied Animal Behaviour Science,* in press.

Gekara, O.J., E.C. Prigge, W.B. Bryan, M. Schettini, E.L. Nestor, & E.C. Townsend (2001) Influence of pasture sward height and concentrate supplementation on intake, digestibility, and grazing time of lactating beef cows. *Journal of Animal Science,* 79, 745-752.

Gibb, M.J., C.A. Huckle, R. Nuthall & A.J. Rook (1999). The effect of physiological state (lactating or dry) and sward surface height on grazing behaviour and intake in dairy cows. *Applied Animal Behaviour Science,* 63, 269-287.

Ginane, C. & Petit, M. (2005). Constraining the time available to graze reinforces heifers' preference for sward of high quality despite low availability. *Applied Animal Behaviour Science* (accepted).

Ginane, C., B. Dumont, M. & Petit (2002a). Short-term choices of cattle vary with relative quality and accessibility of two hays according to an energy gain maximisation hypothesis. *Applied Animal Behaviour Science,* 75, 269-279.

Ginane, C., M. Petit & P. D'Hour (2003). How do grazing heifers choose between maturing reproductive and tall or short vegetative swards? *Applied Animal Behaviour Science,* 83, 15-27.

Ginane C., R. Baumont, J. Lassalas & M. Petit (2002b). Feeding behaviour and intake of heifers fed on hays of various quality, offered alone or in a choice situation. *Animal Research,* 51, 177-188.

Ginane, C., A.J. Duncan, S.A. Young, D.A. Elston & I.J. Gordon (2005). Herbivore diet selection in response to simulated variation in nutrient rewards and plant secondary compounds. *Animal Behaviour*, in press.

Griffiths W.M., J. Hodgson & G.C. Arnold (2003). The influence of sward canopy structure on foraging decision by grazing cattle. II. Regulation of bite depth. *Grass and Forage Science*, 58, 125-137.

Hester, A.J., I.J. Gordon G.J. Baillie & E. Tappin (1999). Foraging behaviour of sheep and red deer within natural heather/grass mosaics. *Journal of Applied Ecology,* 36, 133-146.
Hutchings, N.J. & I.J. Gordon (2001). A dynamic model of herbivore-plant interactions on grasslands. *Ecological Modelling,* 136, 209-222.
Illius, A. W. & I. J. Gordon (1987). The allometry of food intake in grazing ruminants. *Journal of Animal Ecology,*56, 989-999.
Illius, A.W. & I.J. Gordon (1991). Prediction of intake and digestion in ruminants by a model of rumen kinetics integrating animal size and plant characteristics. *Journal of Agricultural Science, Cambridge*, 116, 145-157.
Illius, A.W., I.J. Gordon, J.D. Milne & W. Wright (1995). Costs and benefits of foraging on grasses varying in canopy structure and resistance to defoliation. *Functional Ecology*, 9, 894-903.
Illius, A.W., I.J. Gordon, D.A. Elston & J.D. Milne (1999). Diet selection in goats: a test of intake-rate maximization. *Ecology,* 80, 1008-1018.
Illius, A.W. & J. Hodgson (1996). Progress in understanding the ecology and management of grazing systems. In J. Hodgson & A.W. Illius (eds) The ecology and management of grazing systems. CAB International, Wallingford, 429-458.
Kenney, P.A. & J.L. Black (1984). Factors affecting diet selection by sheep. I. Potential intake rate and acceptability of feed. *Australian Journal of Agricultural Research,*35, 551-563.
Lavorel S. & E. Garnier (2002). Predicting changes in community composition and ecosystem functioning from plant traits: revisiting the Holly Grail. *Functional Ecology*, 16, 545-556.
Lavorel S., S. McIntyre, J. Landsberg & T.D.A. Forbes (1997) Plant functional classifications: from general groups to specific groups based on response to disturbance. *Trends in Ecology and Evolution*, 12, 474-478.
Louault F., V.D. Pillar, J. Aufrère, E. Garnier & J.F. Soussana (2005). Plant traits and functional types in response to reduced disturbance in semi-natural grassland. *Journal of Vegetation Science* (in press)
Marell, A., J.P. Ball & A. Hofgaard (2002). Foraging and movement paths of female reindeer: insights from fractal analysis, correlated random walks, and Lévy flights. *Canadian Journal of Zoology*, 80, 854-865.
Marriott C. A. & P. Carrère (1998). Structure and dynamics of grazed vegetations. *Annales de Zootechnie*, 47, 359-369.
Milne J. (2002). Forage plant characteristics: how to meet animal requirements. In: J.L. Durand, J.C. Emile, Ch. Huyghe & G. Lemaire (eds.). Multi-function grasslands: quality forages, animal products and landscapes. Grassland Science in Europe, 7, 31-36.
Nams, V.O. (2005). Using animal movement paths to measure a response to spatial scales. *Oecologia*, in press.
Newman, J.A., A.J. Parsons, J.H.M. Thornley, P.D. Penning & J.R. Krebs (1995). Optimal diet selection by a generalist grazing herbivore. *Functional Ecology,* 9, 255-268.
Parsons A.J. & B. Dumont (2003). Spatial heterogeneity and grazing processes. *Animal Research*, 52, 161-179.
Parsons, A.J., J.H.M. Thornley, J.A. Newman & P.D. Penning (1994). A mechanistic model of some physical determinants of intake rate and diet selection in a two-species temperate grassland sward. *Functional Ecology*, 8, 187-204.
Parsons, A.J., P. Carrère & S. Schwinning (2000). Dynamics of heterogeneity in a grazed sward. In G. Lemaire, J. Hodgson, A. de Moraes, C. Nabinger & P.C.F. De Carvalho (eds) Grassland Ecophysiology and Grazing Ecology, CAB International, Wallingford, 289-315.
Pastor J., R. Moen & Y. Cohen (1997). Spatial heterogeneities, carrying capacity, and feedback in animal-landscape interactions. *Journal of Mammalogy*, 78, 1040-1052.
Penning, P.D., A.J. Parsons, R.J. Orr & T.T. Treatcher (1991). Intake and behaviour responses by sheep to changes in sward characteristics under continuous stocking. *Grass and Forage Science,* 46, 15-28.
Penning, P.D. & S.M. Rutter (2004). Ingestive behaviour. In P.D. Penning (ed.). Herbage intake handbook. The British Grassland Society, Reading, 151-176.
Prache, S., C. Roguet & M. Petit (1998). How degree of selectivity modifies foraging behaviour of dry ewes on reproductive compared to vegetative sward structure. *Applied Animal Behaviour Science*, 57, 91 – 108.
Prache, S. & J.L. Peyraud (2001). Foraging behaviour and intake in temperate cultivated grasslands. Proceedings of the XIXth International Grassland Congress, 309-319.
Provenza, F.D., (1995). Role of learning in food preferences of ruminants: Greenhalgh and Reid revisited. In: W.V.Engelhardt, S. Leonhard-Marek, G. Breves & D. Giesecke (eds) Ruminant physiology: digestion, metabolism, growth and reproduction. Ferdinand Enke Verlag, 233-247.
Roguet, C., S. Prache & M. Petit (1998). Feeding station behaviour of ewes in response to forage availability and sward phenological stage. *Applied Animal Behaviour Science,* 56, 187-201.
Rook, A.J., A. Harvey, A.J. Parsons, P.D. Penning & R.J. Orr (2002). Effect of long-term changes in relative resource availability on dietary preference of grazing sheep for perennial ryegrass and white clover. *Grass and Forage Science,* 57, 54-60.

Rook, A.J., C.A. Huckle & P.D. Penning (1994). Effects of sward height and concentrate supplementation on the ingestive behaviour of spring-calving dairy cows grazing grass-clover swards. *Applied Animal Behaviour Science,* 40, 101-112.

Rutter S.M., R.J. Orr, N.H. Yarrow & R.A Champion (2004). Dietary preference of dairy heifers grazing ryegrass and white clover, with and without an anti-bloat treatment. *Applied Animal Behaviour Science*, 85, 1-10.

Sauvant, D., R. Baumont & P. Faverdin (1996). Development of a mechanistic model of intake and chewing activities of sheep. *Journal of Animal Science*, 74, 2785 – 2802.

Senft, R.L., M.B. Coughenour, D.W.;Bailey, L.R. Rittenhouse, O.E. Sala & D.M. Swift (1987). Large herbivore foraging and ecological hierarchies. *Bioscience,* 37, 789-799.

Shipley, L.A., D.E. Spalinger, J.E. Gross, N.T. Hobbs & B.A. Wunder (1996). The dynamics and scaling of foraging velocity and encounter rate in mammalian herbivores. *Functional Ecology*, 10, 234-244.

Shiyomi, M., M. Okada, S. Takahashi & Y. Tang (1998). Spatial pattern changes in aboveground plant biomass in a grazing pasture. *Ecological research*, 13, 313-322.

Stephens, D.W. & J.R. Krebs (1986). Foraging Theory. Princeton University Press, Princeton, New Jersey.

Taweel, H.Z., B.M. Tas, J. Dijkstra & S. Tamminga (2004). Intake regulation and grazing behavior of dairy cows under continuous stocking. *Journal of Dairy Science,* 87, 3417-3427.

Tolkamp, B.J., G.C. Emmans, J. Yearsley & I. Kyriazakis (2002). Optimization of short-term animal behaviour and the currency of time. *Animal Behaviour*, 64, 945-953.

Wallis de Vries M.F. & C. Daleboudt (1994). Foraging strategy of cattle in patchy grassland. *Oecologia*, 100, 98-106.

Wang, J. & F.D. Provenza (1996). Food preference and acceptance of novel foods by lambs depend on the composition of the basal diet. *Journal of Animal Science*, 74, 2349-2354.

Ward, D. & D. Saltz (1994). Foraging at different spatial scales; Dorcas gazelles foraging for lilies in the Negev desert. *Ecology*, 75, 48-58.

Wilmshurst, J.F., J.M. Fryxell & R.J. Hudson (1995). Forage quality and patch choice by wapiti (cervus elaphus). *Behavioral Ecology,* 6, 209-217.

Land use history and the build-up and decline of species richness in Scandinavian semi-natural grasslands

O. Eriksson, S.A.O. Cousins and R. Lindborg
Department of Botany, Stockholm University, SE – 106 91 Stockholm, Sweden, Email: ove.eriksson@botan.su.se

Abstract

Scandinavian semi-natural grasslands have an exceptionally high small-scale species richness. In the past, these grasslands covered extensive areas but they have declined drastically during the last century. How species richness of semi-natural grasslands was built up during history, and how species respond to land use change, are discussed. The agricultural expansion from the late Iron Age was associated with increasing grassland extent and spatial predictability, resulting in accumulation of species at small spatial scales. Although few species directly depend on management, the specific composition of these grasslands is a product of hay-making and grazing. Grassland fragmentation initially has small effects on species richness, due to slow extinction of many species. Species loss in grasslands is, however, expected in the coming decades. Restoration efforts may fail due to slow colonization. Effects of landscape configuration may be overlooked, if land use history is not considered, since present-day species richness largely reflects landscape history.

Keywords: colonization, extinction, grassland management, landscape history, regional species dynamics, restoration, species density

Introduction

Changes in habitat conditions and in the occurrence and distribution of habitats in landscapes are among the most important drivers behind species decline and loss at present as well as in future scenarios (Sala *et al.*, 2000). Landscape changes are complex and consist of an array of different processes involving climate, chemical and physical properties of the environment, and changing networks of interactions among organisms. For habitats that are declining, the process is often described as fragmentation, which has become a core issue for research in spatial ecology (e.g. Hanski, 1999) and a primary concern for conservation biology (e.g. Meffe *et al.*, 1997). Fragmentation is manifested both as a decreasing area *per se*, thus increasing the area of surrounding landscape matrix, and an increasing isolation among remaining habitat patches. In addition, fragmentation is usually also associated with other changes, such as deterioration of local habitat conditions and edge effects (Harrison & Bruna, 1999).

Despite the recognition that ecological patterns and processes exhibit scaling relationships in both time and space (e.g. Levin, 1992), and despite extensive studies of, for example, vegetation succession (e.g. Connell & Slatyer, 1977) and effects of wildfires in forests (e.g. Zackrisson, 1977), the temporal dimension of changing landscape structure has not received as much attention as spatial phenomena such as effects of area, isolation, corridors, and edges. However, there is growing evidence that landscape history is of great importance for patterns of species distribution and abundance (e.g. Peterken & Game, 1984; Foster, 1993, 2002; Austrheim *et al.*, 1999; Motzkin *et al.*, 1999; Bruun *et al.*, 2001; Cousins & Eriksson, 2002; Bellemare *et al.*, 2002; Eriksson *et al.*, 2002; Poschlod & WallisDeVries, 2002; Lindborg & Eriksson, 2004). Much of this evidence comes from studies on habitats formed or greatly influenced by human management, such as hay-meadows, pastures and managed woodland, or habitats that have developed after abandonment of management.

In temperate and nemo-boreal regions in Europe, the influence of land use is ubiquitous, extending back several millennia. The remnants of areas with such a long-lasting impact of "traditional" management are often associated with a high biological diversity. In this chapter we will focus on one example of such remnants of traditional management, semi-natural grasslands in Scandinavia. These grasslands contain an astonishing plant species richness at small spatial scales. It is not unusual to find 40-60 (and sometimes even more) different plant species per m^2 (Kull & Zobel, 1991; Eriksson & Eriksson, 1997; Austrheim *et al.*, 1999; Eriksson *et al.*, 2004). As these species-rich grasslands often lie embedded in a heterogeneous landscape, with many types of habitats, e.g. ditches, road verges, stone walls, midfield islets, ponds and forest margins, species diversity at the landscape scale is high, not only for plants, but for organisms such as insects and birds (e.g. Weibull *et al.*, 2000; Söderström *et al.*, 2001). Due to their biological value, semi-natural grasslands are an important concern for conservation programmes in Scandinavia. There are also several other reasons for these grasslands to be maintained, e.g. their aesthetic and cultural values, and their importance for attracting visitors. Moreover, traditional management of grasslands may exploit a developing market for locally produced milk and meat, thus contributing to economic sustainability of pastoral systems in parts of Scandinavia where agriculture is generally declining.

The objectives of this chapter are to (i) present a general hypothesis on how plant species richness of semi-natural grasslands was built up during the period of agricultural management from Late Iron Age up to the modernisation of agriculture, initiated during late 19th century, (ii) present an overview of how plant species richness responds to the ongoing fragmentation and deterioration of remaining semi-natural grasslands, and (iii) discuss some implications for conservation biology of considering land use history in analyses of species richness. The main arguments are based on our own studies performed in semi-natural grasslands in Scandinavia, and we thus primarily restrict the conclusions to this geographical region. However, although it was beyond the scope of this chapter to generalise across broader geographical areas, we consider the suggested hypotheses applicable also to other parts of northern Europe.

The historical build-up of species richness in semi-natural grasslands

The suggested hypothesis on the build-up of the high species richness characteristic of semi-natural grasslands are based on three premises: (a) there existed a large species pool of grassland species in the pre-agricultural landscape, (b) due to dispersal limitation, the colonisation of grassland sites is generally slow, and (c) the managed grasslands are generally open communities, i.e. provided eventual dispersal to a site, and recruitment is the likely outcome.

The first premise, which has been much discussed recently, concerns the origin of the plant species inhabiting semi-natural grasslands, i.e. what were the "natural" habitats for these species before the development of a permanent agriculture, with pastures and infield meadows. Today, no grassland habitat without a long-term history of management has a small-scale species density even close to what is found in semi-natural grasslands. For example, former fields which are presently grazed seldom contain more than 5-15 species per m^2, i.e. less than a third of what is common for semi-natural grasslands. Semi-open forests and woodlands developed on former grasslands that were abandoned during the last century have an even lower density of species. Moreover, approximately 50% of the plant species in semi-natural grasslands are seldom found in habitats which have not been subjected to management in the present-day landscape (Cousins & Eriksson, 2001). It might be tempting to interpret this as an indication of a close association between agricultural management and the migration (or evolution) of species into regions where agriculture developed. However, as discussed in more detail by Eriksson *et al.* (2002), it is likely that the pool of species currently inhabiting semi-natural grasslands already occurred in the pre-agricultural landscape. The previously common notion that this landscape was covered with dense forests (i.e. excluding grassland plants) has been challenged by several authors (e.g. Vera, 2000; Svenning, 2002) who have suggested that there were open habitats harbouring grassland species. Moreover, the time span over which agriculture has existed is short (in the magnitude of some 10^3 years) compared to the average life span (in the magnitude of 10^5-10^6 years) of plant species (Eriksson *et al.*, 2002). Thus, we propose that the species that colonised human-made habitats such as pastures and meadows, existed in the pre-agricultural landscape in marginal habitats such a shores, river-banks, dry grasslands or temporary open sites created by wildfires, or natural grazing.

The second premise concerns the now well-supported notion that plants are generally dispersal-limited (Turnbull *et al.*, 2000; Eriksson & Ehrlén, 2001). Dispersal limitation implies that potentially suitable sites are not necessarily inhabited by all species that may live there. If sufficiently strong, dispersal limitation may cause distribution patterns where occupied and unoccupied sites differ in age. An example is *Thymus serpyllum*, a species that in south-central Sweden has been found positively associated with Iron Age grave fields (Eriksson, 1998). These grave fields are often covered with dry grassland and, since they were generally located in the vicinity of farms or villages, which still basically occur at the same sites, they are likely to have been influenced by management for a long time, probably over a millennium. In contrast, other dry grasslands, which by experimentation were revealed to possess the same qualities for *T. serpyllum* although they often had not harboured natural populations of this species, are likely to be of a more recent origin. This example suggests that dispersal limitation may cause delays in colonisation in the magnitude of centuries. Although *T. serpyllum* is perhaps exceptional regarding the time-scale of colonization, dispersal limitation of plants is ubiquitous in the nemo-boreal landscape. Figure 1 illustrates that regional populations of many investigated plant species in our study area have a large fraction of suitable but unoccupied sites, and that this fraction is not related (linear regression; $r = 0.67$) to the general abundance of the species in the landscape, assessed as the fraction of occupied sites. Thus, dispersal limitation is not something particular for just rare species. A corollary of dispersal limitation is that landscape configuration, basically the size of grassland sites, and their connectivity in the landscape, will influence the degree of dispersal limitation (Eriksson & Ehrlén, 2001; Lindborg & Eriksson 2004). Changing landscape configuration, e.g. increasing grassland area and connectivity, will reduce limitations to dispersal.

The third premise concerns the fate of species after they have dispersed to a potentially suitable grassland site. A common assumption in ecology is that plant communities are "niche-structured" (cf. Hubbell, 2001) implying that colonising species are hindered by the presence of already established species such that increasing species richness would retard colonization. Some experiments performed in grasslands elsewhere support this idea (e.g. Naeem *et al.*, 2000). However, experiments conducted in species-rich Scandinavian grasslands (with over 40 species per m^2) suggest that these communities are basically open to colonisation, and that removal of functional groups has only slight effects on colonisation (Eriksson *et al.*, 2004). This means that the composition of these grasslands may largely reflect the propagule pressure, i.e. the inflow of seed (or generally, diaspores) arriving at the sites.

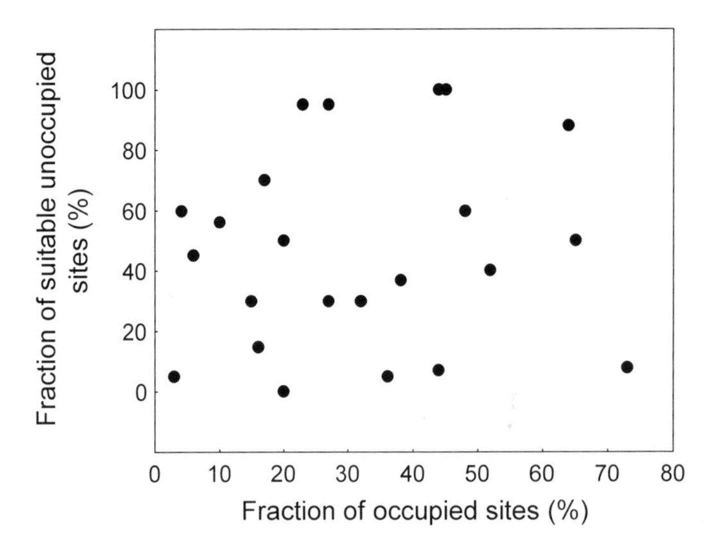

Figure 1 Dispersal limitation in 23 plant species inhabiting semi-natural grasslands or remnant habitats developed after abandonment of grasslands in the province of Södermanland, Sweden. The fraction of suitable unoccupied sites was estimated as the fraction of the unoccupied sites where suitability was detected by recruitment after experimental sowing. The fraction of occupied sites was estimated by surveys of sites subjectively determined as potentially suitable. This fraction is an approximation of abundance of species on a landscape scale. Data are from various sources: Eriksson & Kiviniemi (1999), Kiviniemi & Eriksson (1999), Ehrlén & Eriksson (2000) and O. Eriksson (unpublished data).

Combining these three premises, we suggest that the high species richness of Scandinavian semi-natural grasslands is due to an historical accumulation of species, manifested at small spatial scales. Even such small areas as a few square metres may in fact harbour the majority of species occurring at a semi-natural grassland site covering several hectares (Eriksson & Eriksson, 1997). The development of permanent farms and villages during the late Iron Age and the early Viking Age, which is still reflected in the farm structure in Scandinavia today, implied an increasing area and connectivity of grasslands, and an increase in their spatial predictability. These structural changes in the grassland configuration promoted colonization and establishment of species at existing sites. Over time, in this case a period of over a millennium, an exceptional local-scale species richness developed. Removal of biomass, due to hay-making and grazing (especially if manure was transferred to arable fields) is likely to have reduced productivity, possibly further promoting the invasibility of these grassland communities. Moreover, in the traditional agricultural landscape, seed dispersal was likely aided by moving livestock, hay and equipment (Poschlod & Bonn, 1998), acting to further reduce the effects of dispersal limitation. Thus, a characteristic feature of the semi-natural grassland plant communities in the historical landscape was the ubiquitous presence of most species at even small spatial scales. Although the specific composition and diversity of the plant communities are products of management, individual species now inhabiting semi-natural grasslands may not initially have been migrating to the landscape as a result of the development of permanent agricultural management (such as infield meadows).

This hypothesis for the build-up of species richness during the history of the agricultural landscape in Scandinavia is concordant with the so called "species-pool hypothesis" (Zobel, 1992; Pärtel *et al.*, 1996; Franzén & Eriksson, 2001). It implies that we should expect that the maximum of local species richness was reached during the later phases of the traditional agriculture, i.e. in most of Sweden at the end of 19th century, when it is possible that the limit to local species richness was actually the size of the regional species pool. This prediction is unfortunately difficult to test due to the lack of time-series of species richness. However, the prediction can be expressed as an expected positive relationship between management continuity and local species richness. Indeed, such relationships are found in studies using historical landscape information to assess grassland continuity (Cousins & Eriksson, 2002).

From the early 20th century the area of semi-natural grasslands in Sweden has declined by at least 90%. With the introduction of ley production on fields subjected to the application of artificial fertilisers, semi-natural meadows disappeared rapidly. Remaining semi-natural grasslands were used for grazing. This was also the case for the outland, i.e. grazing outside the infields, a practice that was maintained until the 1940s. Since outland grazing is usually not included in the estimated 90% reduction, the actual loss of grasslands during the last century is probably even higher. After the 1940s a second phase of modernisation of agriculture and forestry resulted in further loss of semi-natural grasslands. During the last decades programmes for restoration of semi-natural grasslands have been initiated, usually by introducing grazing by cattle or sheep. A national goal for biodiversity in Sweden states that no further reduction in semi-natural grassland area should be accepted. However, the economic situation for many farmers, despite subsidies, implies that it is far from certain that this goal will be reached (K. I. Kumm, pers. comm.).

The response of species richness to grassland fragmentation

As a result of the changes in agricultural and forestry practices, the present-day distribution of semi-natural grasslands represents only fragments of the historical distribution (Figure 2). In order to describe the response of plant species to the fragmentation of semi-natural grasslands, we have to add a fourth premise to the three premises used for developing the hypothesis of the build-up of grassland species richness: local extinction rates of established populations are generally slow for perennial species (which constitute a majority of the grassland flora). Results from population matrix simulations and analyses of distribution patterns of selected species suggest that many species may resist extinction for up to 50-100 years after abandonment of grazing (Eriksson & Ehrlén, 2001). This means that the observed distribution of grassland species in a changing landscape may reflect historical habitat conditions rather than present-day conditions. If historical land use is overlooked in spatial analyses, there is a risk of erroneous conclusions about the importance of spatial effects (area and isolation) on species richness. Lindborg & Eriksson (2004) analysed 30 semi-natural grassland sites with respect to relationships between small-scale species richness and grassland connectivity in the present-day landscape, and in the landscapes c. 50 and 100 years ago. There were no such relationships in the present-day landscape, suggesting an apparent lack of spatial landscape effects. However, for both the 50- and 100-year old landscapes grassland connectivity was linked to species richness. The best model was actually for the oldest (100-year old) landscape, where 57% of the variation in present-day species richness of the target grasslands was explained by variation in grassland connectivity within a 2 km radius from the target site.

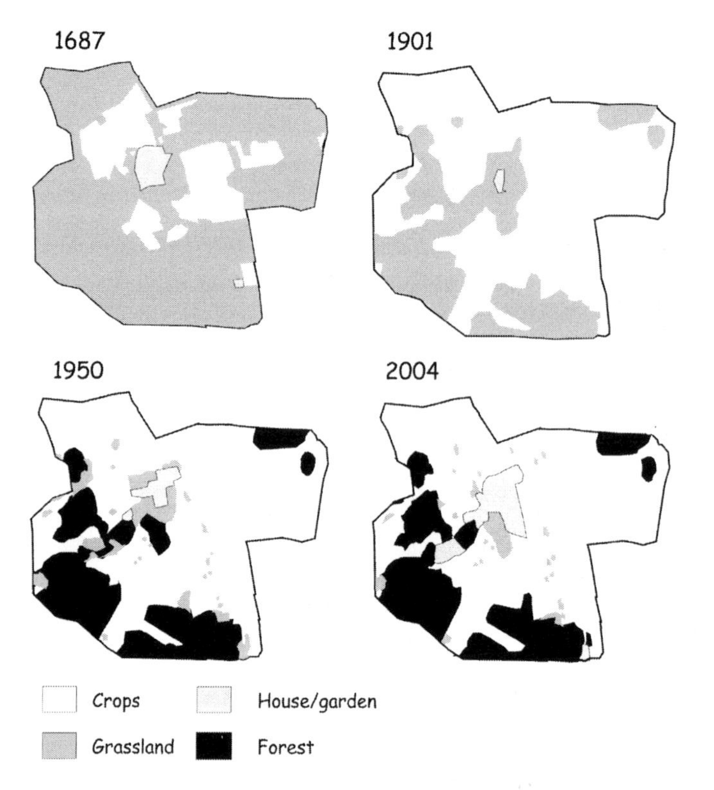

Figure 2 An example of grassland fragmentation. The distribution of semi-natural grasslands in Tösta, Sweden, from 1687 to 2004. The size of the area is c. 2 km^2.

The loss of species after abandonment of management occurs at a different pace for different groups of plants. Short-lived species decline relatively rapidly, and due to the existing isolation of grasslands, spontaneous colonisations balancing local extinctions are unlikely. Although good time-series are scarce, there is reasonably good evidence that typical grassland species, such as the biennial *Gentianella campestris*, decline rapidly as a result of fragmentation of the grasslands (Lennartsson & Svensson, 1996). In a study area in the province of Södermanland, 32 % of populations of *G. campestris* recorded 15 years ago had vanished (O. Eriksson, unpublished data). For long-lived species, in contrast, we still may not actually observe the decline, due to the presence of decreasing but still existing remnant populations. On a landscape scale, the "equilibrium" species richness in the modern landscape may therefore be much lower than the observed species richness. In other words, there may be an extinction debt (Tilman *et al.*, 1994; Hanski & Ovaskainen, 2002). Analyses of land-use change during the last century, combined with population monitoring, will be the best available tool to assess the time scale of these extinction processes. An additional process that is likely to influence species richness in remaining grasslands is colonisation of grassland generalists that dominate the landscape matrix surrounding semi-natural grasslands (Kiviniemi & Eriksson, 2002). This process has not yet been subject to proper study, but it may be that species with a large growth potential under conditions of increasing eutrophication slowly replace species that are less competitive under such conditions. This may cause further species loss even if management is maintained in the semi-natural grasslands.

If species, now considered specialised to semi-natural grasslands, did exist in the pre-agricultural landscape in temporary open habitats, we should ask whether the same distribution could develop if the semi-natural grasslands disappear. Will there be a "dilution process" mirroring the accumulation process that, according to the hypothesis described above, built up the exceptional species richness in semi-natural grasslands? One reason for *not* expecting this to happen is that the habitats that previously harboured grassland species no longer exist, or have been altered (e.g. by eutrophication) to such an extent that they are not any more suitable for most semi-natural grassland plants. On the other hand, sites with grassland conditions can be maintained in environments that in themselves are new, e.g. road verges provide vast areas of mown grassland, although linear in structure and often likely to be influenced by nitrogen input from the traffic. Many typical semi-natural grassland plants are indeed recorded along roads (Eriksson & Kiviniemi, 1999; Cousins & Eriksson, 2001; S. A. O. Cousins, unpublished data).

Implications for conservation biology

We can conclude that studies combining land use history and plant ecology are useful for enhancing understanding of the development of species richness in the traditional agricultural landscape, as well as for analyses of the response of species to ongoing landscape change. We believe there are four specific advantages of incorporating land use history as an aspect of research programmes aimed to promote conservation of semi-natural grasslands.

Knowledge of extinction time-lags improves interpretations of status of declining species

As illustrated by Lindborg & Eriksson (2004), evidence suggests that historical grassland connectivity has a strong effect on present-day patterns of species richness. One mechanism behind this effect is likely to be extinction time-lags, i.e. the inherent capacity of long-lived species to persist at sites despite a negative population growth rate. Also from other regions than Scandinavia, studies of plant species distributions provide indirect evidence of extinction time-lags after local habitat change (e.g. Koerner *et al.*, 1997; Motzkin *et al.*, 1999, 2002; Bellemare *et al.*, 2002; Dupouey *et al.*, 2002). The most important implication of these results is that risk assessments of endangered species associated with traditional agricultural landscapes may fail to recognize the actual status of long-lived species. Monitoring programmes are needed to evaluate the population development of putative remnant populations in order to conclude whether local hotspots of species richness, occurring isolated in a transformed modern landscape, really are sustainable. If such population studies conclude that this is not the case, the temporal dimension of the extinction time lags (assessed from population simulations, or descriptive landscape studies) may provide a time-frame for necessary conservation action.

Knowledge of colonization time lags may serve as basis for decisions on species introductions

In the same way as for extinction time lags, the colonisation time lags inferred from dispersal limitation (e.g. Ehrlén & Eriksson, 2000; Turnbull *et al.*, 2000) contribute to the phenomenon that present-day species richness is not able to catch up with rapid landscape change. This means that restoration programmes focussing on introducing traditional management, for example grazing or mowing on sites where management have been abandoned previously, may fail simply because dispersal limitation is sufficiently strong to hinder spontaneous re-colonisation (at least within a reasonable time-frame). Historical analyses provide an assessment of such a time-frame for dispersal limitation of species, i.e. the time we expect until spontaneous re-colonizations may occur. In turn, coupled with population studies of remaining populations, these assessments provide a basis for evaluating the need for artificial introductions (or re-introductions) of threatened species (van Groenendael *et al.* 1998). Historical landscape analyses may also be a guide for the choice of sites useful for restoration, based on the assumption that a long history of previous management at a site promotes the likelihood of restoration success.

Knowledge of land use history improves assessments of spatial habitat configuration effects

If the distribution of species is far from equilibrium with the present habitat distribution, due to historical effects, an analysis of effects of landscape habitat configuration may yield strongly misleading results. For example, studies that fail to find present-day species-area relationships in fragmented habitats (Eriksson *et al.*, 1995; Kiviniemi & Eriksson, 2002) may underpin false conclusions that landscape structure does not influence species occurrences, and thus provide erroneous recommendations for conservation planning. If historical, but not present-day, landscape effects on species richness are documented, a corollary is that we should expect a continuing long-term decline of species, even if the present-day landscape is maintained. Moreover, landscape effects on local species richness are relevant for the question of whether conservation programmes should focus on a site-scale (i.e. single target sites of traditionally-managed grassland) or at a landscape scale, i.e. take into account also the qualities of the landscape surrounding the target sites. The results from our studies suggest that conservation programmes should be orientated towards a landscape scale.

Knowledge of land-use history improves assessment of combined biological and cultural values of landscapes, and strengthens public support for conservation plans

European conservation biologists have for long recognized that many of the most diverse (and highly valued) habitats are products of human management (e.g. hay-meadows, semi-natural pastures, coppiced woodland), and this view is gaining increased attention in North America (e.g. Foster 2002; Foster *et al.* 2002) and Australia (e.g. Yibarbuk *et al.* 2001; McIntyre 2001). Knowledge of the impact of historical land use for present-day biodiversity, in itself, does not provide guidelines for assigning values to species, habitats or landscapes. However, knowledge of land-use history is helpful in developing historical models for conservation programmes focusing on managed habitats, e.g. hay meadows and pastures. In addition, cultural heritage may add to the perceived values of habitats that are primarily given priorities due to biological reasons, thus promoting the acceptance for conservation actions. Studies in Sweden (Stenseke, 2001; 2004) suggest that cultural values, including the insight that landscapes are shaped by human activities during centuries or millennia, are important factors for the valuation of landscapes, both by stakeholders, residents and visitors. Appreciation of land-use history may also contribute to support traditional cultures and resolve possible conflicts in areas previously regarded as “wilderness”, for example land use by Australian Aborigines (Yibarbuk *et al.*, 2001), and reindeer grazing as a part of traditional management of the Sami culture in the Scandinavian mountains (e.g Austrheim & Eriksson, 2001).

Conclusions

Land-use history of Scandinavian agricultural landscapes is an essential mechanism behind the development of the present-day plant species richness in semi-natural grasslands. Although human impact during the last millennia may not have been the instrinsic driver behind migration of grassland species to Scandinavia, due to the probable existence of grassland habitats in the pre-agricultural landscape, management, such as grazing and mowing, have shaped the composition and small-scale diversity of the plant communities. Today, when concern of species loss is in focus, historical studies contribute important information for analyses of species response to land use change, particularly when estimating time-lags of species colonisation and extinction, and for the assessment of the influence of landscape on local-scale species richness. Moreover, knowledge of land-use history promotes necessary cross-disciplinary research on the valuation of traditional grasslands, and such knowledge enhances public support for conservation programmes.

Acknowledgements

We are grateful to M. Öster for comments on the manuscript, and to the Mistra programme "Management of semi-natural grasslands: economy and biodiversity" and the Swedish research council for environment, agricultural sciences and spatial planning (projects to OE and SC) for financial support.

References

Austrheim, G. & O. Eriksson (2001). Plant species diversity and grazing in the Scandinavian mountains – patterns and processes at different spatial scales. *Ecography*, 24, 683-695.

Austrheim, G., E. G. A. Olsson & E. Grøntvedt (1999). Land-use impact on plant communities in semi-natural sub-alpine grasslands of Budalen, central Norway. *Biological Conservation*, 87, 369-379.

Bellemare, J., G. Motzkin, & D. Foster (2002). Legacies of the agricultural past in the forested present: an assessment of historical land-use effects on rich mesic forests. *Journal of Biogeography*, 29, 1401-1420.

Bruun, H. H., B. Fritzbøger, P. O. Rindel & U. L. Hansen (2001). Plant species richness in grasslands: the relative importance of contemporary environment and land-use history since the Iron Age. *Ecography*, 24, 569-578.

Connell, J. H. & R. O. Slatyer (1977). Mechanisms of succession in natural communities and their role in community stability and organization. *American Naturalist*, 111, 1119-1144.

Cousins, S. A. O. & O. Eriksson (2001). Plant species occurrences in a rural hemiboreal landscape: effects of remnant habitats, site history, topography and soil. *Ecography*, 24, 461-469.

Cousins, S. A. O. & O. Eriksson (2002). The influence of management history and habitat on plant species richness in a rural hemiboreal landscape, Sweden. *Landscape Ecology*, 17, 517-529.

Dupouey, J. L., E. Dambrine, J. D. Laffite & C. Moares (2002). Irreversible impact of past land use on forest soils and biodiversity. *Ecology*, 83, 2978-2984.

Ehrlén, J. & O. Eriksson (2000). Dispersal limitation and patch occupancy in forest herbs. *Ecology*, 81, 1667-1674.

Eriksson, Å. (1998). Regional distribution of *Thymus serpyllum*: management history and dispersal limitation. *Ecography*, 21, 35-43.

Eriksson, Å. & O. Eriksson (1997). Seedling recruitment in semi-natural pastures: the effects of disturbance, seed size, phenology and seed bank. *Nordic Journal of Botany*, 17, 469-482.

Eriksson, Å., O. Eriksson & H. Berglund (1995). Species abundance patterns of plants in Swedish semi-natural pastures. *Ecography*, 18, 310-317.

Eriksson, O. & J. Ehrlén (2001). Landscape fragmentation and the viability of plant populations. In: J. Silvertown & J. Antonovics (eds) Integrating ecology and evolution in a spatial context. Blackwell, Oxford, 157-175.

Eriksson, O. & K. Kiviniemi (1999). Site occupancy, recruitment and extinction thresholds in grassland plants: an experimental study. *Biological Conservation*, 87, 319-325.

Eriksson, O., S. A. O. Cousins & H. H. Bruun (2002). Land-use history and fragmentation of traditionally managed grasslands in Scandinavia. *Journal of Vegetation Science*, 13, 743-748.

Eriksson, O., S. Wikström, Å. Eriksson & R. Lindborg (2004). Species-rich Scandinavian grasslands are inherently open to invasion. *Biological Invasions* (in press).

Foster, D. R. (1993). Land-use history (1730-1990) and vegetation dynamics in central New England, USA. *Journal of Ecology*, 80, 753-772.

Foster, D. R. (2002). Thoreau´s country: a historical-ecological perspective on conservation in the New England landscape. *Journal of Biogeography*, 29, 1537-1555.

Foster, D. R., B. Hall, S. Barry, S. Clayden & T. Parshall (2002). Cultural, environmental and historical controls of vegetation patterns and the modern conservation setting on the island of Martha´s Vineyard, USA. *Journal of Biogeography*, 29, 1381-1400.

Franzén, D. & O. Eriksson (2001). Small-scale patterns of species richness in Swedish semi-natural grasslands: the effects of community species pools. *Ecography*, 24, 505-510.

Hanski, I. (1999). Metapopulation ecology. Oxford University Press, Oxford, 313pp.

Hanski, I. & O. Ovaskainen (2002). Extinction debt at extinction threshold. *Conservation Biology*, 16, 666-673.
Harrison, S. & E. Bruna (1999). Habitat fragmentation and large-scale conservation: what do we know for sure? *Ecography*, 22, 225-232.
Hubbell, S. P. (2001). The unified neutral theory of biodiversity and biogeography. Princeton University Press, Princeton, 375pp.
Kiviniemi, K. & O. Eriksson (1999). Dispersal, recruitment and site occupancy of grassland plants in fragmented habitats. *Oikos*, 86, 241-253.
Kiviniemi, K. & O. Eriksson (2002). Size-related deterioration of semi-natural grassland fragments in Sweden. *Diversity and Distributions*, 8, 21-29.
Koerner, W., J. L. Dupouey, E. Dambrine & M. Benoit (1997). Influence of past land use on the vegetation and soils of present day forest in the Vosges mountains, France. *Journal of Ecology*, 85, 351-358.
Kull, K. & M. Zobel (1991). High species richness in an Estonian wooded meadow. *Journal of Vegetation Science*, 2, 711-714.
Lennartsson, T. & R. Svensson (1996). Patterns in the decline of three species of *Gentianella* (Gentianaceae) in Sweden, illustrating the deterioration of semi-natural grasslands. *Symbolae Botanicae Upsaliensis*, 31, 169-184.
Levin, S. A. (1992). The problem of pattern and scale in ecology. *Ecology*, 73, 1943-1967.
Lindborg, R. & O. Eriksson (2004). Historical landscape connectivity affects present plant species diversity. *Ecology*, 85, 1840-1845.
McIntyre, S. (2001). Biophysical and human influences on plant species richness in grasslands: comparing variegated landscapes in subtropical and temperate regions. *Austral Ecology*, 26, 233-245.
Meffe, G. K., C. R. Carroll and contributors (1997). Principles of conservation biology, 2nd edition Sinauer, Sunderland, 729pp.
Motzkin, G., P. Wilson, D. R. Foster & A. Allen (1999). Vegetation patterns in heterogeneous landscapes: the importance of history and environment. *Journal of Vegetation Science*, 10, 903-920.
Motzkin, G., R. Eberhardt, B. Hall, D. R. Foster, J. Harrod & D. MacDonald (2002). Vegetation variation across Cape Cod, Massachusetts: environmental and historical determinants. *Journal of Biogeography*, 29, 1439-1454.
Naeem, S., J. M. H. Knops, D. Tilman, K. M. Howe, T. Kennedy & S. Gale (2000). Plant diversity increases resistance to invasion in the absence of covarying extrinsic factors. *Oikos*, 91, 97-108.
Pärtel, M. M. Moora & M. Zobel (1996). Variation in species richness within and between calcareous (alvar) grassland stands: the role of core and satellite species. *Plant Ecology*, 157, 205-213.
Peterken, G. F. & M. Game (1984). Historical factors affecting the number and distribution of vascular plant species in the woodlands of central Lincolnshire. *Journal of Ecology*, 72, 155-182.
Poschlod, P. & S. Bonn (1998). Changing dispersal processes in the central European landscape since the last ice age: an explanation for the actual decrease of plant species richness in different habitats? *Acta Botanica Neerlandica*, 47, 27-44.
Poschlod, P. & M. F. WallisDeVries (2002). The historical and socioeconomic perspectives of calcareaous grasslands – lessons from the distant and recent past. *Biological Conservation*, 104, 361-376.
Sala, O. E. & 18 Coauthors (2000). Global biodiversity scenarios for the year 2100. *Science*, 287, 1770-1774.
Söderström, B., B. Svensson, K. Vessby & A. Glimskär (2001). Plants, insects and birds in semi-natural pastures in relation to local habitat and landscape factors. *Biodiversity and Conservation*, 10, 1839-1863.
Stenseke, M. (2001). Landskapets värden: lokala perspektiv och centrala utgångspunkter. *Choros*, 2001:1, 3-106. (In Swedish)
Stenseke, M. (2004). Bönder och naturbetesmarker. Del 1: Bygdeperspektiv. *Choros*, 2004:1, 1-88. (In Swedish)
Svenning, J.-C. (2002). A review of natural vegetation openness in Northwestern Europe. *Biological Conservation*, 104, 133-148.
Tilman, D., R. M. May, C. L. Lehman & M. A. Nowak (1994). Habitat descruction and the extinction debt. *Nature*, 371, 65-66.
Turnbull, L. A., M. J. Crawley & M. Rees (2000). Are plant populations seed-limited? A review of seed sowing experiments. *Oikos*, 88, 225-238.
van Groenendael J., N. J. Ouborg & R. J. J. Hendriks (1998). Criteria for the introduction of plant species. *Acta Botanica Neerlandica*, 47, 3-13.
Vera, F. W. M. (2000). Grazing ecology and forest history. CAB International, Wallingford, 506pp.
Weibull, A.-C., J. Bengtsson & E. Nohlgren (2000). Diversity of butterflies in the agricultural landscape: the role of farming system and landscape heterogeneity. *Ecography*, 23, 743-750.
Yibarbuk, D., P. J. Whitehead, J. Russell-Smith, D. Jackson, C. Godjuwa, A. Fisher. P. Cooke, D. Choquenot & D. M. J. S. Bowman (2001). Fire ecology and Aboriginal land management in central Arnhem Land, northern Australia: a tradition of ecosystem management. *Journal of Biogeography*, 28, 325-343.
Zackrisson, O. (1977). Influence of forest fires on the north Swedish boreal forest. *Oikos*, 29, 22-32.
Zobel, M. (1992). Plant species coexistence – the role of historical, evolutionary and ecological factors. *Oikos*, 65, 314-320.

Recreating pastoralist futures

T.J.P. Lynam
The Institute of Environmental Studies, University of Zimbabwe, P.O. Box MP167 Mount Pleasant, Harare, Zimbabwe. Email: tim.lynam@csiro.au

Abstract

Research experience in southern Africa is used to reflect on key determinants of pastoral futures and how they might need to be addressed. The paper begins with a brief review of what we mean by marginality. A set of observations on key issues defining the option sets for pastoralism in the future is then presented. The first of these is that only a small number of structures or processes actually control the behaviour of social-ecological systems such as pastoralist systems. A second observation is that the future is so uncertain that there is a need to learn to design for robustness across plausible futures. Coupled to this is the observation that a reliable understanding of how we might manage adaptive capacity in pastoral people and communities is needed. Lastly it is suggested that a vital frontier in research is the set of relationships between cognition, emotions and behaviour at the scales of the individual and society.

Keywords: plausible futures, adaptive capacity, cognition, mental models, robustness

Introduction

We peer into the future with great uncertainty. We are familiar with the present and the recent past and so we tend to think of the future as being very much like the present. But the future may be something entirely different. As Allen (1990) noted many years ago, "The future is not what it was." What might the future be like for pastoral systems? What research do we need to be doing now to meaningfully contribute to expanding the options that pastoralists of the future face? The focus of this paper is on outlining some tentative answers to these questions. The answers are posed with the hope that they will stimulate our thinking about pastoral systems in the future and thence guide us to developing better research agendas. The suggestions and ideas presented in the paper are strongly influenced by my work in southern Africa.

I have structured the paper in the following way. Firstly I explore what I understand by the concept of marginality. I suspect that for many researchers marginal means something to do with ecological productivity. It is my contention, however, that marginalisation may come from many other dimensions of the pastoralists' world. Ecological marginality may not be the dominant marginalisation process. I then discuss in turn four ideas that I believe to be of consequence for our collective thinking about pastoral systems in the future. The first of these is the notion that we cannot know the future and therefore need to think about pastoralist systems being robust across all plausible futures. The second idea has to do with the belief that there are at most a handful of a processes or system structures that govern the behaviour of complex adaptive systems (Gunderson and Holling, 2002). We need to understand these and their relationship to the state of the system to understand what the future might be like. The third idea is that of adaptive capacity. The general concept is understood but we need to think through what it is, what impact it has and how it should be measured. Lastly I feel it important to improve our understanding of the way people perceive the world and how individual perceptions are mediated or altered through social processes. The mental models that pastoralists and in fact people at large have and how these influence our behaviour seem crucial determinants of understanding how people, such as pastoralists, interact with their world. In each of these discussions I seek to highlight what I believe to be some of the key unknowns or uncertainties and hence areas where research or investment might well be warranted.

What is marginal to you may not be marginal to me

To people discussing pastoralism or pastoralists, marginal usually mean something related to being on the verge of not being able to sustain a livelihood or being on the verge of slipping into greater poverty. Although we often talk of marginal lands or marginal areas or marginal environments, we need to be cautious in thinking about marginalisation as being a purely ecological or a purely political or a purely economic process; it is more likely to be an inter-related mixture of all three (Blaikie and Brookfield, 1987; PASTORAL, 2003). Even within a given community there may well be differentiation and some people or families may be more economically, politically, socially or ecologically marginal than others (Findlay, 1996; Wisner, 2005). Closely associated with the concept of marginality is the idea of vulnerability; the livelihoods or general well-being of marginal groups or people are vulnerable to perturbation.

What is important to acknowledge is that marginalisation is firstly a relative concept. Marginalisation is conceived in relation to some metric or distribution. We need to be clear what that metric or reference distribution is, who is using it and for what purposes. Secondly marginalisation, in the sense of being on the verge of not being able to sustain an adequate quality of life, can come about through many combinations of demographic, economic, social, political or ecological factors (Blaikie and Brookfield, 1987). Marginalisation may be temporary or a chronic condition.

Marginal groups or individuals may play an important role in some societies and may in fact not perceive themselves as being marginal. Many of the great leaders and thinkers of society have come from what the societies of those times or places have called marginal groups; particularly from a religious perspective individuals, such as Buddha or Muhammad, were marginal in their societies (Armstrong, 2000; 2001). Marginal is measured with respect to a norm or reference and often means on the edge or fringe. For some people or groups this is a choice and having sectors of society living different lifestyles and values can be an important source of innovation and resilience. We need a better understanding, therefore, of the distribution of the marginal condition across a population as well as the impacts of this distribution. Is the modal value within the region that is accepted as marginal or is there just a lower tail of the distribution that is within that region? What is happening to the distribution through time; is the lower tail extending to the marginal or away from it? Is the modal value shifting toward the marginal or away from it?

We also need to understand what it is that makes people marginal. Inadequate production systems, environmental degradation, insecure resource rights, low use of external inputs and credit, lack of public investment in agriculture, remoteness and risky and limited markets have been noted as principal determinants of food insecurity (Scherr, 2003), and these are probably true for most forms of marginalisation, but there are also very strong social and cultural factors that bring about marginalisation. Why does the proportion of people or families in the marginal section of the distribution not self-organise away from that section? What are the processes or forces that keep them in what is essentially an undesirable state? Is it in fact an undesirable state? Unless we have a clear understanding of these processes or forces, we are unlikely to be able to generate solutions that will shift the distributions from the marginal end to the viable end. Even when we do understand the processes and forces, we need to be very clear on what the implications of our interventions will be on the distributions that we faced at the start. Shifting one group of society across a marginal threshold may only result in their replacement by another group. Does the society need a marginal group for it to remain viable? These are hard questions to answer but we need to be honest and open to the nature of the systems we are trying to understand and help people manage.

Implicit in some of these thoughts is the degree to which society writ large has the right to determine who is marginal and thereafter to enforce its perspective on sections of society. Western society has taken on for itself the role of keeper of global values. Whilst I concede that in many instances people perceive themselves to be marginal and would like to change to a better state, it has also been my experience that there are many instances where external agents drive the push away from marginality. We need be very cautious in so doing.

What keeps people in an undesirable state of being? This is the subject of the next section in which I explore the major determinants of a system's behaviour and how these recreate the same conditions unless they themselves are fundamentally altered.

System governing structures and processes

It has been suggested that a handful of key process or structures, that generally change quite slowly, largely determine the behaviour of complex adaptive systems such as pastoral systems (Gunderson and Holling, 2002; Walker *et al.*, 2004). A logical consequence of accepting this proposition is that unless there are fundamental changes in these system-governing structures or processes then the basic dynamics of the system under investigation will not change outside of the behavioural boundaries set by the interactions of these system governing structures or processes. Where these system governing structures or processes are determinants of marginality then they have to be addressed before system level change can occur.

As an example, I reflect on research experiences with colleagues in the semi-arid Zambezi Valley of Zimbabwe. We were working on a project (the Mahuwe Project) to improve the management of vegetation resources in an agro-pastoral system in a semi-arid area of northern Zimbabwe. In a workshop of local leaders and villagers that was designed to identify the factors that prevented the community from achieving its goal of "conserving our natural, grazing and browse resources", the villagers articulated that it was the current allocation of land to crop production, housing and grazing areas, as well as the corrupt processes through which this land allocation was achieved and maintained, that were the major factors preventing them from achieving their vision. We therefore had to restructure the project to firstly address these underlying system constraints. Any attempt to alter vegetation management practices on top of the dysfunctional (from the community perspective) land allocations and land allocation processes would not have been likely to achieve their objectives. We, therefore, went through an eighteen-month process to firstly get the community to accept a re-planning of their area and then reallocation of land as well as the removal of illegitimate village leaders. Only then were we able to begin the original process of addressing resource management issues.

Perhaps we were lucky in being able to address the system governing structures or processes. There are likely to be many more instances where these are outside the control of the people to address. But what do you do when you cannot address the system-governing structure or process? It is these situations that provide the greatest problems for research and management. Solutions are likely to be restricted to mitigation or adaptation; learning to live with undesirable but essentially unchangeable situations.

Robust across plausible futures

We cannot predict the future with any useful degree of certainty. The further into the future we peer, the greater the uncertainty. With several of the major drivers of change in pastoral systems (e.g. climate and markets) expected to change quite markedly but in as yet highly uncertain ways, the future seems even less certain than it was. How should we deal with this situation?

One approach is to analyse the major determinants or drivers of the future and establish which of these are likely to have high impacts and which are likely to be highly uncertain in their impacts or occurrence. The high impact and high uncertainty drivers are those that are most likely to split the future into alternative pathways or trajectories. With these drivers identified the future can be explored through creating storylines of alternative future worlds based on the separation of these drivers along major axes – the axes of discrimination. For example, in the Gorongosa – Marromeu component of the Southern Africa Millennium Ecosystem Assessment (SAfMA-GM), two future scenarios were developed to explore the future of that region of Mozambique (Lynam *et al.*, 2004). The central discriminating axis was a combination of governance; from centralised and corrupt to devolved and locally responsive. Economic investment flows into the region were seen as being linked to the governance state. Scenarios were also developed with local communities (Table 1) and provided a rich opportunity for these groups to explore the robustness of their livelihood systems in each future world (Table 2).

Table 1 Major drivers of change (to 2030) for future scenarios in Vunduzi, Gorongosa District, Mozambique and the relative impact scores for each driver. Scores range from 1 (lowest) to 40 (highest) and are subjective representations of the perceived likely impacts of these drivers (Lynam *et al.*, 2004)

Driver	Relative impact score
Amount and type of armed conflict in Mozambique	40
Amount and timing of rainfall in Vunduzi	25
Relation of government with Vunduzi community	22
Amount of trade in Gorongosa District	20
Amount and type of agricultural commercialisation in Vunduzi	15
Condition of road between Vila Gorongosa and Vunduzi	15
Population of Vunduzi	13
Understanding among people in Mozambique	12
Amounts and types of international projects in Vunduzi	10
Movement of people in and out of Vunduzi	10
Amount of manufactured products available in Vunduzi	9
Numbers of non-locals living in Vunduzi	9
Prices of agricultural produce in Mozambique	9
Amount of greed / covetousness among people of Vunduzi	1

Table 2 Indicators of human wellbeing for the Nhanchururu community under the Patronage scenario. The villagers scored 31 human well being indicators now and in 2015 under the two different scenarios. A score of 10 in 2003 indicated the baseline and from there the villagers used an open ended scoring to identify likely changes under the scenario to 2015. Higher scores generally mean a positive change and scores less than 10 mean a negative change.

HWB Indicator	2003	2015	Explanation
Land for houses	10	10	Land for houses will still be readily available
Land for cultivation	10	10	Poverty or no poverty, there will still be adequate land for cultivation
Houses	10	20	Houses will increase in number as our children establish homes of their own
Water for household use	10	9	Quality of water will not change, but it may become a little drier
Crop production	10	8	Decrease initially ascribed to poor governance and increased drought, subsequently restated as a lack of coordination between government, NGOs and communities
Agricultural equipment	10	6	Under poor economic conditions it will not be easy to purchase agricultural equipment in Vila Gorongosa
Agricultural inputs	10	4	If all crops are sent to Maputo it will become difficult to find seeds locally
Credit/funds for agriculture	10	3	Credit will become less accessible as people will borrow money, have difficulty in paying back and then will not be able to take any more
Knowledge and technology	10	10	Ploughs and tractors have never been used here and are unlikely to be introduced
Livestock	10	2	Livestock such as goats, pigs and dogs will decrease due to conflicts caused with neighbours who will either kill or chase away the offending animals
Hunting of wildlife	10	1	Although only minimal hunting is carried out at present (e.g. cane rats in fields), in future there is likely to be stronger prohibition and control by park rangers
Fishing	10	10	Currently possible to secure a permit to catch fish within the park, and we hope that this will continue
Collection of wild foods	10	10	Currently possible to secure a permit to harvest natural resources within the park, and we hope that this will continue
Grinding grains for food	10	10	The situation will not change – there has never been any grinding mill here and none will come
Cooking of food	10	10	There are so many trees here such that firewood will still be readily available
Sleeping mats	10	14	We will have more children, so there will have to be more sleeping mats
Household items	10	12	Young people will come with new ideas and improve on what we use at present
Access and transport	10	10	The road will remain in its current poor state
Selling of crops	10	13	Increase in crops will be due only to the population increase, and not from any other factors
Selling of livestock	10	1	Livestock numbers will be strongly reduced, so there will be few to sell
Selling of natural resources	10	12	This will increase due to population increase
Purchases from shops	10	10	There are no shops here at present and they will not come
Local employment	10	11	Limited increase due to the increase in population which will create additional opportunities for local employment
Formal employment	10	10	There have never been any formal employment opportunities here and they will not come. Some may go to seek employment outside, but opportunities are restricted as people prefer those who can write nicely. Our life is based on agriculture here.
Education status	10	11	School is satisfactory at present, and will continue. The teacher is a problem, but we will manage to replace him, and this will improve the education
Health status	10	10	There are no formal health facilities in the village at present. When we are sick we must go to Vila, and this will continue.
Status of traditions	10	10	These will continue without change
Status of government regulations	10	10	These will continue without change
New activities	10	13	People say that this area is good for tobacco – maybe tobacco farmers will come and then we can start to grow tobacco too
Social differentiation	10	15	Some families will be able to increase crop production and sales, but others will not – this will lead to greater differences between families than at present
Village population	10	17	Population will increase due to natural growth: Firstly, the age of marriages is decreasing (from 20/21 years before to 18/19 years now) and, secondly, polygamy is increasing. We do not expect any families to leave Nhanchururu, but some will come from outside to settle here. This is already happening, and we will accept them, and show them where to put their houses and where to make their fields

Examining livelihood systems in the framework of future scenarios (Peterson *et al.*, 2003) can provide opportunities to explore what adaptations (social, institutional or technical) would be required to maintain (or enhance) specific human wellbeing objectives in the alternative futures. But what is important is the potential to explore these adaptations in a relatively low risk, simulated environment.

The capacity of individuals or communities to adapt to changes, either proactively in anticipation of changes or reactively in response to changes, is an important determinant of the resilience of people and communities. This is the focus of the next section of the paper.

Understanding and enhancing adaptive capacity

Changes occur and people adapt through mechanisms such as altering the rules governing resource use, altering or developing new technologies or changing consumption. In some situations human adaptations improve overall wellbeing but in many others human wellbeing decreases. Why is this? Why are people able to adapt without loss of wellbeing in some situations but not in others? Why are people in some situations not able to adapt at all? Understanding the determinants of adaptation, at both an individual and a community level, is clearly of importance to our ability to deliver research that might enhance the ability of people to make wellbeing-enhancing adaptations and to avoid wellbeing decreasing adaptations.

In this section I explore, largely from a conceptual perspective, some aspects of human adaptation. We need to be clear what it is we mean by adaptive capacity. In his classic treatise "Adaptation and natural selection", George Williams (1966) considered adaptation to be "an aspect of a phenotype (structure, behaviour, physiology or mind) that was designed by natural selection to serve a specific function" (Symons, 1989). Williams (1966) was very clear that adaptations must be considered in relation to the environmental factors that they were the consequence of. From this evolutionary perspective with enough time, adaptive capacity could be considered to be almost infinite. This is not particularly useful when talking of pastoralist societies that are currently in existence. However, it does tell us that we have to express adaptive capacity in relation to some response time. In the context of current pastoral peoples or societies, we need to be clear whether we are referring to the adaptations of current (extant) individuals or populations or we referring to their societies. We can clearly adopt a much longer timescale with the latter. In this paper I will use the term, adaptive capacity, to refer to the capacity of individuals or society to generate responses to changed circumstances that are intended to sustain or to improve their wellbeing. These might, for example, be behavioural, institutional, technological or consumptive responses.

It is important to recognise that this is not purely a positive set of responses. When there is no other recourse, people have to reduce their consumption to damaging levels (i.e. resulting in malnourishment) in order to continue. This is an adaptation. When people talk of adaptive capacity, I get the sense that they do not mean this response of last resort.

What then are the determinants of this capacity to adapt? I suggest that there are two major determinants; exposure history together with the society's or individual's memory of the history and resources as measured along axes of human, social, natural, financial and physical capital.

Throughout their existence societies and individuals experience a great variety of change situations. This is their exposure history. If we express these change situations as frequencies of change events of different magnitudes, we might expect the great majority of events or processes to be on the left end of the x-axis (the small events end) with only a relatively few large magnitude events or processes. For most people and societies their experiences of change will be largely in the range of small-scale events or processes. Only very occasionally will they have to face (adapt to) large-scale events. The more common and thence by definition largely small change events will become part of their normal routines – intra-season drought in southern Africa, for example. Adaptations to these change situations, such as moving livestock to distant grazing areas, will be largely institutionalised. The more common a change event the more learning the individual or society will have had a chance to do and the more likely they will have the adaptive capacity to respond to these situations. Experience provides learning opportunities. The more difficult situations are those that are rare and thence likely to be very much larger events. Where there is no previous experience, of either the same or similar class of change events, then I expect the capacity to adapt to be much less well developed. However, experiential history may also wear down the capacity of a community or individual to adapt (Hobfoll *et al.*, 1995). With repeated stresses, individuals or societies may not have the time to recover and thence may be even more vulnerable to change events that would not usually make much difference to them.

In part this last point reflects a second important determinant of adaptive capacity; the availability of resources to buffer the individual or society from changes that have emerged. Resources provide for alternative pathways for recovery or adaptation from a change situation. When grazing shortages impact Australian outback graziers, they lease grazing elsewhere, load their livestock onto trucks and move them to the new grazing. Southern African agro-pastoralists do much the same thing but walk the animals to new locations. As noted earlier, repeated high magnitude disturbances will erode the resources (social as well as financial or natural) available to an individual or society. Where there are no resources, there can be little adaptation other than altering consumption.

This brief outline of adaptive capacity leaves me with many more questions than I am able to answer. What, for example, are the links between local (individual) and social adaptation or adaptive capacities? How do we measure adaptive capacity? What do we understand by the more complex concept of transformative capacity or the capacity of societies or communities to transform themselves into something entirely new (Walker *et al.*, 2004)? Perhaps most important of all is the question of how we might go about enhancing human adaptive capacities? Clearly the majority of opportunities available to us are associated with the provision of resources. Of these it would appear that knowledge and social capital in the form of networks, governance, trust, the capacity to visualise alternatives and plan their implementation are key. These were important aspects of the success of the Mahuwe Project mentioned earlier and in many respects are associated with cognition – how people understand or conceive of the world and work with these conceptions to manage the world and their place in it. This is the subject of the next section of the paper.

Cognition, emotion and mental models

People have an extraordinarily well-developed ability to conceptualise, in quite abstract ways, the world they seek to manipulate or manage in order to live and satisfy their goals and aspirations. These mental models (Johnson-Laird, 1988) of how the world works, and their attitudes and beliefs as to what is important and relevant in the world, appear to be a major determinant of human behaviour. This is a large field of research that spans a number of disciplines that I cannot adequately represent here. I would, however, like to outline what I believe should be a major area of enquiry in relation to pastoral systems. This is the area of the relationship between human mental representations of the world and human behaviour. It is my thesis that without a reasonably reliable understanding of this domain we stand little chance of understanding human decision-making or human behaviour at either an individual or collective scale. I furthermore suggest that human cognitive and emotional processes underpin much of what I have discussed so far in this paper and we, therefore, need a greatly improved applied understanding of these processes in order to guide the emergence of successfully adaptive policies and management options for pastoral societies.

Virtually all human interactions with the world are mediated through our cognitive and emotional systems (Johnson-Laird, 1988; Goleman, 1996). We construct, manipulate and update mental (simulated) representations of the world and from these draw conclusions on what actions to take, stimulate the physiological, physical and mental processes needed to take actions as well as monitor progress and adapt our actions to improve performance. Our emotions appear to play crucial roles in this process through acting as the equivalent of programme interrupts to halt or switch cognitive processing (Johnson-Laird, 1988), or fast and parallel processing systems that bypass the slower cognitive processes to achieve rapid results when these are needed (Le Doux, 1989; 1998). This rapidly developing field of research that integrates psychology, neural and brain research and human behaviour has a number of important lessons for research and policy in any field of human endeavour. We have a collective responsibility to bring the findings of this exciting frontier into our mental models of how pastoralists view the world. We need to understand the relationships between these mental representations and human behaviour. As importantly, we need to understand the relationships between individual mental models and social constructions of reality and action. Rather in the sense of the thinking of Kuhn (1970) on scientific paradigms, I believe that societies maintain collective mental models of how the world works that incorporate attitudes, beliefs and values. These collective mental representations act as attractors for the thinking of people living within that society. They usually change only very slowly but are capable of very rapid changes across paradigmatic thresholds such as when societies gear up for war.

People do not necessarily alter their behaviour when provided with information on the benefits of different behaviours. Participatory research methods have been successful because they involve working with people in ways that, I believe, develop revised cognitive representations of the world. The mental models or at least small parts of them are re-coded. I believe that the acknowledged successes of model-building for altering the understanding of modellers are due in large to the same processes. But I suggest that we need a better understanding of how this happens and how stable these revised representations are. We also need to better understand what combinations of information, participatory learning and action will yield the most reliable and useful cognitive and emotional systems for pastoralists dealing with the many changes they face in the coming decades. But it is not only the pastoralist mental models that need to be understood and altered. We, as scientists as well as elected representatives of society, need to rework our cognitive and emotional processing systems. Are we sure that we are in the best attractor of cognitive representations?

Conclusions

I have suggested that pastoralists form part of a larger society that, when viewed at some scales, we are also a part of. As such they face similar problems and issues that many of us do, albeit at different times and perhaps without the buffering that our relative wealth across several scales of measurement provides us. I have identified a handful of concepts which I believe are of significance to our thinking about pastoralists and therefore that compel us to act in particular ways or to adopt particular attitudes or beliefs. The first of these is that for many systems there is only a small number of what I call system governing structures or processes that fundamentally control the behaviour of the system. We need to pay attention to these because unless we alter them the same system behaviour will be recreated. To fundamentally alter the way a system behaves, we need to address these system-governing structures or processes. Where we cannot alter these system-governing structures or processes, and they are creating undesirable behaviours, we can only hope to mitigate the consequences of these behaviours.

A second observation made was that, because we cannot know the future (and neither can most pastoralists), we need to start thinking about what interventions are likely to be robust across a plausible range of alternative futures. We should examine the likely outcomes of any intervention across a number of futures. Many pastoral societies are on the edge of social and economic viability. They are not in a position of taking great risks. Few of them have large capital stocks to buffer them when interventions go wrong. We need therefore to think of them in terms of resilience (*sensu* Holling, 1973) rather than production.

The third observation that I made was that, if we accept the conclusion that we need to design for robustness across plausible alternative futures then a crucial determinant of that robustness will be the capacity of people and the society in which they exist to adapt to change. I believe that there is a major gap in our understanding in relation to our ability to measure and improve the adaptive capacities of pastoralist peoples and societies.

Finally I have suggested that much of the research we do, the policy formulated and the behaviour of pastoralists themselves is determined by a complex web of cognitive and emotional models of the world. I argue that we need a far better applied understanding of these models and their relationships to individual and collective behaviour. This latter strikes me as one of the major research challenges in the coming decade.

A common thread throughout this paper is the unknown relationships between individuals and the collective. Science seems to have hit a wall in its ability to deal with scale. Although great strides have been made in thinking about scale, we are far from having a workable understanding of the relationships among small-scale and large-scale (in both a spatial and temporal sense) structures or processes. This is research frontier that beckons enticingly, hydra-like.

Acknowledgements

The Mahuwe Project was funded by the United Kingdom Department for International Development (DFID) for the benefit of developing countries (R7432 Livestock Production Programme). The views expressed are not necessarily those of DFID. The SAfMA-GM Project was funded by a grant from NORAD to the Millennium Ecosystem Assessment.

References

Allen, P. M. (1990). Why the Future Is Not What It Was. Futures, 22, 55-570.
Armstrong, K. (2000). Islam. A Short History. Random House, New York, USA.
Armstrong, K. (2001). Buddha. Penguin, New York, USA.
Blaikie, P. & H. Brookfield (1987). Land Degradation and Society. Longman, Harlow, Essex, UK.

Findlay, A.M. (1996). Population and Environment in Arid regions. International Union for the Scientific Study of Population (IUSSP), Policy and research paper Number 10, Paris, France.
Goleman, D. (1996). Vital Lies Simple Truths. The Psychology of Self Deception. Simon and Schuster, New York, USA.
Gunderson, L. & C.S. Holling (eds) (2002). Panarchy: Understanding Transformations in Human and Ecological Systems. Island Press, Washington, DC, USA.
Hobfoll, S.E., S. Briggs & J. Wells (1995). Community Stress and Resources: Actions and Reactions. In: S.E. Hobfell & M.W. de Vries (eds.) Extreme Stress and Communities: Impact and Intervention. Kluwer Academic Publishers, Maastricht, The Netherlands.
Holling, C.S. (1973). Resilience and stability of ecological systems. *Annual Review of Ecology and Systematics*, 4, 1-23
Johnson-Laird, P. (1988). The Computer and the Mind. An Introduction to Cognitive Science. Fontana Press, London, UK.
Kuhn, T.S. (1970). The Structure of Scientific Revolutions. Chicago University Press, Chicago, USA.
LeDoux, J. (1998). The Emotional Brain. Simon and Schuster, New York, USA.
LeDoux, J. (1989). Cognitive-Emotional Interactions in the Brain. *Cognition and Emotion*, 3, 267-289.
Lynam, T.J.P., B. Reichelt, R. Owen, A. Sitoe, R. Cunliffe & R. Zolho (2004). Human Well-Being and Ecosystem Services: An Assessment of their Linkages in the Gorongosa – Marromeu Region of Sofala Province, Mozambique to 2015. CBC Publishing, Bath, UK.
PASTORAL (2003). Trends and Threats to the Viability of European Pastoral Systems. PASTORAL Project Information Note 5. http://www1.sac.ac.uk/envsci/external/Pastoral/webpages/project_description.htm
Peterson, G. D., T. D. Beard Jr., B. E. Beisner, E. M. Bennett, S. R. Carpenter, G. S. Cumming, C. L. Dent, and T. D. Havlicek (2003). Assessing future ecosystem services: a case study of the Northern Highlands Lake District, Wisconsin. *Conservation Ecology*, 7, 1. [online] URL: http://www.consecol.org/vol7/iss3/art1
Scherr, S. J. (2003). Halving Global Hunger. Background Paper of the Task Force 2 on Hunger. Millennium Project, UNDP, New York.
Symons, D. (1989). A Critique of Darwinian Anthropology. *Ethology and Sociobiology*, 10, 131-144.
Walker, B.H., C. S. Holling, S. R. Carpenter, & A. Kinzig (2004). Resilience, adaptability and transformability in social–ecological systems. *Ecology and Society*, 9, 5. [online] URL: http://www.ecologyandsociety.org/vol9/iss2/art5
Williams, G.C. (1966). Adaptation and Natural Selection. Princeton University Press, Princeton, USA.
Wisner, B. (2005). Tracking Vulnerability: History, Use, Potential and Limitations of a Concept. Invited Keynote Address, SIDA & Stockholm University Research Conference on Structures of Vulnerability: Mobilisation and Research, Stockholm, 12-14 January, 2005.

Challenges and opportunities for sustainable rangeland pastoral systems in the Edwards Plateau of Texas

J.W. Walker[1], J.L. Johnson[2] and C.A. Taylor, Jr[1]
[1]*Texas Agricultural Experiment Station, 7887 U.S. Highway 87 N., San Angelo, TX 76901 USA, jw-walker@tamu.edu*
[2]*Texas Cooperative Extension, 7887 U.S. Highway 87 N., San Angelo, TX 76901 USA.*

Abstract

This paper focuses on pastoral systems in an area of west-central Texas known as the Edwards Plateau. These rangelands have a combination of grass, forb and browse species and are used primarily for combinations of grazing by cattle, sheep, goats and wildlife. A major ecological challenge is woody plant encroachment. Stocking rate is the major factor affecting sustainability and historically this area was heavily grazed. Today the stocking rate is half or less of its historical peak. Species of livestock has shifted from predominantly small ruminant to cattle. About 70 % of pastoralists use some sort of rotational grazing system. On average ranches lose equity from livestock but appreciating land values make up for this loss. Rangelands are still a good investment for those that can afford them. Land ownership is changing and is creating land fragmentation problems and a need for targeted educational provision.

Keywords: economic, livestock, wildlife, grazing management

Introduction

This paper provides a brief description and history of an area in west-central Texas known as the Edwards Plateau. Our objective is to describe past and current conditions in order to provide recommendations for a sustainable future. As part of this effort, we will describe the principles of grazing management that are the basis for sustainable use of this resource area and the economic principles that influence management decisions of pastoralists. The focus will exceed the narrow definition of pastoral, i.e., devoted to or based on livestock production, and encompass the broader definition of this term, and encompasses all aspects pertaining to rural livelihoods in rangeland-based production systems. The reason for this broader definition is that in Texas sustainable pastoral systems require consideration of all potential products of the rangeland resource including forage production for livestock and wildlife, recreation, ecosystem services and existence values. To reflect the diversity of motivations of people responsible for managing Texas rangelands they will be referred to by a variety of synonyms, e.g. pastoralists, ranchers and landowners.

Texas consists of ten natural regions that are defined on the basis of the interaction of geology, soils, physiography and climate (Gould *et al.*, 1960). The Edwards Plateau is an area of approximately 10 million hectares in west-central Texas that is bounded on the east and south by the Balcones Escarpment, an area know as the "Texas Hill Country." The Stockton Plateau forms the western border. The northern boundary is less distinct and blends with other natural areas known as the High Plains, Rolling Plains and Cross Timbers and Prairies. Soils are usually shallow with a wide range of surface textures. They are underlain by limestone or caliche on the Plateau proper and by granite in the Central Basin. Elevations range from slightly less than 300 m to more than 900 m. The surface is rough and well drained, being dissected by several river systems.

Annual rainfall ranges from about 350 mm in the west to more than 800 mm in the east increasing by about 1 mm for each km from west to east. On the average, there are more years with below-average than above-average rainfall. Droughts have occurred in the area frequently, but the most severe drought on record was from 1950 to 1957. Seasonal rainfall patterns are bimodal with May and September peaks. This pattern shifts to a late summer high on the western edge of the Stockton Plateau.

The original vegetation was grassland or open savannah-type plains with tree or brushy species found along rocky slopes and stream bottoms. Tallgrasses, such as cane bluestem (*Bothriochloa barbinodis* var. *boabinodis*), big bluestem (*Andropogon gerardii* Vitman), Indiangrass (*Sorghastrum nutans* [L.] Nash), little bluestem (*Schizachyrium scoparium* [Michx.] Nash), and switchgrass (*Panicum virgatum* L.) are still common along rocky outcrops and protected areas having a good soil moisture conteny. These tallgrasses have been replaced on shallow xeric sites by midgrasses and shortgrasses such as sideoats grama (*Bouteloua curtipendula* [Michx.] Torr.), buffalograss (*Buchloe dactyloides* [Nutt.] Engelm.), and Texas grama (*Bouteloua rigidiseta* [Steud.] A. S. Hitchc.). The western part of the area comprises the semi-arid Stockton Plateau, which is more arid and supports shortgrass to midgrass mixed vegetation. Common woody species are live oak (*Quercus virginiana* Mill.), sand shin oak (*Q. havardii* Rydb.), post oak (*Q. stellata* Wang.), mesquite (*Prosopis glandulosa* Torr.), and juniper (*Juniperus ashei* Bucholz. and *J. pinchotii* Sudw.). As a result of high grazing pressures and a reduction of fire, the density of woody vegetation has increased dramatically compared to its density prior to European settlement. Encroachment of juniper on shallow soils and slopes, and mesquite on deeper clay loam, is one of the major threats to sustainable use of this area.

The Edwards Plateau is about 98% rangeland with cultivation largely confined to deeper soils in valley bottoms. Rangelands have a combination of grass, forb (i.e., herbaceous dicot) and browse species and are used primarily for mixed livestock (combinations of cattle, sheep, and goats) and wildlife production. This area is the major wool- and mohair- producing region in the United States, providing about 98 % of the nation's mohair and 20 % of the lamb and wool production. Because of abundant grass the area was known as "Stockman's Paradise" when it was first settled in ca. 1880. It also supports the largest deer population in North America. Exotic big-game ranching is becoming important, and axis, sika, and fallow deer and blackbuck antelope are increasing in number. Native white-tailed deer are abundant over much of the area and serve as a valuable source of income for ranchers.

Uses of Edwards Plateau rangelands have been in transition for about 40 years. Prior to the 1950s, livestock production was the primary use of this resource. During the 1960s, hunting for fees began in earnest and its importance has steadily increased to present, somewhat offsetting the loss of income from livestock. Revenues from hunting have an advantage compared to livestock revenues in that they are affected less by droughts. Other trends that occurred in this area starting around the decade of the 1960s were an increase in the number of deer and an explosion in brush. The deer population in Texas benefited from an increase in woody plants and eradication of the screwworm which was complete by 1966. Recently, large ranches are being subdivided and purchased primarily by people without agricultural experience.

Before discussing pastoral systems that are sustainable in west Texas, it is first necessary to define the goals of the system and examine the constraints to sustainability. Long-term sustainability has two dimensions: ecological and economic. The first dimension is self-evident and we use the definition of rangeland health, i.e., the degree of integrity of the soil and ecological processes that are most important in sustaining the capacity of rangelands to satisfy values and produce commodities (National Research Council, 1994). It implies that the soil and its ability to capture and retain precipitation have not been unduly compromised. The second dimension, i.e. economic sustainability, is often ignored by ecologists but is equally important. As Ainesworth (1989) stated, "the only sustainable agriculture is profitable agriculture." However, for many landowners profit often takes many non-market forms such as the ability to provide a desired lifestyle or improve the resource, and this will be discussed in greater detail later.

A major ecological challenge not only of the Edwards Plateau but most of Texas, and arid lands worldwide, is woody plant (i.e., brush) encroachment (Archer, 1989). In Texas, this is primarily a result of overgrazing, which reduces competition from herbaceous species as well as a reduced fire frequency because of a reduction in fine fuels. The two most problematic species in this area are mesquite and redberry juniper. Left untreated both the canopy cover of mesquite (Ansley *et al.* 2001) and redberry juniper (Ueckert *et al.* 2001) increases by about 1% annually. Carrying capacity for domestic livestock significantly and rapidly declines when the canopy cover exceeds 20 % to 30 % for redberry juniper and mesquite respectively. Up to a point increased brush cover improves the habitat for white-tailed deer, the most economically important wildlife in this area, eventually it can reach densities that reduce their carrying capacity as well (Teer 1996). Current conditions have decreased the frequency of naturally-occurring fires below the level necessary to keep brush below an economic threshold thus requiring inputs, i.e., prescribed fire, herbicides or mechanical treatments, to keep the woody vegetation below an economic threshold so that these pastoral systems remain productive. This creates a problem because the cost of brush control often exceeds the economic return that results from reduced woody plant cover. Thus one of the most important principles of sustainable grazing management is management that facilitates low-cost alternatives for brush management.

Grazing management

There are three components of grazing that must be managed, namely stocking rate, species of livestock and distribution of livestock. A fourth component, namely, season of grazing which is important in more temperate climates, is less important in Texas because these ranges are grazed year-round. When domestic livestock are grazed, decisions relative to livestock species, timing of grazing and stocking rate are made, either consciously or unconsciously. Improving grazing distribution, however, normally requires the investment in capital improvements such as fencing or water development.

Stocking rate is the most important principle of grazing management and historically was high. Grazing in the Edwards Plateau began in earnest around 1880. Fencing was introduced with the invention of barbed wire in 1873 and by 1900 all of the ranches had been fenced. Based on census data since 1919, total livestock numbers in the Edwards Plateau peaked in 1940 and declined until 1970 when they stabilized until a drought that began in 1997 caused a further reduction in livestock numbers (Figure 1). However, other sources (Merrill, 1959; Smeins *et al.*, 1979) suggest stocking rates around 1900 may have been as much as 3 times higher than the peak reported for 1940. What is clear regardless of the source, is that stocking rates routinely exceeded the sustainable carrying capacity throughout the history of grazing by domestic livestock and in the early years of the twentieth century stocking rates may have been extreme. The reduction in livestock numbers was caused by a reduction in the number of sheep and goats. Thus the proportion of forage demand shifted from 85 % from small ruminants down to 34 % by 2000.

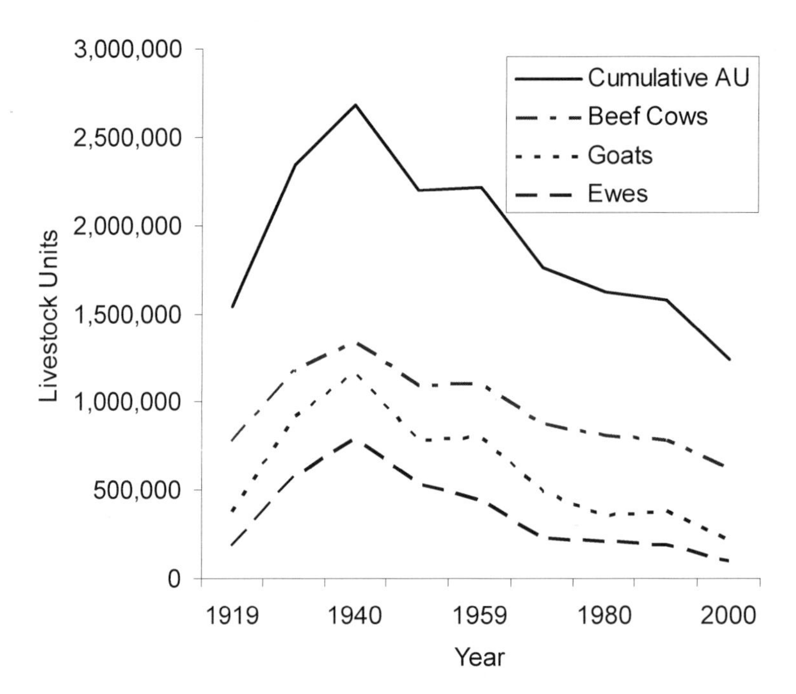

Figure 1 Changes in livestock units of grazing by breeding females of different livestock species commonly found in the Edwards Plateau, Texas. Data prior to 1970 are from U.S. Census Bureau and then from U.S. Department of Agriculture, National Agriculture Statistic Service.

Early overgrazing of this area was a result of early settlers with no experience of pastoral systems in semi-arid environments. The first pastoralists in this area had neither an appreciation for the certainty of drought nor an understanding of the importance of leaving residual forage to maintain plant vigour. Long-term overgrazing resulted in desertification of this region because of soil loss and a successional shift from tallgrasses and midgrasses to shallow-rooted shortgrasses. Another result of overgrazing was a reduction in fire frequency which in conjunction with reduced vigour of herbaceous plants allowed for encroachment of woody plants, particularly juniper. Because juniper can intercept and evaporate up to 47% of the precipitation that falls on it (Owens and Lyons, 2003), this further exacerbated the desertification cycle.

Much research has been conducted in the last 50 years that has shown the effect of stocking rate on livestock production and economic returns (Walker and Hodkinson 1999). These results clearly demonstrate that moderate to light grazing will result in greater economic returns and less risk than heavy grazing. Nonetheless there is still a strong tendency of ranchers in this area to overgraze. This is a result of several factors. One is the failure of pastoralists to anticipate and respond to the inevitability of drought. Another has been government programmes such as emergency feed programmes and price support for fibre production (i.e., wool and mohair) which encourages overgrazing because, when prices are favourable, fibre can be produced at stocking rates that are too high to produce meat profitably. Finally, 62% of Texas rangelands are leased usually at a fixed rate per unit area.

Such arrangements provide little or no incentive for lessees to adjust stocking rates for climatic variation or to preserve the long-term ecological condition and productivity of the resource. Flexible long-term grazing leases based on the number of animals grazed are necessary to provide lessees incentives to stock appropriately.

Paradoxically, juniper encroachment, which occurred as a result of high grazing pressures, may ultimately provide the impetus for pastoralists to graze more conservatively. Prescribed fire is the most cost-effective method for controlling juniper. Effective burning programmes depend upon adequate amounts of fine fuel, which can only be accumulated with moderate to light stocking rates. Thus, while ranchers may be reluctant to reduce stocking rates for the purpose of maintaining the vigour of the herbaceous vegetation, they may be willing to defer grazing to accumulate fuel to control juniper.

Species of livestock is the second most important component of grazing management. Unlike other parts of the U.S., the Edwards Plateau has a tradition of grazing cattle, sheep and goats together. However this was not always the case. During the early settlement of this area (ca. 1880), when there was still competition for unclaimed grazing lands, there was competition between sheep producers and cattle producers (McSwain, 1996). For instance, legislation passed in 1881 made it illegal to graze sheep on land belonging to another without the owner's permission, but cattle and horses were exempted from this law. As available land was bought, fenced and ownership established, conflicts eventually subsided. Around 1923 the proportion of sheep and goats began to increase as their purchasing power relative to cattle increased (Gabbard *et al.*, 1930). In time as a result of the profitability of sheep in contrast to hard times in the cattle industry, cattle producers in the Edwards Plateau raised sheep and goats (McSwain, 1996).

This tradition continued until recent times when a combination of low fibre prices, loss of government support programmes, and greater management requirements of sheep and goats has resulted in a severe reduction in grazing by small ruminants (Figure 1). Matching livestock grazing preference with the botanical composition of available forage increases carrying capacity because grazing pressure is distributed more evenly across the available forage (Walker, 1994). Furthermore, grazing with cattle, sheep and goats will help keep less desirable forage from replacing preferred species. For instance Taylor and Fulendorf (2003) reported an increase of 16 large juniper trees (i.e. > 2 m tall) per hectare for each year following the cessation of goat grazing. At about the same time as sheep and goat numbers were declining demand for fee hunting was increasing and replacing the income that formerly had come from small ruminants. However, wildlife managers generally discourage pastoralists whose primary interest is deer production from grazing goats because they believe that deer and goats compete for forage resources. This is unfortunate because there is little data to support this claim while the ability of goats to help manage juniper encroachment is well documented.

The final component of grazing management is livestock distribution. Poor spatial distribution is the cause of many problems associated with grazing livestock (Holechek *et al.*, 1989). Cross-fencing in conjunction with grazing systems is used to manage the spatial and temporal distribution of livestock grazing pressure. Species of livestock will affect spatial distribution as well. There is good evidence that any increased livestock production as a result of rotational grazing is caused by improved distribution, which has the effect of increasing carrying capacity (Hart *et al.*, 1993) rather than altered foraging behaviour (Gammon & Roberts, 1978) or increasing primary production (Heitschmidt *et al.*, 1987). Carrying capacity in large pastures with poor livestock distribution is lower than on similar areas with good distribution. Poor distribution may buffer the effect of changes in stocking rate on livestock performance compared to the classic response curve where animals are well distributed (Stafford Smith, 1996). Grazing systems in conjunction with proper stocking rates can help maintain or improve rangeland health.

Much of the research on grazing systems was conducted in the Edwards Plateau. The Merrill 3-herd, 4-pasture grazing system was developed by Dr. Leo Merrill, the long time superintendent of the Sonora Experiment Station (Merrill 1954). This system is still widely used in Texas. As indicated by the name it involves grouping livestock into 3 herds and rotating them through 4 pastures. In this system each pasture is grazed for 12 months, then rested for 4 months. The rest period comes at a different season in each succeeding 16-months grazing cycle. Thus over a 4-year period each pasture receives a 4-month rest during a different season. Adoption of this grazing system was enhanced because it is simple, and increases both individual animal performance and range condition. Interest and research on single herd multi-pasture grazing systems began in the 1970s. Some of these systems, such as high-intensity, low-frequency, which had one herd and eight or fewer paddocks, were useful for improving range condition at the cost of reduced animal performance. Others, such as short-duration grazing, in which one herd rotated rapidly through eight or more paddocks, were investigated for their potential to increase carrying capacity but did not meet these expectations. About 70 % of livestock producers use some sort of rotational grazing system (Hanselka *et al.* 1990). One of the greatest advantages of rotational grazing is that it provides management flexibility to implement other range improvement practices. As previously mentioned, encroachment of woody plants is one of the greatest threats to sustainable rangeland utilization in the Edwards Plateau. Management of woody plants, using prescribed fire, or mechanical removal, requires the ability to provide periods of rest for the accumulation of fine fuel in the case of fire and rest for the regrowth of herbaceous vegetation for both methods. This is only possible if some type of rotational grazing system is in place.

Financial management

Profit, defined as increasing equity, does not appear to be a primary motivation for most pastoral systems in Texas. On average, between 1991 and 2001, Texas ranches lost $33 US per cow annually if appreciating land value is excluded (McGrann, 2003). Ninety-one % of cow-calf operations have less than 100 cows and are part-time operations. Pastoral enterprises can operate with a long-term net loss because they are subsidized with non-ranch income. Ignoring incentives that result from appreciation in land values fails to provide a complete picture of the factors that motivate pastoralists. During the same time-period that ranches had negative returns from cow-calf production (1991 – 2001) real (i.e., deflated) Texas rural land prices in the Edwards Plateau more than doubled in value for a gain of $83 US /ha in real terms (Gilliland *et al.*, 2004). This rapid appreciation was caused by recreation-based demand rather than agriculture production potential. Based on an average stocking rate of 7 ha per animal unit (AU) the annual appreciation of land values in the Edwards Plateau was 345$ US/AU or about ten times the average net loss per cow reported by McGrann (2003). Thus for many Texas pastoralists owning land is an insurance policy whose premiums are the cost of raising livestock. Insurance is a better description than investment for the motivation for owning land because the latter implies the willingness to sell is a financial decision, when in fact most pastoralists are motivated to sell land only in emergency financial situations when other assets are not available. Agriculture tax valuation, which is based on the production value as opposed to the market value of land, is another incentive for raising livestock even at a financial loss. In 2001 the average agriculture valuation for rangeland in the Edwards Plateau was $38 US compared to a market value of $692 US. These values provide the basis for property taxes.

Although, cash flow is still an important consideration of most pastoralists, their primary motivations often are non-financial goals, such as preserving heritage, maintaining ownership of the land and/or improving the ecological condition of the resource. Ranching provides an opportunity to pursue these goals and simultaneously maintain a desired lifestyle and a large degree of security. Blank (2002) describes this phenomenon as living poor and dying rich.

While increasing equity may not be the primary motivating factor, cash flow or gross profit often is because it affects standard of living and the ability to achieve non-financial goals described previously. Standardized Performance Analysis (SPA) can help identify areas to manage cost and minimize equity use for pastoralists. While on average Texas ranches lost equity some were able to make a profit. More specifically one-half of the ranches had at an annual loss of $152 US per cow, while the other half had a net income of $84 US per cow (McGrann, 2003). Operations with a positive net income were characterized by weaning 8% more live weight per cow joined and more importantly having a 31 % lower production cost per cow. On average only those ranches with a cow herd size greater than 500 head were profitable. These large ranches accounted for less than 1% of the 133,000 cow-calf operations in Texas and only 14 % of the Texas beef cow inventory. Based on SPA results, all herds, regardless of size, can lower production cost by controlling grazing and feed cost, which account for 40% of total costs. This can be done by an appropriate stocking rate combined with a controlled breeding season that matches forage production with livestock nutrient requirements.

Nature tourism has been promoted as a way to increase economic activity in rural areas. While there are many potential outdoor recreational opportunities currently hunting fees have the greatest potential for increasing equity. Because of the growing importance of hunting fees for ranching enterprises, many operators make land and livestock management decisions based on the presumed effect on wildlife. However, such decisions do not appear to affect the economic reality. Based on a comprehensive investigation of ranches in the Edwards Plateau, hunting income accounts for 14 % of gross agricultural receipts. This agrees closely with respondent's estimate that 19 % of total ranch income came from hunting. Nonetheless the importance of hunting is increasing relative to livestock. Between 1993 and 2000 the average lease value of land for hunting increased by 30 % compared to a 10 %increase in livestock grazing fees. Currently, in the Edwards Plateau average lease rates are $10.11 US/ha for livestock grazing compared to $8.69 US/ha for lease hunting.

Although income from hunting provides less than 20 %of net ranch income, the effect of range management practices on wildlife and hunting is probably the deciding factor for many of the management decisions. As discussed above, land appreciation is the primary source of wealth accumulation for a ranching enterprise and this source of wealth is driven by recreational demands for land. Therefore range improvement practices, e.g., brush removal, are conducted only to the extent that they will improve wildlife habitat rather than to the degree that would be optimal for livestock production. Secondly, although for management decisions it is useful to analyze pastoral operations by their individual enterprises, ultimately ranching is an integrated economic activity that on the whole is profitable or not. Thus hunting income allows ranch enterprises to operate at a profit. Within this context livestock production is a commodity business, i.e., livestock producers are price-takers. As such, to make this segment of a ranching operation profitable ranchers must manage production costs. In contrast, hunting enterprises, which are experientially-based, have a greater opportunity to add value to the hunting experience. As such, these experiences, i.e. hunting opportunities, are priced according to what the market will bear. This price has proven to be less cyclical than livestock prices and allows for ranch operators to capitalize on their existing infrastructures and/or willingness to provide additional amenities (e.g., meals, lodging and guide service). This allows ranchers to convert their time and labour to ranch equity according to their personal preference to trade services and ranch amenities for financial reward.

Land ownership is rapidly changing in the Edwards Plateau as urban dwellers see this region of the state as a good financial investment that allows them to enjoy recreational activities. Because of these financial and quality of life values, the numbers of land sales are increasing dramatically and tract size is decreasing Gilliland *et al.* (2003). From 1990 to 2003, we estimate that 6-8 % of the land in the Edwards Plateau has been purchased by people with no experience in land management or ecological processes.

The median size of the tracts purchased by these new landowners was 93 ha and has decreased by 5 ha annually. This change in ownership has potential positive and negative aspects. On the positive side these new landowners are well-educated, have good financial resources and are motivated to make improvements on their investment. On the negative side fragmentation of large ranches is considered the greatest single threat to wildlife habitat (Wilkins *et al.*, 2003). Furthermore, most of the new owners lack the knowledge to manage these resources properly and many have unrealistic expectation of the production potential of these semi-arid lands. Thus without the knowledge and appreciation for the capacity of this area, there is a potential that new landowners may repeat the mistakes of the original European settlers in this region. This points to a great need for targeted educational provision that provide these new owners with the skills necessary to become responsible land stewards.

Because of this educational need, five years ago the Academy for Ranch Management (ARM) was established at the Texas Agricultural Experiment Station at Sonora, Texas, to provide training for new ranch owners including those that may be one or more generations removed from living on the ranch. Most of the ARM students own rangeland in the Edwards Plateau. Almost all of the students are genuinely interested in restoring their ranch to a better ecological condition and improve habitat for wildlife. Courses include training in prescribed fire, rangeland ecology, plant identification, rangeland restoration, animal nutrition, grazing management and improving wildlife habitat.

Conclusions

The greatest threats to sustainable pastoral systems in the Edwards Plateau natural resource region of Texas are encroachment of woody vegetation, rangeland degradation resulting from overgrazing, and the lack of cash flow required to control brush and maintain ranch infrastructure. Cash flow can be improved by managing the cost of livestock production, grazing cattle, sheep, and goats in combination, and hunting fees. Pastoralists that do not use all available options to increase cash flow can still accumulate wealth through land appreciation but the health of the rangeland will likely suffer if financial resources are not used to make necessary range improvements. Many large ranches have been divided into smaller holdings and purchased by people without an agricultural background. New landowners need to be educated about the effect overgrazing can have on the ecological health of their property and how properly structured leases can help prevent this problem.

References

Ainesworth, E. (1989). LISA men have called you. *Farm Journal*, 113, 1.

Ansley, R.J., B. Wu, & B. Kramp (2001). Observation: long-term increases in mesquite canopy cover in north Texas. *Journal of Range Management,*54,171-176.

Archer, S. (1989). Have southern Texas savannas been converted to woodlands in recent history? *American Naturalist*, 134, 545-561.

Blank, S.C. (2002). The end of agriculture in the American portfolio. Quorum, Books,Westport, 218pp.

Gabbard, L.P., C.A. Bonnen & J.N. Tate. (1930). Planning the ranch for greater profit: a study of the physical and economic factors affecting organization and management of ranches in the Edwards Plateau grazing area. Texas Agricultural Experiment Station, College Station, Bull 413. 45pp.

Gammon, D. M. & B. R. Roberts (1978). Patterns of defoliation during continuous and rotational grazing of the Matopos Sandveld of Rhodesia. 1. Selectivity of grazing. *Rhodesian Journal of Agricultural Research*, 16, 117-31.

Gilliland, C.E., J. Robertson and H. Cover (2004). Texas rural land prices, 2003. Real Estate Center at Texas A&M University. College Station, Technical Report 1676, 7pp.

Gould, F.W., G.O. Hoffman & C.A. Rechenthin (1960). Vegetational areas of Texsa. Texas Agricultural Extension Service, College Station, L-492.

Hanselka, C.W., A. McGinty, B.S. Rector, R.C. Rowan & L.D. White (1990). Grazing and brush management on Texas rangelands an analysis of management decisions. Texas Agricultural Extension Service Report.

Hart, R.H., J. Bissio, M.J. Samuel & J.W. Waggoner, Jr. (1993). Grazing systems, pasture size, and cattle grazing behavior, distribution and gains. *Journal of Range Management*, 46, 81-87.

Heitschmidt, R. K., S. L. Dowhower & J.W. Walker (1987). 14- vs. 42-Paddock rotational grazing: above ground biomass dynamics, forage production, and harvest efficiency. *Journal of Range Management*, 40, 216-223.

Holechek, J.L., R.D. Pieper & C.H. Herbel (1989). Range Management Principles and Practices. Regents/Prentice Hall, Englewood Cliffs, 501 pp.

McGrann, J. (2003). Standardized performance analysis – SPA – for southwest herds – 1991 – 2001, Texas Cooperative Extension, College Station, 6 pp.

McSwain, R. (1996). Texas sheep and goat raisers' association: a history of service to the industry. Anchor Publishing, San Angelo, 288pp.

Merrill, L.B. (1954). A variation of deferred-rotation grazing for use under southwest range conditions. *Journal of Range Management*, 7, 152-154.

Merrill, L.B. (1959). Heavy grazing lowers range carrying capacity. Texas Agricultural Progress Report, 5(2),18 1pp.

National Research Council (1994). Rangeland health: new methods to classify, inventory, and monitor rangeland. National Academy Press. Washington, D.C. 180 pp.

Owens, K. W. & R. K. Lyons (2003). Evaporation and interception water loss from juniper communities on the Edwards Aquifer Recharge area. Final Report to San Antonio Water System., Texas Agricultural Experiment Station. Uvalde

Smeins, F., S. Fuhlendorf & C.A. Taylor, Jr. (1979). Environmental and land use changes: a long-term perspective. In: 1997 Juniper Symposium. Texas Agricultural Experimental Station. Sonora Technical Report 97-1. 1-3 – 1-21.

Stafford Smith, M. (1996). Management of rangelands: paradigms at their limits. In: J. Hodgson & A.W Illius (eds), The Ecology and Management of Grazing Systems, CAB International, Wallingford, 325-357.

Taylor, C.A., Jr. & S.D. Fuhlendorf. (2003). Contribution of goats to the sustainability of Edwards Plateau rangelands. Texas Agriculture Experiment Station. Sonora Technical Report 03-1. 24 pp.

Teer, J.G. (1996). The white-tailed deer: natural history and management. In: P.R. Krausman, (ed.) Rangeland wildlife. The Society for Range Management, Denver. CO, 193–210.

Ueckert D.N., R.A. Phillips, J.L. Petersen, X.B. Wu,& D.F. Waldron (2001). Redberry juniper canopy cover dynamics on western Texas rangelands. *Journal of Range Management*, 54,603-610.

Walker, J.W. & K.C. Hodkinson. (1999). Grazing management: new technologies for old problems. Proceedings of VI International Rangeland Congress, People and Rangelands Building the Future. Volume 1, 424-430.

Walker, J.W. (1994). Multispecies grazing: the ecological advantage. *Sheep Research Journal.* Special Issue, 52-64.

Wilkins, N., A. Hays, D. Kubenka, D. Steinbach, W. Grant, E. Gonzalez, M. Kjelland & J. Shackelford (2003). Texas rural lands trends and conservation implications for the 21st century. Texas Cooperative Extension, College Station, Bulletin B-6134, 26pp.

Working within constraints: managing African savannas for animal production and biodiversity

J.T. du Toit
Department of Forest, Range and Wildlife Sciences, Utah State University, UT 84322, USA,
E-mail: jtdutoit@zoology.up.ac.za

Abstract

The mean density of livestock biomass on African rangelands now greatly exceeds that of indigenous large herbivores, although livestock cannot fully substitute for wildlife with respect to co-evolved ecosystem processes involving herbivory. The dominance of livestock in semi-arid rangelands is largely due to water provision, which uncouples livestock population dynamics from the rainfall-driven trajectories followed by indigenous ungulate species in wildlife areas. Ecological sustainability cannot be achieved with a few exotic species maintained at unprecedented biomass densities in savanna ecosystems, which are evolutionarily adapted for species-rich communities of ungulates of a wise range of sizes. Integrating wildlife and livestock in multi-species animal production systems offers a partial solution. Community-based ecotourism can effectively augment pastoralism in semi-arid rangelands but, as rainfall increases, the opportunity costs become too high. Direct payment approaches show promise for offsetting opportunity costs, although major obstacles remain in the form of political corruption and obstructive practices by national governments.

Keywords: rangeland ecology, pastoral systems, tropical conservation, livestock-wildlife interface

Introduction

African pastoralists maintain a tenacious hold on their livestock traditions despite the ravages of droughts, parasites, infectious diseases and predators. With the human population of Africa (presently ~885 million people, of which ~65% conduct rural livelihoods) growing at the world-leading rate of 2.4% per annum (PRB, 2004), it is inevitable that livestock pressure on rangelands must increase even further. Across the pastoral lands of the African savanna biome and its arid ecotones, livestock species already dominate the standing crop of ungulate biomass. In southern African countries the average national ratio of biomass of domestic herbivores:wild herbivores is in excess of 6:1, while the regional ratio of cattle:elephant biomass is about 16:1 (Cumming & Cumming, 2003). The high value of this ratio cannot be simply explained by over-hunting of indigenous species, and is remarkable given that domestic herbivores were introduced to the Horn of Africa some 8,000 years ago and reached southern Africa only about 2,000 years ago (Denbow & Wilmsen, 1986). Without the advantages of ecosystem-specific evolutionary adaptations, as have accumulated in Africa's diverse indigenous ungulate fauna, how did a few introduced livestock species achieve such dominance? Or is this dominance unstable? And what is the future for African rangelands that are becoming increasingly marginal for pastoralism due to combinations of overstocking and global change? These are the main questions that are addressed in this chapter, in which it is also suggested how some current trends in the science and management of tropical conservation might show promise for securing the natural capital of Africa's rangelands as well as the livelihoods of its pastoralists.

Overview of the problem

Subsistence pastoralists in the semi-arid regions of Africa include the world's poorest and most politically marginalized people, who strive to improve their livelihoods under conditions of extreme uncertainty. Widespread and continuous risks of war, forced displacement, social upheaval and institutional, and economic, collapse are compounded by illiteracy, pandemic diseases (especially malaria and AIDS) and climatic unpredictability. Under such conditions it is impossible for pastoralists to adopt the long-term view espoused by advocates of 'sustainable development'. It is also very difficult for aid agencies from the developed world to achieve incremental progress with the management of rangeland ecosystems in such countries because of political corruption and the disintegration of formal structures of governance. Ironically, at present, tropical countries typically have inherently high levels of biodiversity but low governance scores, and indicators of conservation success (or failure) are correlated with national governance scores (Smith *et al.*, 2004). Pastoral systems in African countries will thus continue to lose natural capital as long as pastoralists subsist without the self-organizing institutional structures needed for effective environmental management (du Toit *et al.*, 2004).

Nevertheless, the long-term implications of biodiversity loss have little direct relevance to most subsistence pastoralists in Africa (for reasons already outlined), who are understandably more concerned with how many of their animals they can bring through each drought period. In that respect they are astoundingly successful, to the extent that Oesterheld *et al.* (1992) suggested that extensive pastoral management in African savannas has achieved a similar effect to that in South America, where animal husbandry has increased the overall herbivore carrying capacity of the system at a regional scale. After detailed analysis, however, Fritz & Duncan (1994) refuted this for African savannas where indigenous large herbivore assemblages include mega-herbivores (i.e. with a body mass >1000 kg), which cannot be fully substituted by cattle (Fritz *et al.,* 2002; du Toit & Fritz, 2003). Nevertheless, considering buffaloes (*Syncerus caffer*) as the indigenous African 'equivalent' of cattle, Owen-Smith & Cumming (1993) compared the average metabolic biomass density attained by buffaloes in large protected areas against that for cattle in regions of Africa free of tsetse flies (*Glossina* spp, vectors of *Trypanosoma* parasites). They found that cattle outweigh buffaloes, and indeed any other single species of indigenous grazing ruminant in a protected savanna ecosystem, by a factor of three. This is mostly as a result of water provision; the component of animal husbandry that has provided the greatest boost to livestock carrying capacity in African pastoral systems.

Water provision

Semi-arid grazing systems exhibit high spatial and temporal variability in animal numbers, and this is because density-dependent resource limitation operates episodically in key resource areas where the accessibility of food is restricted during dry seasons and drought periods (Illius & O'Connor, 1999). The crucial factor here is resource accessibility rather than mean resource abundance across the landscape, since grazing ungulates in the dry season are central-place foragers, having to return regularly to a water source. Foraging radius is determined by energetic relationships involving body size and gait, and indigenous ungulates in the 200-500 kg range of live weight can achieve a dry-season foraging radius of ~20 km (Pennycuick, 1979). For cattle the radius is considerably less due to their comparatively slow gait, which is why semi-arid pastoral systems are characterized by discrete concentrations of livestock occurring in overgrazed and trampled piospheres around water points. If, however, through the construction of dams, wells and boreholes, the mean nearest neighbour distance among water points can be reduced to ~30 km, then virtually all of the standing crop of dry season forage in the landscape becomes accessible to cattle. Given the dependence of animal numbers on dry season food resources in semi-arid pastoral systems (Illius & O'Connor, 1999), it follows that a dramatic increase in the accessibility of these resources will result in a similarly dramatic increase in animal numbers. While pastoralists and politicians celebrate this result, and international development agencies maintain a penchant for introducing water-points, because *inter alia* they are tangible, popular, quick to construct and convenient for media exposure, the long-term ecological implications of a hyper-inflated grazer biomass require more serious thought.

Comparative studies on the processes, by which extensive subsistence pastoralism influences rangeland dynamics, are remarkably scarce, and there is a particular need for controlled experimentation. For example, the ways in which an African savanna landscape changes through time under a regime of subsistence pastoralism should be compared with a set of "control" landscapes with similar vegetation, soils and rainfall but with intact indigenous herbivore communities. From the few comparative studies conducted so far, it appears that the single most important finding concerns the use of groundwater as drinking water, which profoundly influences the dynamics of the plant-herbivore interaction. Walker *et al.* (1987) analyzed demographic data for a range of indigenous large herbivore species in relation to a severe and extended drought in southern Africa during 1982-1984. They found that, although the drought caused widespread declines in wildlife populations, these were particularly catastrophic (80-90% in extreme cases) for grazing ungulates in wildlife areas that had the most closely spaced artificial water points, e.g. Klaserie in South Africa and Tuli in Botswana. In areas where permanent water was more dispersed, e.g. Kruger in South Africa, or where managers had heavily culled the pre-drought populations, e.g. Umfolozi in South Africa, the demographic responses to the drought were much less pronounced. The implication is that where the mean nearest-neighbour distance between water points is artificially reduced, the accessibility to key resources is greatly increased in pre-drought conditions. This causes depletion of "reserve" grazing areas, which lie beyond the normal foraging radius and are not heavily grazed but can be accessed by very hungry animals. If pre-drought grazer densities are allowed to increase in response to the increased accessibility of key resources, then the population crashes caused by droughts are much steeper and deeper than in rangelands with more widely dispersed water points.

Another important finding from Chivi in Zimbabwe was that the number of cattle owned per village in semi-arid communal lands varies widely from decade to decade in response to the cumulative departure of annual rainfall figures from the long-term mean (Sayer & Campbell, 2004). Changes in cumulative rainfall control groundwater, which is abstracted through wells and boreholes at water points, and the collapse of Chivi cattle herds in the 1990s was attributed to water-point failure following a prolonged rainfall deficit from 1981 to 1992. In wildlife areas (e.g. Kruger) it is also apparent that indigenous grazing ungulate populations decline in response to long-term rainfall deficits (Owen-Smith & Ogutu, 2003), but the controlling factor there is availability of green grass in the dry season rather than access to drinking water (Dunham *et al.*, 2004).

Water provision in semi-arid African rangelands has the effect of uncoupling livestock population dynamics from the rainfall-driven trajectories followed by indigenous ungulate populations. In a run of normal or wet years the artificial dispersion of water points allows livestock biomass density to exceed that expected from an indigenous wildlife community.

But during droughts the livestock numbers crash due to both the depletion of "reserve" areas of key resources and the drying of water points, which causes an exponential decline in the accessible food supply. Recovery of livestock herds to pre-drought sizes is slow because the extent of each population crash can be staggering: in Chivi the cattle population in 1993 was just one quarter of what it had been three years earlier (based on data in Sayer & Campbell, 2004). Furthermore, the return of herds to their former grazing areas depends on the resurrection of water points and therefore on the rate of groundwater recharge, but groundwater profiles are ultimately determined by the underlying geology, so recovery time differs widely from one water point to the next. Finally, catchment hydrology can be altered during droughts by changes in land cover, causing unfavorable changes in *inter alia* runoff, infiltration and river sedimentation. The net effect is that livestock numbers in semi-arid pastoral systems tend to oscillate more widely and less predictably than for wildlife in conserved systems. It comes as no surprise, therefore, that in a retrospective analysis of an 11-year data set from Namibia, Burke (2004*a*) found no statistical relationships between livestock numbers and either annual rainfall or water-point numbers. Quite simply, livestock numbers cannot increase in direct response to rainfall at the end of a dry spell, and counting water points on a map is meaningless without detailed records of which water points were functional over what periods. Moreover, the relationship between livestock density and water point density asymptotes when the mean nearest-neighbour distance (D) between functioning water points is less than twice the foraging radius (R) of the dominant livestock species (i.e. when $D < 2R$).

Ecosystem effects

African savannas are especially heterogeneous at the ungulate habitat scale, and the uniquely high diversity of African ungulates is directly linked to vegetation heterogeneity across multiple spatial and temporal scales (du Toit & Cumming, 1999). The regime of natural herbivory is thus patchily distributed, the species-rich community of indigenous large herbivores (typically ~30 species in any one ecosystem) represents a body size range that spans three orders of magnitude (<10 kg to >1,000 kg), and it includes two main trophic guilds (grazers and browsers), each with a diverse set of co-evolved herbivore-plant interactions: for reviews, see McNaughton & Georgiadis (1986) and du Toit (2003). In sharp contrast, the strategy of extensive livestock production is to remove indigenous large herbivores, and, through water provision, maximize the proportion of rangeland available to livestock, thereby imposing a herbivory regime of high biomass and low diversity (<5 species in any one system). By effectively reducing the heterogeneity of a semi-arid ecosystem in this way, and substituting the indigenous herbivore community for an ecologically mismatched assemblage of livestock, it is inevitable that rangeland degradation will follow, i.e. reduced production of ecological goods and services: see Illius & O'Connor (1999) for a discussion on the definition of rangeland degradation.

To begin with, there are some purely physical implications associated with eradicating a diverse and size-structured indigenous ungulate community and substituting it with a livestock assemblage of much lower species diversity. Across ungulate species of increasing body mass (M), hoof area increases in direct proportion with body mass (αM^{1}) but stride length increases more gradually ($\alpha M^{0.26}$), which means that in each step taken a cow exerts the same pressure per cm^2 of soil as a gazelle, but in walking the same distance the cow causes greater soil compaction than the gazelle (Cumming & Cumming, 2003). For semi-arid rangelands, the implications become apparent if herbivore biomass density is held constant and the trampling impact of a pastoral system against that of an indigenous wildlife community is compared. Cattle typically dominate pastoral systems in southern and eastern Africa, while indigenous communities are typically comprised of numerous species of differing size of which only some are water-dependent. The pastoral system has a far greater impact on the soil, not only because of the greater trampling impact per unit animal biomass, but also the extremely high trampling intensity along stock trails that radiate from water points. Trampling causes soil compaction, which impedes water infiltration and seed germination, and some stock trails become the erosion gullies that characterize degraded rangelands.

From the perspective of a savanna ecosystem, removing the indigenous assemblage of large herbivore species implies the termination of multiple co-evolved interactions, especially with indigenous plant species. Some of these interactions, involving seed dispersal, nutrient cycling, mediation of inter-specific plant competition and maintenance of vegetation structure, are taken over by livestock species, which become surrogates for their indigenous counterparts, but the extent to which this occurs requires investigation. It is likely that the greatest losses of co-evolved interactions occur at the extremes of the body size range, with the functional properties of small habitat-specific antelopes from the tribes Antilopini and Neotragini, and mega-herbivores, i.e. hippopotamus, rhinoceros, elephant and giraffe, being particularly non-substitutable (du Toit & Cumming, 1999). Empirical evidence from African savannas on the ecosystem-level effects of the switch from wildlife to cattle is still scarce and equivocal (Richardson-Kageler, 2003). Nevertheless, evidence from rangelands on other continents, e.g. Australia, indicates an association between the introduction of livestock and the local extinction of certain plant species, some of which have functional properties of importance for maintaining ecosystem resilience (Walker *et al.*, 1999).

Global change and unpredictability

Subsistence pastoralists in African savannas depend on traditional knowledge for maintaining their herds but the world is changing faster than traditional knowledge can adapt. Scientific knowledge is not, however, filling the gaps, mainly because of the challenges of maintaining an effective extension service in rural Africa. Furthermore, while scientists seem fairly confident in predicting that global warming will increase the frequency and severity of droughts in African rangelands, there are many other changes that are expected, yet cannot be predicted. For example, the rising atmospheric concentration of CO_2 is expected to influence rangeland plants to the extent that changes should occur in the quantity and quality of food available to livestock. Higher C: N ratios are expected in foliage with increased concentrations of structural tissues and C-based secondary chemicals, which reduce plant digestibility and palatability for large herbivores. Indeed, from experimental studies it is clear that C: N ratios do generally become elevated in plants growing in the CO_2 concentrations predicted for future scenarios but allocations of C to structural tissue and secondary chemistry are unpredictable (Diaz *et al.*, 1998; Penuelas & Estiarte, 1998). It thus remains unclear how such responses might influence large herbivores and translate into ecosystem-level effects at larger spatial and temporal scales.

The inherent problem of unpredictability in African rangelands is likely to become exacerbated by the globalisation of trade, which can exert powerful and unexpected forces on local markets. In this respect a lesson for African savannas can be learned from another biome - the rainforests of West Africa. There the human population uses fish as a primary source of protein, but European fishing fleets operating off the West African coast have reduced oceanic fish stocks and at the same time developed a local dependence on the fishing industry for food and jobs. Brashares *et al.* (2004) have shown that, with a dramatic decline in the availability of fish in local markets in Ghana, alternative protein sources from animal production are inadequate to meet demands because the dependence on fish had obstructed the development of a local self-organising livestock industry. The protein deficit is now being made up by bush meat, which is extracted from the forest in alarmingly unsustainable quantities. In this example, the global market for one protein source, fish, have prevented, and then unexpectedly presented, opportunities for the local markets of livestock, as a sustainable protein source, at a scale of demand that cannot be met. The price is now being paid, locally, in biodiversity.

In another example, subsistence pastoralists have seized an opportunity presented by the globalization of trade but ironically the outcome of that is also a reduction in natural capital. With the current boom in global markets of organically-grown coffee, pastoralists in and around the village of Hangala, in southern India, are exporting cow dung for use as manure on coffee plantations. This lucrative source of income has caused pastoralists to increase their herds and make greater use of their grazing rights in the adjacent Bandipur National Park, but the intensified herbivory and almost complete removal of dung is posing a serious threat to ecosystem function (Madhusudan, 2005). In short, the preferences of coffee-buying people in the developed world have caused a traditional subsistence pastoral system in India to 'flip' into a commercial nutrient-mining operation of a scale that was completely unexpected by local conservation authorities. A similar transition could occur in pastoral systems within trading distance of the coffee-producing regions of East Africa.

Ways forward?

The preceding review begs two obvious questions. Firstly, are subsistence pastoral systems, as influenced by human population growth, poverty, weak institutional structures, socio-political instability and globalisation, driving the communal rangelands of Africa into scenarios of irretrievable degradation? Secondly, have any new approaches emerged to improve the effectiveness of conservation funds that filter through the hierarchies of international development agencies and non-government organisations, African national governments and traditional pastoral communities? The first question is left for future environmental historians to answer but an attempt is now made at answering the second question. In doing so, it must be pointed out that to prevent any loss of human livelihood options for future generations, a goal of sustainable development, the herbivory regime on African rangelands has to be restored to something resembling the indigenous, naturally selected state. This might be considered idealistic but it is assumed there is adequate interest among decision makers to justify a brief discussion on conserving ecosystem processes while also improving livelihoods in African savannas.

Although animal husbandry can increase livestock biomass densities well above those maintained by any indigenous bovids in conserved ecosystems (Owen-Smith & Cumming, 1993), the advantages of an intact wildlife community, that includes mega-herbivores, will generally outweigh those of any livestock assemblage in the same ecosystem (du Toit & Fritz, 2003). This applies particularly in moist-dystrophic savannas, e.g. in the miombo vegetation type of south-central Africa, where the percentage contribution of elephant biomass to total herbivore biomass reaches ~60% in regions of comparatively low soil nutrient status that receive ~1000 mm of annual rainfall (Fritz *et al.*, 2002). The reason is that the standing crop of plants in such regions is of high abundance but low quality, being fibrous, woody and chemically defended, and elephants, with their voluminous guts and low mass-specific metabolic demands, tolerate such food better than smaller herbivores can (Bell, 1982). Due to greater efficiency of resource use, therefore, a multi-species animal production system that retains elephants will outperform a livestock-dominated pastoral system in terms of annual meat production per unit area in moist-dystrophic savannas, e.g. in central and north-western Zimbabwe, northern Botswana, western Zambia and eastern Angola. The problem is that subsistence pastoralists cannot manage elephants as they do livestock, and cannot easily harvest and distribute the meat due to logistical constraints and external objections to the killing of elephants. Furthermore, the direct costs of retaining elephants in a settled area, including risks of injury and death to people, damage to crops, granaries and water supplies, adds to the opportunity costs incurred by foregoing land use options that are incompatible with elephants, and these are usually higher than the sustainable elephant-derived benefits that may be relied upon to augment household economies. This problem of managing elephants outside of wildlife reserves exemplifies the general tendency for opportunity costs associated with conserving biodiversity to increase along the arid-to-humid gradient in tropical ecosystems (Burke, 2004b; du Toit, 2004). Quite simply, as rainfall increases so does productivity, and so does the range of opportunities for exploiting it.

In Africa's semi-arid rangelands, however, the opportunity costs associated with retaining wildlife are comparatively low because the alternatives are limited. There is thus considerable scope for developing community-based ecotourism (Kiss, 2004) as an adjunct to pastoralism, and some promising results are emerging from this approach in low-rainfall countries like Namibia (Burke, 2004*b*). The obvious implication for pastoral systems in rangelands with higher rainfall is that the adoption of a conservation-oriented approach to land use will require financial subsidies to offset the opportunity costs. Deriving these subsidies and channeling them to the communities that need them is a major challenge, especially since political corruption presents a significant obstacle to conservation activities in African countries (Smith *et al.*, 2003). Nevertheless, current thinking within major international funding agencies, such as the World Bank, is tending towards the 'direct payments' approach to funding conservation interventions in developing countries (Ferraro & Kiss, 2002). In principle, direct payments could be provided to semi-arid pastoralists in return for reaching contractually arranged short-term goals, such as destocking village herds, performing reclamation work on erosion gullies, closing water points, removing snares and cutting out alien invading shrubs. In the longer term, negotiated payments in cash or kind could be earned when monitoring surveys detect population increases in key wildlife species, reduced sediment loads in rivers or improved rangeland condition. The financial implications to donors are obviously vast. However, if such payments can indeed be channeled directly to pastoralists living in rangelands of high conservation value, thereby avoiding the costs of corruption and overheads associated with conventionally funded projects that operate through government agencies, then conservation gains per unit cost will be greatly improved.

Conclusions

The livelihoods of Africa's pastoralists depend on the flows of ecosystem goods and services from Africa's rangelands, and so the conservation of savanna ecosystem processes simply has to be the central feature of all major rangeland development interventions. Meat, milk, dung, blood, hides and draught power are all examples of goods and services that rangelands can provide through livestock, while the pastoral lifestyle is a key component of the social-ecological systems of African savannas. However, to achieve a sustainable balance requires an integration of wildlife and livestock management, which in turn requires subsidies in the form of direct payments on a sliding scale based on the opportunity costs associated with wildlife. Opportunity costs of biodiversity conservation are generally lower in the more arid rangelands, where community-based ecotourism can significantly augment the household economies of traditional pastoralists. In the more humid rangelands, direct payments from global funds are increasingly required for the goals of ecological sustainability to be realistically met. Contractually negotiated direct payments should not be viewed as 'handouts', since tropical biodiversity is actually of greater value to the global community than it is to local pastoralists (Balmford & Whitten, 2003). Nevertheless, we cannot delude ourselves over the enormous obstacles presented by political corruption and weak or obstructive governance systems in the countries of which Africa's pastoralists are citizens. The only way forward is for rangeland ecologists and managers to interact more effectively with economists, sociologists and political scientists to quantify and demonstrate the benefits of sustainable rangelands to everyone, ranging from subsistence pastoralists to government leaders (du Toit *et al.*, 2004).

Acknowledgements

This chapter was written while the author was in the Mammal Research Institute of the University of Pretoria.

References

Balmford, A. & T. Whitten (2003). Who should pay for tropical conservation, and how should the costs be met? *Oryx,* 37, 238-250.

Bell, R.H.V. (1982). The effect of soil nutrient availability on community structure in African savannas. In: B.J. Huntley & B.H. Walker (eds) Ecology of Tropical Savannas. Springer-Verlag, Berlin, 193-216.

Brashares, J.S., P. Arcese, M.K. Sam, P.B. Coppolillo, A.R.E. Sinclair & A. Balmford (2004). Bushmeat hunting, wildlife declines, and fish supply in West Africa. *Science,* 306, 1180-1183.

Burke, A. (2004a). Range management systems in arid Namibia – what can livestock numbers tell us? *Journal of Arid Environments,* 59, 387-408.

Burke, A. (2004b). Conserving tropical biodiversity: the arid end of the scale. *Trends in Ecology and Evolution,* 19, 225-226.

Cumming, D.H.M. & G.S. Cumming (2003). Ungulate community structure and ecological processes: body size, hoof area and trampling in African savannas. *Oecologia*, 134, 560-568.

Denbow, J.R. & E.N. Wilmsen (1986). Advent and cause of pastoralism in the Kalahari. *Science,* 234, 1509-1515.

Diaz, S., L.H. Fraser, J.P. Grime & V. Falczuk (1998). The impact of elevated CO_2 on plant-herbivore interactions: experimental evidence of moderating effects at the community level. *Oecologia*, 117, 177-186.

Dunham, K.M., E.F. Robertson & C.C. Grant (2004). Rainfall and the decline of a rare antelope, the tsessebe (*Damaliscus lunatus lunatus*), in Kruger National Park, South Africa. *Biological Conservation*, 117, 83-94.

du Toit, J.T. (2003). Large herbivores and savanna heterogeneity. In: J.T. du Toit, K.H. Rogers & H.C. Biggs (eds) The Kruger experience: ecology and management of savanna heterogeneity. Island Press, Washington, DC, 292-309.

du Toit, J.T. (2004). Response to Burke. Conserving tropical biodiversity: the arid end of the scale. *Trends in Ecology and Evolution,* 19, 226.

du Toit, J.T. & D.H.M. Cumming (1999). Functional significance of ungulate diversity in African savannas and the ecological implications of the spread of pastoralism. *Biodiversity and Conservation,* 8, 1643-1661.

du Toit, J.T. & H. Fritz (2003). Size matters: body size diversity and resource use efficiency in ungulate guilds. In: N. Allsopp, A.R. Palmer, S.J. Milton, K.P. Kirkman, G.I.H. Kerley, C.R. Hurt & C.J. Brown (eds) Proceedings of the VIIth International Rangelands Congress, July/August 2003, Durban, South Africa. Document Transformation Technologies, Irene, 499-501.

du Toit, J.T., B.H. Walker & B.M. Campbell (2004). Conserving tropical nature: current challenges for ecologists. *Trends in Ecology and Evolution,* 19, 12-17.

Ferraro, P.J. & A. Kiss (2002). Direct payments to conserve biodiversity. *Science*, 298, 1718-1719.

Fritz, H. & P. Duncan (1994). On the carrying capacity for large ungulates of African savanna ecosystems. *Proceedings of the Royal Society, London, B,* 256, 77-82.

Fritz, H., P. Duncan, I.J. Gordon & A.W. Illius (2002). Megaherbivores influence trophic guilds structure in African ungulate communities. *Oecologia*, 131, 620-625.

Illius, A.W. & T.G. O'Connor (1999). On the relevance of nonequilibrium grazing concepts to arid and semiarid grazing systems. *Ecological Applications,* 9, 798-813

Kiss, A. (2004). Is community-based ecotourism a good use of biodiversity conservation funds? *Trends in Ecology and Evolution,* 19, 232-237.

Madhusudan, M.D. (2005). The global village: linkages between international coffee markets and grazing by livestock in a South Indian wildlife reserve. *Conservation Biology*, 19, 411-420.

McNaughton, S.J. & N.J. Georgiadis (1986). Ecology of African grazing and browsing mammals. *Annual Review of Ecology and Systematics,* 17, 39-65.

Oesterheld, M., O.E. Sala & S.J. McNaughton (1992). Effect of animal husbandry on herbivore-carrying capacity at a regional scale. *Nature,* 356, 234-236.

Owen-Smith, N. & D.H.M. Cumming (1993). Comparative foraging strategies of grazing ungulates in African savanna grasslands. *Proceedings of the XVII International Rangelands Congress, New Zealand*, 691-698.

Owen-Smith, N. & J. Ogutu (2003). Rainfall influences on ungulate population dynamics. In: J.T. du Toit, K.H. Rogers & H.C. Biggs (eds) The Kruger experience: ecology and management of savanna heterogeneity. Island Press, Washington, DC, 310-331.

Pennycuick, C.J. (1979). Energy costs of locomotion and the concept of the "foraging radius". In: A.R.E. Sinclair and M. Norton-Griffiths (eds) Serengeti - dynamics of an ecosystem. University of Chicago Press, Chicago, 164-184.

Penuelas, J. & M. Estiarte (1998). Can elevated CO_2 affect secondary metabolism and ecosystem function? *Trends in Ecology and Evolution*, 13, 20-24.

PRB (2004). 2004 World Population Data Sheet. Population Reference Bureau, Washington, DC, 16pp.

Richardson-Kageler, S.J. (2003). Large mammalian herbivores and woody plant species diversity in Zimbabwe. *Biodiversity and Conservation,* 12, 703-715.

Sayer, J. & B. Campbell (2004). The science of sustainable development. Local livelihoods and the global environment. Cambridge University Press, Cambridge, 268pp.

Smith, R.J., R.D. Muir, M.J. Walpole, A. Balmford & N. Leader-Williams (2004). Governance and the loss of biodiversity. *Nature*, 426, 67-70.

Walker, B.H., R.H. Emslie, R.N. Owen-Smith & R.J. Scholes (1987). To cull or not to cull: lessons from a southern African drought. *Journal of Applied Ecology,* 24, 381-401.

Walker, B.H., A. Kinzig & J. Langridgen (1999). Plant attribute diversity, resilience, and ecosystem function: The nature and significance of dominant and minor species. *Ecosystems*, 2, 1-20.

Range-based livestock production in Turkmenistan

R.H. Behnke and G. Davidson
Macaulay Institute, Craigiebuckler, Aberdeen AB15 8QH, UK,
Email: Kerven_behnke@compuserve.com

Abstract

Turkmenistan retains a centralized system of livestock production in which many critical assets are owned by the state. Though technically in the temperate zone, the country's climate is harsh and unstable. Groundwater resources are unevenly distributed, leaving many potential grazing areas seasonally inaccessible due to lack of drinking water for livestock. This paper summarizes the results of a three-year study of rangelands, livestock production, flock economics and land tenure at two study sites, one in central and the other in eastern Turkmenistan. The results of this study suggest that pastoral communities in Turkmenistan have coped remarkably well with the institutional changes that followed the demise of the Soviet Union, and with the country's persistently unstable climate and scarce natural resources.

Keywords: Turkmenistan, Central Asia, pastoralism, grazing systems, agricultural reform

Introduction

Turkmenistan has preserved more of its Soviet agricultural legacy than almost any other part of the former USSR. In Turkmenistan the state owns almost all agricultural land and maintains large collective farms that supply critical commodities, such as wheat, cotton and meat, in response to state production targets and procurement orders. Following the dissolution of the Soviet Union and market reforms in China, centralized agricultural regimes of this kind are increasingly rare. The object of this paper is to describe how this system works in the pastoral sector of Turkmenistan.

The paper argues that the system works because, contrary to expectations, it is effectively decentralized. Despite state controls, herders have considerable freedom to fashion husbandry systems that are adapted to their individual needs and local resource availability. The result is a remarkably constant level of livestock output irrespective of variation in natural resource endowments and herd sizes. There is also evidence that the system supports equitable livestock distributions, with small herds growing more consistently than large ones over the study period.

Administrative organization of the livestock sector

The present organization of the livestock sector resulted from reforms following presidential decrees in the 1990s.

In 1994 and 1995 presidential decrees transformed the Soviet collective (*kolkhoz*) and state (*sovkhoz*) farms into farmer associations or *dihan birlishik.* The new associations took on the assets of the old Soviet farms and adopted the old farm boundaries. What did change after the Soviet era was the way agricultural production was organized inside these farms. Arable farm land was no longer worked collectively, but was subdivided and leased to individual families. Instead of a salary, these farmers now sold the produce of their lease holdings, either at controlled prices to the government or on the open market (Lerman and Brooks, 2001). In the pastoral sector, leasehold contracts pertained not to land plots but to herds of state-owned animals, which became the responsibility of individual shepherd families. Like leasehold farmers, these shepherds no longer received a salary from the state or the collective farm. They were instead entitled to a proportion of the offspring of the herd under their care, in return for bearing the costs of herd maintenance and assuming all risks if animals died or were lost. Unlike cotton and wheat marketing, which was characterized by both input and output price distortions, shepherds and the state transacted their business on a barter basis, each side taking its income in live animals. Shepherds were free to sell their produce on the open market, and neither the state nor the collective farms provided shepherds with subsidized services or inputs.

The adoption and standardization of the contract leasing (*arinda*) system took several years to work out. In the late 1990s herding contracts were not uniform, and the share of a herd or flock's offspring that belonged to the shepherd differed according to agro-ecological conditions or by administrative region (see Lunch, 2004, for the period up to 1999). In 1999 officials briefly considered paying shepherds a salary calculated on the value of their share of flock output, rather than in live animals. In the late 1990s it was also unclear who should provide inputs like supplementary fodder and veterinary services. Some collective farm managers said that provision of these services for state-owned flocks should be their responsibility, but few of the shepherds keeping state-owned animals received enough inputs from the collective farms and most depended on their own resources.

By 2000 there was in place a uniform national system of livestock leasehold. Shepherds with breeding flocks were the single most common type of contractor in the livestock sector. The terms of their contracts assumed a 95% lambing rate with half of the lamb crop going to the shepherd and half to the association. For example, for a flock of 1000 ewes, presumed lamb production would be 950 with, at weaning, 475 head going to the shepherd and a similar number to the farmer association, with the association having first claim to female animals. The shepherd bore all risks. Inputs, such as fodder, veterinary services or water transport by tanker truck, could be purchased by the shepherd from the association, and payment deferred until the end of summer when lambs were counted, separated and accounts settled. The shepherd was also free to obtain these inputs on the open market. The shepherds were responsible for shearing and kept all wool, and were entitled to slaughter a set number of animals for home consumption and to receive advances on their 'wages' prior to weaning.

By 2000 three different kinds of large farms owned livestock - specialized livestock farms, arable farms in which livestock keeping was an ancillary activity, and district-wide 'shareholder stock associations'. Transformed into farmer associations, the specialized livestock and arable farms were the institutional descendants of Soviet-era collective and state farms. The shareholder stock associations were new and were created by presidential decree in 1999 to address the problems of keeping animals on farms that were involved predominately in irrigated farming. In the middle to late 1990s, livestock populations on these arable farms had declined. To stem these losses, it was decided to take state-owned livestock away from arable farms, which were only marginally interested in pastoral activities and which had scattered livestock holdings that were difficult for the central authorities to supervise. Livestock collected from these farms were pooled into a single operation that managed all the state-owned livestock and pastures in a district (*etrap*), the lowest level in the national administrative system.

Shareholder stock associations were the largest operations in the livestock sector. In Mary Province, for example, there were nine shareholder associations, each covering a single district of the province, with holdings averaging 84,000 head of sheep per association in 2003. The shareholder stock association for the District of Bayram Ali north and west of Mary town was typical. It was formed in 1999 from about 70,000 sheep and 246,000 ha of pastures appropriated from the district's ten wheat and cotton farms. By 2003 the association had increased its holdings to 84,000 head, kept in 85 separate flocks, but had supplied its shareholders - the farms from which it had initially taken stock and land - with only 1000 head. The low dividends paid to the shareholding farms followed an explicit government policy to minimize animal sales and slaughter in order to expand the size of the national flock. In winter, association flocks grazed pastures about 100 km from Bayram Ali town; in summer the flocks moved to distant pastures in the mountains and foothills near the Afghan border. Pastures near the Karakum canal were not used by state-owned animals and were available for use by herds and flocks owned privately by people living along the canal.

Methods and field sites

Agro-ecological and socio-economic research was carried out at two sites - one on a collective farm in Mary Province (*wilayat*) in eastern Turkmenistan, and the other in the pastoral portion of the District (*etrap*) of Gokdepe, in Ahal Province close to the capital city of Ashgabat. Work at these sites included a livestock census and survey of livestock husbandry practices, in-depth interviews with shepherds, farm managers and district-level officials, and the analysis of statistical data available from state organizations.

The field site in Mary Province included all of the Ravnina village *dihan birleshek*, or farmer association, located in Baydram Ali District. Ravnina village is located about 100 km to the north and east of Mary city along a paved road and with good rail links to the city. About 260-270 families - roughly 1850 people - lived in the association's territory - including the population herding in the desert, living in hamlets at stops along the railroad line, and in the central village. The farm, which receives about 140 mm of precipitation per year, consists almost exclusively of desert pastures, is 346,000 ha in size and is traversed by 45 km of the Karakum canal. Ravnina *dihan birleshek* was a specialized livestock production farm and only a few families on the farm engaged in any cultivation aside from irrigated backyard gardening. In 2004 the farm kept about 26,000 head of state-owned sheep in 34 flocks averaging slightly less than 800 head per flock. In 2004 families on the farm privately owned about 7000 head of sheep and goats, 100 camels and a couple of hundred cattle.

Fifty-five randomly selected shepherds, keeping both private and state-owned animals, were interviewed in Ravnina in 2003 and 2004. A standard questionnaire was used to collect information on herd composition and size, herd movement patterns over the last year and the use of fodder. Intensive open-ended interviews on a wide range of subjects related to livestock-keeping were conducted with selected shepherds and farm staff. Officials responsible for livestock were also interviewed in Bayram Ali District, where Ravnina is located.

The second study area consisted of the pasture areas that make up the northern two-thirds of Gokdepe District. Gokdepe town, the administrative center of the District, lies about 50 km west of the national capital of Ashgabat on paved roads along the Karakum Canal. The pastures belonging to the district stretch about 150 km to the north of the canal into the Karakum desert. At the time of the study, eleven collective farms with their headquarters and main settlements along the canal held northern pastures. All of these collective farms were primarily engaged in arable agriculture, but held some state-owned sheep and camels under the care of shepherds permanently resident in the pasture areas. Pasture areas and settlements north of the canal were accessible only by unpaved desert tracks.

Using the same questionnaire that had been developed for Ravnina, ninety-two interviews with randomly selected state and private shepherds were conducted along a north-south transect that began at the northern fringe of the district's settled zone and ran north to the northern boundary of the district. The transect included 20 settlements ranging in size from a single family to just under 40 families. The pastures covered by the transect supported approximately 21,000 head of sheep and goats, 2000 camels and, in the most southern settlement on the fringes of the cultivated zone, 130 cattle. Intensive open-ended interviews were held with the managers of collective farms, district-level government staff and shepherds between 1999 and 2004.

To estimate flock performance twenty sample flocks were also selected in each study area. The sample flocks were chosen to reflect the distribution of sheep and goats in flocks of different sizes in the study communities. The flocks were visited approximately every three months from August 2001 for eighteen months. During each visit live weights were recorded on a sample of up to 30 sheep in each flock, in the morning prior to the animals going out to graze. For flocks of less than 30, all animals were weighed. For those of more than 30, a representative sample was monitored. For mixed flocks of sheep and goats of over 50 animals, the species were chosen roughly in proportion to the species in the whole flock. To ease identification monitored animals were ear-tagged.

A specialized livestock farm: Ravnina

At the time of the Soviet Union, Ravnina village and farm was one of two specialized livestock farms in the district of Bayram Ali. When the district-wide shareholder stock associations were formed in 1999, specialized pastoral operations, like Ravnina, were permitted to keep their animals and their independent identity.

In the late Soviet period Ravnina state farm employed around 300 people. By 2004 the association employed about 30 people, half in the farm's engineering section as drivers, watchmen, mechanics and pump operators, and the other half consisting of managers and office staff - the director, accountants, economists, veterinarians and secretaries. Thirty-four shepherds kept state-owned association sheep on contract. Aside from the farm itself, there were roughly another 100 salaried employees living in the village and working at the local school, health post and on the railroad.

Ravnina village is a station on the railroad line between Mary and Charjev and owes its existence to the railroad. In 1882-3 the railroad arrived and people started to settle in the vicinity. But the village was not founded until 1927 when the Soviet authorities confiscated livestock from rich owners in the neighboring province of Lebab on the other side of the Amu Darya River, and resettled both animals and shepherds in the new village. Initially Ravnina was a department within a neighbouring state farm, but it became an independent state farm in 1932 and was subsequently upgraded to a Karakul sheep breeding station in 1966. In 1963 the village was supplied with piped water from the Karakum canal. In 1996 the farm was re-established as a farmer association.

The pastures operated by the farm are bisected by the Karakum canal. Ravnina village and roughly a quarter of its pastures lie north of the canal. These northern pastures were occupied primarily by privately-owned livestock belonging to village residents. The remaining pastures south of the canal contained about 40 wells constructed between 1932 and 2000 and varying in depth from about 20 to over 100 metres. Shepherds keeping association animals, usually a flock of 500-900 head resident year-round at a single well, occupied these pastures.

Mean annual rainfall in Ravnina is about 143 mm. *Haloxylon aphyllum, H. persicum, Calligonum setosum* and *C. divaricatus* dominate the vegetation north of the canal and around the village itself. The vegetation of the rolling dune country in the pastures to the south of the farm is dominated by *Calligonum eriopodum, Ephedra strobilacea, Salsola richteri* and *Astragalus unifoliolatus*. Dry matter yields varied from a low of 250 kg/ha near the village to 650 kg/ha on the southern rangelands (Gintzburger *et al*., 2005).

Figure 1 shows the numbers of state-owned sheep on the farm from 1940-2004. Several phases in the farm's development can be detected from these figures:

- From 1940 to the early 1960s there was a steady increase in small ruminant numbers, which peaked at 67,000 head in 1962. There is little evidence that the digging of new wells led to these increases in stock numbers. Instead, the grazed area which was accessible to water was relatively constant over the 1950s and the stocking rate in this area increased.
- 1969 was a disastrous year for which no flock size figure is available. Old farm managers recall that in 1969 after losing about 27,000 animals the farm had about 23,000 head. The reason for the mortality was a severe winter, which proved a turning point in the farm's management strategy. Thereafter it focussed on fodder collection as a buffer against winter weather. Also, in the 1970s fodder collection was mechanized, which substantially increased the amount that could be harvested. Adequate fodder provision was calculated to be 150 kg per head of livestock. Despite these precautions, total flock size never again equalled that of the early 1960s. When total numbers began to increase in the late 1990s, they were again reduced by a severe winter.
- When records resumed again in 1970, there was a steady two-decade-long increase in flock size from about 30,000 to around 40,000 head. During this period, extreme weather events - either good or bad years - had no consistent or visible impact on stocking levels. This result conforms to the opinions of experienced shepherds who assert that there is no reason for poor years to become disasters if precautions have been taken to collect sufficient winter fodder. During this period, growth in sheep numbers was contained by a high offtake - roughly 5000 head annually for meat to the government, 5000 head as breeding stock for other farms, and 10-15,000 karakul lamb pelts per year.
- 1999 was another year for which no records were kept. Up to 14,000 sheep may have died in that year. When record-keeping resumed again in 2000 the flock was down to 20,000 head, whereas it had stood at 45,000 in 1998. Heavy snowfalls occurred in late winter when the sheep had already moved to fresh pasture and would not return to eating the dry standing material which was all that was available after snow covered the ground for a week to 10 days. The weather was, therefore, a genuine problem, but it need not have been a disaster according to most shepherds. This was a period of transition to the current *arinda* system of contract flock management. At this point the collective farm was responsible for fodder provision but in fact had collected very little, and shepherds were being erratically paid. The poor weather revealed underlying institutional problems.
- After the crash in 1999, sheep were in good condition because of the decline in their numbers, and the contract herding system paid shepherds well and there was no confusion over the responsibilities of shepherds and farm managers. Flock numbers were again rising and abandoned wells were being re-opened to accommodate newly created flocks.

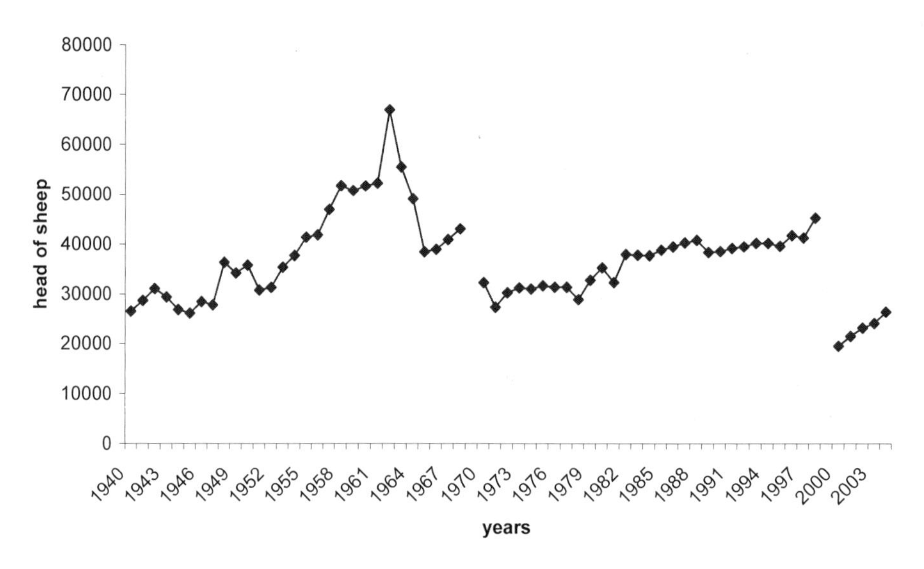

Figure 1 State-owned sheep numbers in Ravnina, 1940-2004

Table 1 documents two geographically separate systems of livestock production on Ravnina Farm. Pastures south of the canal were occupied by over 30 flocks of state-owned sheep combined with the private animals owned or cared for by the shepherds looking after the state animals, a total of about 26,000 sheep. Sampled herds averaged over 900 sheep equivalents and generally occupied a single well each. In contrast, with the exception of one recently established government flock and the farm's collective camel herd, all animals north of the canal were privately owned. These private flocks were small but increased in average size as one moved further from the central village. In the village itself, average holdings were 38 sheep equivalents typically consisting of about 30 head of sheep and goats and a small number of dairy cows for household milk consumption. Herds in outlying hamlets were substantially larger (at 70 sheep equivalents) and those based in isolated farmsteads or mobile camps were larger still (at 227 sheep equivalents per holding). In total, about 7000 sheep equivalents were kept north of the canal.

The government flocks south of the canal were managed very differently than the private flocks to the north. The contrast is starkest when village-based flocks are compared with those at wells in the southern desert. Per head, village animals received about fifteen times more feed supplements than the desert animals - 78.9 versus 5.3 fodder units per sheep equivalent, respectively (Table 1). This supplementary feed was also used very differently in the two locations. In the village, supplementation was a regular feature of animal diets, and all animals in village flocks were supplemented for nearly five months of the year. In the desert, feed supplementation was reserved for emergencies when snow prevented grazing or was given selectively to weak, pregnant or lactating animals for less than two months per year. On average, desert rations consisted of camel thorn (*Alhagi persarum*) for bulk plus one additional feed item of higher quality, such as alfalfa, a feed concentrate or grain; village rations were based on camel thorn and two additional high-quality feed items.

Table 1 Herd size, forage availability and fodder use in Ravnina: 2003-4

	Wells[1]	Outside village[1]	Hamlet[1]	Main village[1]	s.e.d.[2]	P value
Number of sampled herds	15	9	4	27	-	-
Sheep equivalents per sampled herd[3]	971	247	70	38	-	-
Small ruminants per sampled herd	792[4]	227	52	30	-	-
Pasture production (kg DM/ha/year)[5]	448	415	415	379	-	-
Stocking rate (ha/sheep equivalent)	4.5[5]	7.1	9.6	5.0	-	-
Pasture production (kg DM/year/sheep equivalent)	2016	2946	3984	1895	-	-
Fodder units/sheep equivalent[6]	5.3	13.2	20.6	78.9	9.5	<0.001
Fodder cost/sheep equivalent (in manat)[7]	5943	4108	2958	30441	5521.5	<0.001
Cost/fodder unit (in manta)[7]	1046	140	82	353	290	<0.001
Number of kinds of fodder used[8]	1.9	1.8	2.0	2.9	0.47	<0.01
Flocks in which all animals received fodder	7%	33%	100%	100%	0.13	<0.001
Months of regular winter feeding	1.8	1.7	3.0	4.8	1.23	<0.001

Notes:
[1]Wells = Association wells in pastures south of the Karakum canal; Outside village = households occupying isolated farmsteads north of the Karakum canal; Hamlet = settlements of 6-12 households at railroad stops along the line; Village = the central village of Ravnina
[2]Standard Error of the Difference
[3]Estimated live weights of livestock species were converted into sheep equivalent units by a multiplication factor based on the estimated mean live weights derived from FAO (1989). 1 SEU or LU was deemed to be equivalent to a 45 kg Karakul ewe. 1 camel = 4.6 stock units; 1 cattle = 3.6 stock units.
[4]Shepherds were allowed to keep their private animals at the desert wells with the state flock. On average 136 of the sheep and goats in each flock were privately owned.
[5]Estimates based on Gintzburger *et al.* (2005).
[6]One Soviet Fodder Unit is equivalent to the total nutritive value of 1 kg of dry oats (Zhambakin, 1995). The following conversions were used for other types of fodder: 1 kg of camel thorn = 0.3 of a fodder unit (fu); 1 kg alfalfa = 0.45 fu; 1 kg maize = 1.24 fu; 1kg of wheat = 1.16 fu; 1 kg cottonseed residues = 0.66 fu; 1 kg wheat bran = 0.71 fu; 1 kg crushed straw = 0.21 fu; 1 kg maize stems = 0.15 fu; 1 kg natural grasses = 0.3 fu; 1 kg of kombicorn (feed concentrate) = 0.71 fu.
[7]$1.00 USD = 22,000 Turkmenistan manat at the informal exchange rate in 2003-4.
[8]Typical types of fodder are listed in note 6.

Transportation costs explain part of the difference between northern and southern feeding regimes. Cost per fodder unit could be up to ten times higher in the southern desert than in the northern areas where supplies could be obtained cheaply from nearby farming communities or urban markets. Village shepherds also complained about poor quality and over-used village pastures, which forced them to purchase feed supplements to compensate for poor natural grazing. Estimates of pasture production (annual plant and ephemerals biomass) revealed a steady north to south gradient of rising productivity, with the least productive pastures north of the village and the most productive around the southern-most wells (Gintzburger *et al.*, 2005). However, stocking rates around the village were very similar to those at the wells, largely because village flocks walked further to their pastures. As a consequence, estimated pasture production per stock unit was roughly similar around both the wells and the village, which does not explain why shepherds complained about poor grazing conditions in the vicinity of the village. The most likely explanation is that village flocks had to walk further to their pastures, and that these pastures provided a flush of productivity in spring followed by dearth in summer and winter due to a relative absence of perennial and woody vegetation.

Despite the differences in feeding regime, there were no significant differences in the live weights of adult female sheep between those from the wells and those in village flocks 48.0

kg vs. 48.4 kg; s.e.d. 0.58). There was a significant difference ($P<0.01$) in goat live weights between those from the wells and those in village flocks (38.6 kg vs. 41.9 kg; s.e.d. 0.98).

Gokdepe district

Ahal Province, which includes the national capitol of Ashgabat, was exempted from the reorganization that created shareholder stock associations in the late 1990s. Unlike the rest of the country, in Ahal Province arable farms continue to own and manage livestock as an adjunct to irrigated farming. Gokdepe District illustrates this arrangement.

In 2003 Gokdepe District contained fourteen farmer associations situated along the Karakum canal, primarily involved in wheat and cotton production. These farms owned a total of over 23,000 sheep and goats and 3500 camels, kept on over 4000 km^2 of rangeland. The bulk of this grazing land lay within a rectangle roughly 30 km wide in an east-west direction that stretched from the canal north into the Karakum desert for about 150 kilometres. Farms tended to own between two and four discontinuous blocks of grazing land, and to have herds scattered throughout the northern part of the district.

Rainfall is higher at the southern than in the northern desert pastures - 140 mm per annum in the south versus 110 mm in the north. Groundwater is also more plentiful in the south, as waste water from crop irrigation is channelled into canals that feed marshes and lakes in the desert. Although parasite-infested, this waste water is abundant in the southern sand-clay desert and freely available for watering stock, in contrast to the limited supplies of well-water available elsewhere.

To the north, vegetation in the sand desert is dominated *by Haloxylon persicum, Carex physodes, Ephedra stroboliacea* and *Aristida pennata*. Average DM yields are 211-239 kg/ha with an available fodder portion of 88-107 kg/ha (Khanchaev *et al.*, 2004). Vegetation in the sand-clay desert at the south end of the transect has been modified by grazing and by the removal of *H. persicum* for fuel-wood. Dominant species are *Calligonum rubens, Salsola richteri*, and *Carex physodes*. Average total dry mass production is 187 kg/ha per year, with an available fodder portion of 99 kg/ha/year (Khanchaev *et al.*, 2004). Mean production figures are, however, deceptive. Over a three-year period that included both drought and good rainfall years (2001 to 2003), DM yields varied threefold at sampling sites along the transect (Khanchaev, 2005).

Human settlements and livestock population levels along this transect are directly correlated with water availability. Areas that offer more water and better quality water have attracted more settlers and more livestock. In the far south, where water was freely available, residents owned about 19 sheep equivalents for every km^2 of pasture accessible from their settlements (Table 2). In the far north where water was scarce and of poor quality, local residents kept many fewer animals relative to the pasture area available to them - 7.6 sheep equivalents per km^2. In two middle grazing zones, both water availability and stock densities were intermediate.

When feed is in short supply, the herder has several options: to move the animals to the feed, to move the feed to the animals, or to move water where it is needed. The husbandry practices described in Table 2, i.e. nomadism, fodder provision, water transport, and changes in herd composition, therefore compensate for the aggregation of stock around plentiful water supplies. In zones of heavy stock concentration, herds and flocks spent less time in the immediate vicinity of the village, thereby reducing stocking rates around settlements. Fodder provision further closed the gap between the amounts of natural forage available from lightly- versus heavily-used pasture areas, and trucked water opened up fresh pastures in areas where natural water supplies did not exist. Finally, the choice of herd species influenced local stocking densities and feed availability. Camels roam widely around settlements on a daily or weekly basis, but become attached to home ranges that are incompatible with long-distance seasonal migration. Sheep are the opposite, with restricted daily movement but the capacity to migrate long distances.

At the south end of the transect, water for livestock was abundant, stocking densities were high, and it was the supply of natural forage that limited further expansion in herd numbers. Local herders responded by using fodder on a regular basis, trucking water, migrating seasonally, and by specializing in sheep that were adapted to long-distance migration. At the opposite extreme was the area at the far northern end of the transect, a cluster of isolated wells deep in the Karakum desert. In this area livestock numbers were restricted by the small quantity of poor quality, saline water that was available. Herders in this zone pushed their wells hard, maintaining more animals per working well than communities elsewhere along the transect. But the density of animals owned by residents was low relative to available grazing, and aside from keeping many camels, herders employed none of the husbandry practices that were used elsewhere to improve feed availability.

Evidence suggests that shepherds were remarkably successful in adapting their husbandry practices to equalize livestock output despite variable resource availability. A total of 1353 small ruminants were weighed quarterly for a year with the sample divided into three groups: flocks based in the southern sand-clay desert and migratory, those based in the southern sand-clay desert but resident year-round, and those based in the southern sand desert. Despite the differences in location and husbandry practices, there were no significant differences in adult sheep weights (43.3 kg vs. 43.9 kg; s.e.d. 0.45) between flocks based in the southern sand-clay desert and migratory, those based in the sand-clay desert but resident all year round and those based in the northern sand desert.

Table 2 Water and feed availability and husbandry practices by herding households in Gokdepe District, Turkmenistan 2002-3

Grazing zone	Southern sand-clay desert	Northern sand-clay desert	Southern sand desert	Northern sand desert
Water availability	Abundant, fresh surface water	Fresh well water	Fresh and saline well water	Saline well water and runoff
Grazing area around settlements (km^2)	377	710	707	974
Resident sheep equivalents per working well	2830	283	293	974
Stocking rate (sheep equivalents per working well	146	175	169	200
Total sheep equivalents owned by residents/km^2	19.3	11.2	12.9	7.6
Stocking rate (sheep equivalents/ km^2)	10.9	6.9	6.8	7.6
Proportion of sheep and goats	0.95 (n=22)	0.56 (n=9)	0.52 (n=12)	0.59 (n=13)
Proportion of camels	0.05 (n=10)	0.44 (n=10)	0.48 (n=15)	0.41 (n=7)
Proportion of migratory flocks	0.68	0	0	0
Proportion of flocks using trucked water	0.68	0.89	0.58	0
Proportion of sheep regularly receiving fodder	0.23	0	0.08	0
Proportion of camels regularly receiving fodder	1.00	0	0.08	0

Equity and taxation

Twenty flocks in Ravnina and twenty in the Goktepe study area were sampled four times annually in 2001-2002. Over this period, 25 flocks grew in size and 15 became smaller, and the propensity to either expand or contract was correlated with their initial size (Table 3).

Table 3 Growth and decline in size of flocks

Initial flock size	Decrease in flock size (%)	Increase in flock size (%)
1-150 sheep equivalents	13	87
150+ sheep equivalents	71	29

Based on a single year of observations in an extremely variable climate, these results are inconclusive but suggest that greater differentiation in herd wealth was not occurring in the communities studied during the short time they were monitored.

The egalitarian ethos of rural Turkmen undoubtedly played a part in sustaining small herds. A contributing factor may also be the way that the contract (*arinda*) herding system taxed the pastoral sector. Shepherds, herding for the state, surrendered half of the offspring of their flocks, equivalent to an income tax rate of 50%. Private flocks were not taxed. The actual tax rate for the pastoral sector as a whole therefore depended on the balance of private versus state animals. In the Gokdepe sample, sheep holdings were evenly divided between private and state animals, giving an average pastoral income taxation rate of about 25% of animal offtake, with shepherds liable for herding expenses but the beneficiaries of dairy and fibre production. In Ravnina, where state-owned animals constituted about 80% of the holdings, the comparable taxation rate was just under 40%.

Conceived of as a taxation system, contract herding does not apportion the tax burden evenly. While the state may have claimed up to half of the income of contract herders, private herders paid nothing. The all-or-nothing nature of the *de facto* taxation system encouraged the growth of small private herds. As a general rule, the pastures adjacent to large agricultural settlements were set aside for grazing by private animals, while more productive but distant pastures were occupied by state-owned herds. This system worked well for the private owners of a few dozen sheep and goats, who wanted to keep their animals around the village, did not need extensive pastures for their flocks and could afford to offset poor grazing by providing feed supplements for a small number of animals. The system was less advantageous for large private flock owners who wanted secure access to extensive pastures. This access could only be obtained by caring for state animals, which gave the shepherd the privilege of pasturing private and family-owned animals alongside the state-owed flock. In this way the *de facto* tax burden fell disproportionately on larger herd owners or on extended kin groups that had to herd state animals in order to secure better grazing rights.

Conclusion

The large agricultural enterprises, state ownership of livestock, and contract herding system of independent Turkmenistan have a long regional history. Referring to Mongolia, Humphrey and Sneath (1999) observed that:

In some respects the change from a 'feudal' to a collective organizational form was a less radical change than the one currently underway as the government attempts the transition to a market economy. In both 'feudal' and collective periods there were centralized, commandist politico-economic units that regulated residence, the use of pasture, and extracted a surplus through right to livestock.... Like the feudal lords and the monasteries before them, the collectives organized movement, single-species herds, and allocated pasture.

Humphrey and Sneath (1999) were arguing that hierarchical institutional forms persisted in the transition from feudalism to socialism in Mongolia. Much the same point can be made for the transition from socialism to state-dominated capitalism in contemporary Turkmenistan.

The dominant role of the state in arable farming has been judged as, at best, a mixed success in Turkmenistan (Lerman and Brooks, 2001). State involvement in the pastoral sector has, thus far, been more successful. The contract herding system was uniform, reasonably transparent and deemed by most shepherds to be a fair payment system. It avoided price distortions by paying shepherds in live animals and provided material incentives for those who exceeded their contractual obligations. On a day-to-day basis most decision-making had been delegated to the shepherds to devise husbandry systems and to obtain the inputs that they needed in their particular circumstances. These arrangements resulted in remarkably constant levels of productive performance despite differences in herd sizes and local variations in pasture and water resources. The state has remained ultimately in control through its ownership and command over the allocation of natural resources.

Acknowledgements

Research in Turkmenistan was supported by the European Commission Inco-Copernicus RTD Project ICA2-CT-2000-10015 Desertification and Regeneration: Modelling the Impact of Market Reform on Central Asian Rangeland (DARCA), Macaulay Institute, Aberdeen, U.K.

References

FAO (1989) Animal Genetic Resources of the USSR. N.G. Dimitriev and L.K. Ernst (eds.) Food and Agricultural Organisation Animal Production and Health Paper 65, FAO, Rome. 517pp

Gintzburger, G., S Saidi and V. Soti (2005). The Ravnina rangelands in Turkmenistan: current vegetation condition and utilisation. Unpublished report, the DARCA project. CIRAD-ECONAP, Montpellier, 75pp.

Humphrey, C. and D. Sneath (1999). The end of nomadism? Society, state and the environment in Inner Asia. Duke University Press, Durham, 353pp.

Khanchaev, K. (2005). The results of vegetation monitoring in Goktepe. Unpublished report, the DARCA project. Macaulay Institute, Aberdeen, 11pp.

Khanchaev, K., C. Kerven and I.A. Wright (2004). The limits of the land: pasture and water conditions. In: C. Kerven (ed.) Prospects for pastoralism in Kazakstan and Turkmenistan: from state farms to private flocks. RoutledgeCurzon, London, 194-209.

Lerman, Z. and K.Brooks (2001). Turkmenistan: an assessment of leasehold-based farm restructuring. World Bank Technical Paper No. 500. Europe and Central Asia Environmentally and Socially Sustainable Development Series, Washington, D.C., 68pp.

Lunch, C. (2004). Shepherds and the state: effects of decollectivisation on livestock management. In: C. Kerven (ed.) Prospects for pastoralism in Kazakstan and Turkmenistan: from state farms to private flocks. RoutledgeCurzon, London, 171-193.

Zhambakin, Z. A. (1995). Pastbisha Kazakhstana (Pastures of Kazakstan). Kainar, Almaty.

Section 1

Biological constraints on pastoral systems in marginal environments

A new perennial legume to combat dryland salinity in south-western Australia

L.W. Bell, M.A. Ewing, M. Ryan, S.J. Bennett and G.A. Moore
CRC for Plant-based Management of Dryland Salinity & The University of Western Australia, Nedlands Western Australia A 6100, Australia, Email: lbell@agric.uwa.edu.au

Keywords: perennial pastures, canary clover, alfalfa, recharge control, water use

Introduction Dryland salinity has devastated large tracts of productive land in Australia. This has resulted from the clearing of native perennial vegetation and its replacement with annual crops and pastures. As annual plants are shallow rooted and only use water during their winter-spring growing season, unutilised rainwater leaks into groundwater tables which rise and bring stored salt to the soil surface. The adoption of deep rooted perennial pasture plants that increase the water use can help to manage dryland salinity whilst maintaining productivity. However, new plants are needed as few perennial pasture options currently exist. Preliminary research into the potential of hairy canary clover (*Dorycnium hirsutum* (L.) Ser.) to increase water use is presented.

Materials and methods Six replicate plots (2.5 m x 4 m) of *Dorycnium hirsutum*, lucerne cv. Sceptre (*Medicago sativa* L.), annual burr medic cv. Santiago (*Medicago polymorpha*) and a bare ground control, were established in 2002 at Merredin in the Western Australian wheat-belt, a low rainfall (315 mm annual mean) site with a mediterranean climate. Soil moisture content was monitored under these pastures using a neutron moisture meter from September 2002 to April 2004 at approximately 3 week intervals.

Results Establishment was slow in 2002 with an extremely dry winter growing season. During the first summer lucerne dried the soil more than the annual medic, but little difference was observed in soil water between *D. hirsutum* and the annual medic (Figure 1). Both annual and perennial species dried the soil during spring 2003, but rainfall during summer increased soil water under the annual medic while both perennials maintained a drier soil profile. *D. hirsutum* and lucerne dried the soil to a depth of 180 cm in the 2003/04 summer increasing the soil moisture deficit to 80 mm greater than that recorded under the annual medic.

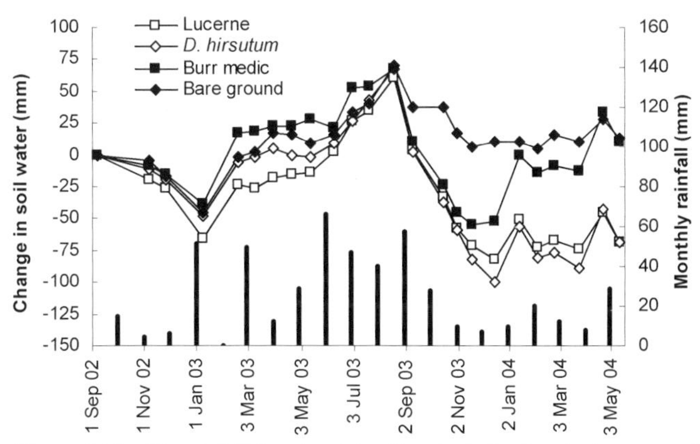

Figure 1 Monthly rainfall (bars) and Δ soil water (< 3 m) under bare ground, annual medic, lucerne and *D. hirsutum* pastures from September 2002 to April 2004 at Merredin, Western Australia

Conclusions Both lucerne and *D. hirsutum* dried the soil profile more than annual medic, thereby creating an additional buffer of 80 mm that would need to be exceeded before drainage could occur. Similar studies of lucerne have shown it to be effective at reducing recharge (Latta *et al.*, 2001 & 2002; Ward *et al.*, 2001). Water use of *D. hirsutum* was equal to lucerne in the second year. This was surprising given the different breeding history and adaptation of these species to a semi-arid mediterranean environment. We conclude that *D. hirsutum* shows some promise for reducing groundwater recharge and helping to manage dryland salinity in farming systems of the Western Australian wheat-belt.

References

Latta R.A., L.J. Blacklow & P.S.Cocks (2001). Comparative soil water, pasture production, and crop yields in phase farming systems with lucerne and annual pasture in Western Australia. *Australian Journal of Agricultural Research*, 52, 295-303.

Latta R.A., P.S. Cocks & C. Matthews (2002). Lucerne pastures to sustain agricultural production in south-western Australia. *Agricultural Water Management*, 53, 99-109.

Ward P.R., F.X. Dunin & S.F. Micin (2001). Water balance of annual and perennial pastures on a duplex soil in a Mediterranean environment. *Australian Journal of Agricultural Research*, 52, 203-209.

Diversity and variation in nutritive value of plants growing on 2 saline sites in south-western Australia

H.C. Norman, R.A. Dynes and D.G. Masters
CSIRO Livestock Industries, Private Bag 5, Wembley, WA 6913, Australia, Email: Hayley.Norman@csiro.au

Keywords: dryland salinity, biodiversity, nutritive value, saltbush, feeding value

Introduction In south-western Australia 10% or 1.8 million ha of the farmed area is affected by dryland salinity and a further 6 million ha are at risk of salinity (NLWRA, 2001). Animal production from saltbush (*Atriplex* spp.)-based pasture systems represents the most likely large-scale opportunity for productive use of saline land in the short to medium term. Feeding saltbush-based pastures as a maintenance feed during the prolonged autumn feed gap typical in Mediterranean-type climates maximises their economic value. The aim of this study was to explore the diversity and nutritive value of plants that typically persist in saltbush-based saltland pastures.

Materials and methods Two highly saline pastures were chosen at Meckering and Tammin in Western Australia. The Meckering site (19 ha) was situated 130 km east of Perth in the 400 mm annual rainfall zone. The Tammin site (12 ha) was situated 180 km east of Perth in the 325 mm annual rainfall zone. Saltbush was established at both sites more than 10 years before this study and no other plants species had been deliberately introduced. Plant diversity and quality were determined along 3 transects within each plot in autumn 2001. Diversity was measured through the Botanal technique (Mannetje & Haydock, 1963), ranking only the herbaceous component. All species were sampled for laboratory analysis of digestibility (pepsin-cellulase digestibility of the organic matter in the dry matter (P-CDOMD, Norman *et al.*, 2004), pepsin-cellulase organic matter digestibility (P-COMD), ash, acid-detergent fibre (ADF %OM) and crude protein (CP) concentrations.

Results The Meckering site contained 31 herbaceous plant species, of which 26 were volunteers and 7 were native to Australia. The Tammin site contained 24 plant species of which 19 were volunteers and 8 were native to Australia. Both sites contained greater botanical diversity than is found in adjacent non-saline areas. Of the feed on offer at the Tammin site, 60% was derived from halophytic shrubs, 38% grasses and 2% from forbs and legumes. Feed on offer at the Meckering site consisted of 43% halophytes, 52% grasses, 4% forbs and 1% legumes. The legumes provided the best quality biomass with both high high digestibility and CP content. There was considerable variation in P-CDOMD within the grasses. Many had sufficient energy for maintenance of an adult sheep however CP content was low. Some forbs and all halophytes accumulated excess salts. Although many of these plants have high P-COMD values, intake is likely to be restricted by salt and, therefore, sheep are unlikely to maintain themselves on these species alone. The halophytes provided a good source of CP.

Table 1 Mean nutritive values of plants collected from 2 saline sites in autumn

	#	P-CDOMD %	OMD %	CP %	ADF %OM	Ash %
Legumes	4	61.8 (0.6)	66.8 (1.7)	14.9 (1.2)	27.3 (2.9)	8.9 (2.0)
Grasses	16	56.8 (1.1)	62.2 (1.1)	4.4 (0.3)	27.8 (1.4)	9.1 (1.1)
Halophytes	16	48.0 (0.9)	64.2 (1.2)	9.2 (0.6)	14.4 (0.9)	25.5 (1.0)
Forbs	5	58.2 (2.1)	72.9 (7.1)	6.1 (0.5)	22.5 (4.4)	17.8 (6.2)

Numbers in parentheses are the s.e. of the means

Conclusions Mature saltbush pastures in Western Australia are rich in diversity. The variation in nutritive value within and between these species is significant. Legumes have the highest feeding value but are least salt-tolerant with a low biomass contribution. Halophytes and grasses were the largest contributors to biomass but the nutritive values of grasses (low digestibility and CP content) and halophytes (high ash content) suggest that sheep need a combination of both to maintain live weight. Given the opportunity to select, mature sheep should maintain live weight during autumn while grazing. The variation in nutritive value implies opportunities to improve the feeding value of saltbush-based pastures through manipulation of species composition and agronomic selection of high value species. A mixed sward of shrubs, grasses, legumes and forbs appears to be the most productive option, since species of different salt tolerance can occupy niches in these very heterogeneous areas. This mix in turn provides the best opportunity for sheep to select for energy, CP and salt.

References

Mannetje L.'t, & K.P. Haydock (1963). The dry-weight-rank method for the botanical analysis of pasture. *Journal of the British Grasslands Society,* 18, 268-275.

NLWRA (2001). Australian Dryland Salinity Assessment 2000: extent, processes, monitoring and management options, National Land and Water Resources Audit. Canberra.

The long road to developing native herbaceous summer forage legume ecotypes

J.P. Muir, T.J. Butler and W.R. Ocumpaugh
Texas Agricultural Experimental Station, 1229 North U.S. Highway 281, Stephenville, Texas 76401, USA, Email: j-muir@tamu.edu

Keywords: pasture, prairie restoration, range reseeding

Introduction Only a handful of well-adapted herbaceous summer forage legumes are currently marketed for drier regions of North America and even fewer are true natives. There is a growing demand for native germplasm in the region as a new generation of landowner attempts to return grasslands to a semblance of their original species and diversity. The objective of this paper is to describe preliminary research results of a grasslands team collecting, studying and promulgating native leguminous germplasm in Texas.

Materials and methods Initial efforts focused on screening commercial varieties considered even remotely native to Texas. Subsequent evaluation of locally collected germplasm has provided some idea of herbage and seed production potential and will now focus on anti-quality factors, seed harvest and establishment questions.

Results A wide variety in herbage and seed production has been observed in this germplasm (Table 1), often dependent on climate, cultivation and genetic potential. Crude protein (CP) and acid-detergent fibre concentrations also vary considerably. Data from Foster *et al.* (2004) indicated high rates of CP disappearance *in sacco* but unpublished data on condensed tannins and elevated levels of lignin indicate limitations to use as a ruminant forage. BeeWild, a *Desmanthus bicornutus* mix, is the only recent release. Further studies include establishment and performance grazing and in competition with native bunch grasses.

Table 1 Seed and forage production and forage crude protein (CP) and acid-detergent fibre (ADF) concentrations of native herbaceous legumes in Texas, USA (average values reported in the literature)

Herbaceous legume name	Seed production (kg/ha/yr)	Herbage production (kg/ha/yr)	CP conc. (g/kg DM)	ADF conc. (g/kg DM)
Acacia angustissima	120	5,000	170	200
Desmanthus bicornutus	800	6,500	200	200
Desmanthus illinoensis	800	4,560	170	230
Desmanthus leptolobus	780	2,200	220	200
Desmanthus velutinus	560	1,700	170	230
Desmanthus acuminatus		800	200	210
Desmodium nuttallii		4,100	140	350
Indigofera miniata	180	2,890	150	280
Lespedeza stuevei	210	1,200	120	300
Lespedeza procumbens	50	2,780	120	270
Neptunia pubescens		2,200	160	290
Neptunia lutea		5,000	180	270
Rhynchosia americana	130	1,700	140	250
Rhynchosia senna var. *texana*		700	160	250
Strophostyles helvula	650	7,500	130	320
Strophostyles leiosperma	750	7,500	160	310

References

Foster, J.L., J.P. Muir, W.C Ellis & M.F. McFarland (2004). A nutritive evaluation of two native north Texas legumes (*Strophostyles*) for goat diet. American Society of Animal Science Abstracts, 26 July, 2004, St. Louis, MO.

Mortality model for a perennial grass in Australian semi-arid wooded grasslands grazed by sheep

K.C. Hodgkinson[1] and W.J. Muller[2]
[1]*CSIRO Sustainable Ecosystems, Canberra, ACT 2911, Australia, Email: ken.hodgkinson@csiro.au*
[2]*CSIRO Mathematics and Information Sciences, Canberra ACT 2600 Australia*

Keywords: grasslands, grazing, mortality

Introduction Grazing of sheep in marginal semi-arid environments is risky because grazing appears to predispose grass plants, especially palatable species, to sudden death (Hodgkinson, 1994; 1995). These early observations were based on a preliminary analysis of perennial grass survival in a single drought and supported the concept of tactical grazing proposed by Westoby *et al.* (1989) as a preferred management. Later this idea was developed by suggesting the existence of critical thresholds for perennial grass survival, which when crossed, collapses grass populations (Hodgkinson, 1994). Here we examine the relationship between mortality of a palatable perennial grass, *Thyridolepis mitchelliana,* and a number of variables measured during a 10-year period.

Materials and methods Seven paddocks, 4 to 15 ha in area, were each continuously grazed by six Merino wether sheep from 1986 to 1996. The landscapes within each paddock have been described by Tongway and Ludwig (1990) and are typical of banded mulga (*Acacia aneura*) woodland. In each of 5 zones within the 7 paddocks at least fifteen $1 m^2$ quadrats were randomly located. At 3- or 4-monthly intervals each grass plant was examined to see if it were alive and a number of plant and climate variables were measured. Logistic step-wise regression was used to develop the model.

Results On those occasions when plants died, the variables given in Table 1 accounted for a significant amount of the variation considered in a step-wise model. The height of plants (reflecting degree of grazing) accounted for most variation in mortality and then rain/evaporation variables were the next fitted. The only significant interaction term was zone x plant height. All variables had negative coefficients indicating that mortality decreased as variables increased. The relationships derived from the model are shown in Figure 1.

Table 1 Deviance of measured variables ($P<0.001$, %) in a step-wise regression accounting for mortality of *T. mitchelliana* over a 10-year period. Other variables, NS

Variable	Deviance
plant height	124
rain/evap. 3-6 mo	64
rain/evap. 0-3 mo	39
zone x plant height	25
zone	20
basal diameter	5

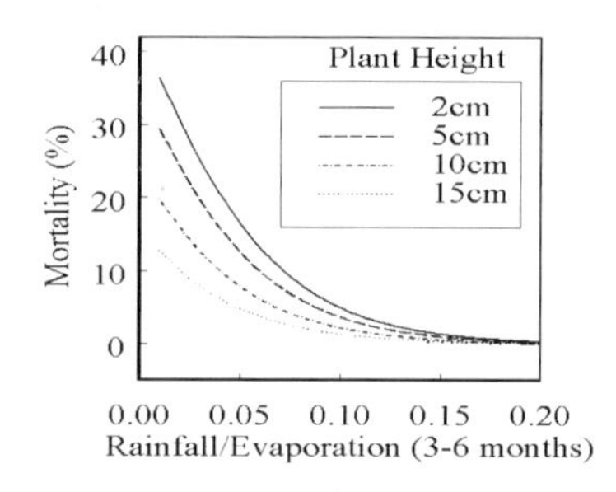

Figure 1 Modelled mortality for different heights and rain/evaporation in the preceding 3-6 months

Conclusions In this wooded grassland, the co-occurrence of drought and heavy grazing predisposes the palatable grasses to higher mortality than if the plants were not grazed. Grazing intensity, as indicated by the average height of grazed plants, had a significant negative effect on plant mortality and the effect increases with the severity of the drought period. This model supports the concept of "death traps" for perennial grasses; the trap is set by grazing and sprung by drought. The relationships can inform managers about when to consider reducing stock numbers on the basis of an approaching drought.

References

Hodgkinson, K. C. 1994. Tactical grazing can help maintain stability of semi-arid wooded grasslands. *Proceedings XVII International Grassland Congress, Palmerston North, New Zealand*, 75-76.

Tongway, D. J. & J. A. Ludwig 1990. Vegetation and soil patterning in semi-arid mulga lands of Eastern Australia. *Australian Journal of Ecology,* 15, 23-34.

Westoby, M., B. Walker & I. Noy-Meir. 1989. Opportunistic management for rangelands not at equilibrium. *Journal of Range Management,* 42, 266-274.

Selecting grassland species for saline environments

M.E. Rogers, A.D. Craig, T.D. Colmer, R. Munns, S.J. Hughes, P.M. Evans, P.G.H. Nichols, R. Snowball, D. Henry, J. Deretic, B. Dear and M. Ewing

Cooperative Research Centre for Plant-Based Management of Dryland Salinity, Perth, Western Australia 6009, Australia, Email: MaryJane.Rogers@dpi.vic.gov.au

Keywords: genetic diversity, plant salt tolerance, soil salinity

Introduction In Australia, around 5.7 million hectares of agricultural land are currently affected by dryland salinity or at risk from shallow water tables and this figure is expected to increase over the next 50 years (LWRA, 2001). Most improved grassland species cannot tolerate the combined effects of salt and waterlogging and, therefore, the productivity of sown grasslands in salt-affected areas is low. However, there is potential to overcome the lack of suitably adapted fodder species by introducing new, salt and waterlogging-tolerant species and by diversifying the gene pool of proven species. Potential species include exotic, naturalised and native Australian grass, legumes, herb and shrub species that are halophytes and non-halophytes. A collaborative national project in southern Australia commenced in 2004 with the objective of evaluating a range of forage species for saline environments.

Materials and methods The project involves glasshouse and field research. Forage germplasm is being acquired from Australian and International Genetic Resource Centres and by direct collection from centres of natural diversity for salt and waterlogging tolerance. Plant material will be evaluated for salt and waterlogging tolerance under glasshouse conditions before promising species are assessed in the field and validation phases undertaken in saline environments. It is envisaged that some priority plant material will be available and recommended to primary producers by year 6 of the project.

Results and discussion Table 1 lists the priority plant genera that have been identified with potential for salt and waterlogging tolerance and that will be evaluated in this research project (Rogers *et al.*, 2004). Initially, priority will be given to the development of superior legume cultivars, since generally these are less tolerant than their companion grasses, yet are considered "drivers" of the system, being the providers of nitrogen and forage of high nutritive value. Within this project, species will be ranked according to forage nutritive value, biomass, ground cover potential, seasonality, ease of establishment, seeding potential, persistence (eg. perenniality, palatability, drought tolerance, crown exposure etc) and for their potential weediness. Species will also be recognised for their role in areas where there is lateral and/or mosaic variation in the salt/waterlogging profile, and where a mixture of species with a range of salt tolerance levels may give the most productive option.

Table 1 High priority legumes, grasses, herbs and shrubs with salt or waterlogging tolerance

Plant category	Genera
Legumes	*Astragalus, Ceratoides, Glycyrrhiza, Hedysarum, Lotus, Medicago, Melilotus, Swainsona, Trifolium, Trigonella, Viminaria*
Grasses	*Aeluropus, Chloris, Cynodon, Dactyloctenium, Distichlis, Enteropogon, Eragrostis, Festuca, Lachnogrostis, Leptochloa, Paspalum, Pennisetum, Porteresia, Puccinellia, Saccharum, Sporobolus, Stenotaphrum, Thinopyrum, Zoysia,*
Herbs	*Cichorium, Plantago, Ptilotus*
Shrubs	*Acanthus, Atriplex, Chenopodium, Maireana, Minuria, Rhagodia*

Conclusion Finding new forage species that are adapted to saline and periodically waterlogged land will provide new options to manage dryland salinity within Australia and internationally.

References

National Land and Water Resources Audit (2001) Australian dryland salinity assessment 2000. Extent, impacts, processes, monitoring and management options. National Land and Water Resources Audit, Canberra. 129pp.

Rogers, M.E., A.D.Craig, S.J. Bennet, C.V. Malcolm, A.J. Brown, W.S. Semple, T.D. Colmer, P.M. Evans, S.J. Hughes, R. Munns, P.G.H. Nichols, G. Sweeney, B.S. Dear & M. Ewing (2004). Fodder plants for the salt-affected areas of southern Australia. Scoping Document 2 - CRC Plant Based Management of Dryland Salinity. ISBN 1740521021. 36pp.

Grazing animal production systems and grazing land characteristics in a semi-arid region of Greece

I. Hadjigeorgiou[1], G. Economou[1], D. Lolis[2], N. Moustakas[1] and G. Zervas[1]
[1]*Agricultural University of Athens, 75 Iera Odos, Athens 118 55, Greece, E-mail:ihadjig@aua.gr*
[2]*Directorate for Rural Development, Prefecture of Larisa, Larisa, Greece.*

Keywords: grazing lands, semi-arid areas, livestock systems

Introduction Rough grazing in Greece cover about 40% of the total land area, is publicly owned and managed extensively (Hadjigeorgiou *et al.*, 2002). The Prefecture of Larisa is in the centre of Greece, and has 212,000 ha of rough grazing land, with a variable topography ranging from sea level up to 3,000 m a.s.l. This area is utilized by a total population of 135,000 LU (mainly sheep, goats and some suckler cows), which consumes annually an appreciable fraction of their total nutrient requirements from rough grazing.

Materials and methods Grazing land characteristics and the grazing animal production systems were studied during a two-year period in the above region. Forty exclusion cages were erected in 4 representative areas to harvest herbage twice yearly and soil samples were collected. Herbage samples were analyzed both botanically and chemically. Nutrition balance sheets, on a yearly basis, were constructed according to information provided by twenty farmers regarding the numbers of animals farmed, the quantities of meat and milk produced and the amount of homegrown or purchased supplementary feeds fed to the animals indoors.

Results Soils are low in contents of organic matter (4.5%, s.e. 0.49), $CaCO_3$ (5.1%, s.e. 0.85), and the basic nutrients (0.24% N, s.e. 0.02; 17.4 μg P g^{-1}, s.e. 4.05; 282 μg K g^{-1}, s.e. 23.3). Climate is characterized by low rainfalls (~450 mm $year^{-1}$), high temperatures (mean annual temperature 15.5°C) and a dry summer (June to September). Herbaceous vegetation is dominated by a multitude of annual species (42 were identified) and characterized by a short growth period in spring (February to May); therefore total herbage productivity is low (c. 350 g DM/m^2 per year). Herbage nutritional quality is similarly low. Mean Crude Protein (CP) content (s.e.of mean) was 69 (6.2) g/ kg dry matter (DM) and mean Crude fibre content (s.e. of mean) was 292 (7.8) g /kg DM) and these variables not significantly different between the areas studied. The average farm raised 58 LU, which grazed for most of the year (300 days) and which were fed supplementary roughages (26% of DM requirements) and concentrates (37% of DM requirements). However, although 37% of their nutritional requirements in DM terms were covered through grazing, only by 27% of their metabolisable energy requirements and 19.5% of CP requirements were met by grazing.

Conclusions Rough grazing lands are an important element in herbivore farming systems in Greece. It appears that they have a further potential for improved herbage production, both quantitatively and qualitatively, but traditional management practices prevent the optimal use of resources.

References

Hadjigeorgiou, I., F. Vallerand, K. Tsimpoukas & G. Zervas, 2002. The socio-economics of sheep and goat farming in Greece and the implications for future rural development. Options Mediterraneennes, Series B, 39, 83-93.

The productivity of coastal meadows in Finland

R. Nevalainen, A. Huuskonen, S. Jaakola, J. Kiljala and E. Joki-Tokola
MTT Agrifood Research Finland, North Ostrobothnia Research Station, Tutkimusasemantie 15, FIN-92400 Ruukki, Finland, Email: sari.jaakola@mtt.fi

Keywords: coastal meadows, forage quality, dry matter yield

Introduction The coastal meadows of Finland have gained a new interest as a summer pasture for cattle. These habitats have great historical, aesthetic and biological value (Pessa & Anttila, 2000). Typical features of the coastal meadows are the varying vegetation zones and wet, sometimes waterlogged, soils. The meadows are important nesting and feeding habitats for many water birds. When grazing ceases, reeds, trees and shrubs take over and the area loses its openness. Lately the amount of grasslands and pastures has drastically declined all over Europe. In Finland, the area of semi-natural biotopes has decreased to 1% of what it had been at the beginning of the twentieth century (Pitkänen & Tiainen, 2001). The goal of this study was to determine the yield and nutritional value of grass herbage in the meadows.

Materials and methods The data was collected from four meadows located on the shore of the Baltic Sea. Each meadow was 60-250 ha and 20-120 cows with their calves grazed the areas for 70-120 days per year. There were two fenced areas unavailable to cows on each meadow of which the vegetation was cut every five weeks in 2003 and 2004 from 25 cm x 50 cm areas. Eight grass samples were taken from each meadow. Samples were collected weekly from the beginning of June to the middle of July. The regrowth samples were taken from vegetation cut a month previously. The soil samples were taken in 2003 from each sample area at the beginning of the growing season.

Results The soil samples showed that the soil was low in nutrients. The pH varied between 4.4 and 5.5 and the phosphorus content was 45 mg/kg. The plant samples indicated a low level (121 g/kg dry matter (DM)) of crude protein, which decreased as the summer progressed (103 g/kg DM in the regrowth sample). Mean *in vitro* DM digestibility of the herbage was 0.69 (0.64 in the regrowth sample). Neutral-detergent fibre concentrations slightly increased towards the end of the summer being 557 g/kg DM and 641 g/kg DM in the regrowth sample. The DM content of the plants was high (325 g/kg). The development of DM (Figure 1) increased with time but drastically decreased in the regrowth sample. Concentrations of Cu, K, P and Ca in herbage were low overall. Fe, Na and Mn concentrations in herbage were remarkably high.

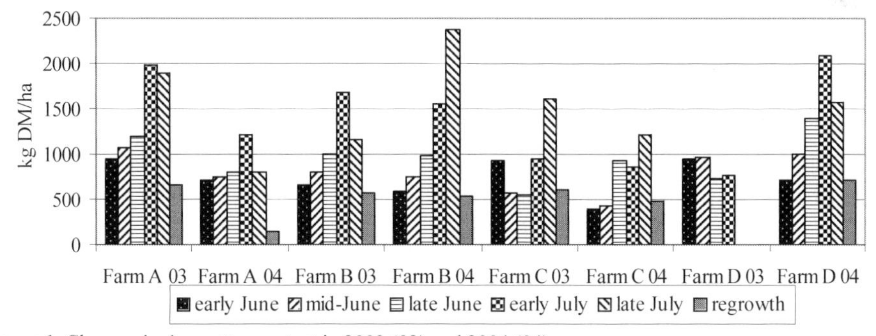

Figure 1 Changes in dry matter content in 2003 (03) and 2004 (04)

Conclusions Coastal meadows are a scarce habitat for vegetation in Finland. The plant species have low crude protein and high Fe and Na concentrations. The production of DM is generally less than half that of cultivated pasture. The quality of herbage decreases after early summer. Forage quantity and quality in late summer do not necessarily meet the feed requirements of growing calves. Nevertheless, the importance of these meadows remains high. The species diversity of vegetation is high, especially in meadows which have a long history of grazing or mowing. There is a need to continue the traditional husbandry practices to manage the landscape.

References

Pessa, J. & I. Anttila (2000). Conservation of habitats and species in wetlands. A case of Liminganlahti. LIFE Nature-project in Finland. The Finnish Environment 443. North Ostrobothnia Regional Environment Centre, Oulu, 108 pp.

Pitkänen, M. & J. Tiainen (eds) 2001. Biodiversity of agricultural landscapes in Finland. BirdLife Finland Conservation Series no 3. 93 pp.

Ear emergence of different grass species under Finnish growing conditions

M. Niskanen[1], O. Niemeläinen[2] and L. Jauhiainen[3]
[1]*MTT Agrifood Research Finland, South Ostrobothnia Research Station, Alapäätie 104, 61400 Ylistaro, Finland, E-mail:markku.niskanen@mtt.fi,* [2]*MTT Agrifood Research Finland, Plant Production Research, Crop Science, 31600 Jokioinen, Finland,* [3]*MTT Agrifood Research Finland, Research Services, 31600, Jokioinen, Finland*

Keywords: ear emergence, timothy, meadow fescue, cocksfoot, tall fescue

Introduction Timothy is the most commonly cultivated grass species in Finland. Swards cultivated for silage of hay are of pure timothy or timothy is the dominant species of the mixture. Successful timing of harvesting of the primary growth is very critical in Northern latitudes where the stand develops very rapidly at daylengths of above 18 hours with a daily mean temperature close to 20 °C. The primary growth has to be harvested within a very short period to obtain a yield of high and uniform quality. Short harvesting periods require high capacity harvest machinery which lead to high costs. If it would be possible to prolong the harvesting over a longer period of time, it would decrease the risk of bad weather conditions and reduce machine costs through better utilization of harvesting capacity. Sowing different grass species makes it possible to extend the harvesting period. The aim of this study was to investigate timing of ear emergence of different grass species in comparison with timothy. The suitable harvesting time for silage in Finland is closely related to ear emergence of the stand.

Material and methods The emerging ears of different grass species were obtained from the Finnish official variety testing database conducted at 15 various sites in Finland in 1976-2001. Ear emergence data of standard varieties of five grass species were used: timothy cv. Iki, meadow fescue cv. Boris, tall fescue cv. Retu, cocksfoot cv. Haka and perennial ryegrass cv. Riikka. The statistical model used took account of the species and varieties that had been in trials during the same years and had been grown at the same trial sites. In the variety experiments the ear emergence is specified to have take place when approximately 5% of the ears/panicles of the stand have fully emerged from leaf sheaths. The harvest time recommendation for meadow and tall fescue, cocksfoot and perennial ryegrass is to harvest when 5-10 % of the ears/panicles have emerged from leaf sheaths. Detailed information about official variety testing protocols can be found at http://www2.mtt.fi/atu/epo/lajikekoe/koeohje.html.

Results and discussion Time to ear emergence among different species is shown in Table 1. Timothy ears emerged on average on 16 June. Cocksfoot ears emerged on average eight days earlier than timothy. Perennial ryegrass came into ear three days later than timothy. There were no statistically significant differences between ear emergence for timothy and tall fescue.

Table 1 The average time to ear emergence of different grass species in official variety trials in Finland counting from 1 May. Statistic significances are in comparison with timothy

Timothy	Meadow fescue	Tall fescue	Cocksfoot	Perennial ryegrass
47	44***	47 ns	39***	50***

***, $P<0.001$, ns, not significant

Results indicate that, when it is necessary to extend the harvesting period of primary growth over a longer period of time, the use of different species represent a good possibility. In silage production the timing of harvesting is more important than in haymaking, where timing of harvesting is more based on weather conditions than on quality changes of the stand. In haymaking, extension of the harvesting period could alleviate the weather risk as well as facilitate easier management of the labour-intensive harvesting. The results suggest that in timothy dominant stands the harvesting period could be extended by using cocksfoot- or meadow fescue-dominant swards to facilitate earlier harvesting and by using perennial ryegrass-dominant swards to facilitate later harvesting than that of timothy-dominant swards. The validity of the results obtained has, however, to be tested by using data on quality changes in the studied species in the primary growth. Unfortunately the quality data available at the variety-testing databank is currently not adequate to facilitate such a study.

Effects of sowing date and phosphorus fertiliser application on winter survival of lucerne cv. Aohan in the northern semi-arid region of China

Z.L. Wang[1], Q.Zh. Sun[1], Y.W. Wang[2], Zh.Y. Li[1] and Sh.F. Zhao[3]

[1]*Grassland Research Institute, Chinese Academy of Agricultural Sciences, Huhhot, 010010, Email: wangzongli@sina.com,* [2]*Department of Grassland Science, College of Animal Science and Technology, China Agricultural University, Beijing, 100094,* [3]*Grassland Service Station of Linxi County, Linxi County, 024550*

Keywords: *Medicago sativa*, fertilisation, semi-arid region, winter survival

Introduction In the northern semi-arid region of China, winter survival is always a limiting factor for lucerne production, because low temperatures and a dry climate in winter (Zhou *et al.*, 1993; Ma, 2000; Sun & Gui, 2001; Sun *et al.*, 2003). An experiment was conducted to find an appropriate sowing date and P application rate in order to improve lucerne winter survival.

Materials and methods The study was conducted in Linxi county, Inner-Mongolia. The experimental design was a randomised block with 3 replications, which included sowing time, and fertiliser application rates. Each plot measured 5.0 m ×2.0 m. Sowing of lucerne cv. Aohan occurred on 28 May,1997 and at eight dates in 1998. Two levels of P fertiliser were applied with 15 kg/ha N on 30 May, in 1997 and on 25 June, in 1998, respectively.

Results Sowing date influenced winter survival significantly (Table 1). Shoot number per plant, buds per plant before winter and winter survival decreased at dates later than early June. P and N fertilisers applied at sowing increased winter survival in both years and winter survival increased with P application rate (Table 2). Compared to the control treatment, 75.0 kg P_2O_5/ha + 15 N kg/ha increased lucerne winter survival by 15.1% and 45.9% in 1997 and 1998, respectively.

Table 1 Responses of seeding dates in 1998 on winter survival of alfalfa

Sowing date	Shoot numbers per plant	Buds per plant	Winter survival rate (%) *
18 May	3.5-4.0	5.0-5.7	98.6
30 May	3.5-4.0	5.0-6.3	95.5
5 June	3.0-3.5	4.5-6.0	91.8
12 June	1.0-2.0	2.0-2.5	63.7
25 June	1.0-1.5	1.5-2.2	47.2
4 July	1.0	1.0-2.3	38.2
16 July	1.0	0.0-1.5	18.4
27 July	1.0	0.0-1.0	6.5

*Effects of seeding date on winter survival rate were significant ($p<0.01$)

Table 2 Effects of fertiliser on lucerne winter survival

Sowing date	Fertiliser rates (P_2O_5+N kg/ha)	Plant crown diameter (m)	Winter survival rate (%)
30 May, 1997	0	0.47	81.5
30 May, 1997	37.5 + 15.0	0.53	89.6
30 May, 1997	75.0 + 15.0	0.72	96.6
25 June, 1998	0	0.26	47.2
25 June, 1998	37.5 + 15.0	0.51	78.3
25 June, 1998	75.0 + 15.0	0.67	93.1

Conclusion In the northern semi-arid region of China, seeding date of lucerne cv. Aohan should not be later than early June. The application of 75.0kg P_2O_5/ha + 15 kg N /ha at the time of sowing is recommended in order to increase winter survival.

References

Ma, Zhiguang (2000). The technology of alfalfa industrialization in northern semiarid region of China. ICET2000-Session 6: Technology innovation and sustainable agriculture pp 531-534.

Sun, Qizhong, X. Y. Hao & Y. Q. Wang (2003). The study of alfalfa on winter survival. The secondary convention of alfalfa development in China, 34, 37.

Sun, Qizhong & R. Gui (2001). The freezing injury and preventing methods of Aohan alfalfa in the Chifeng Region. *Acta Agri-Culturae Boreali-Sinica*, 16,136-142.

Zhou, Xingmin, Y. L. Feng & D. J. Cai. (1993). Study on the introduction of cold-resistant alfalfa cultivars in the northern cold region of china. *Animal Husbandry and Veterinarian,* 11, 4-7.

Reasons for the premature decline in *Astragalus adsurgens* stands in Kerqin sandy land

Q.Zh. Sun[1], Z.L. Wang[1], J.G. Han[2], Y.W. Wang[2] and G.R. Liu[3]

[1]*Grassland Research Institute, Chinese Academy of Agricultural Sciences, Huhhot, 010010, China, Email: Sunqz@126.com,* [2]*Department of Grassland Science, College of Animal Science and Technology, China Agricultural University, Beijing, 100094,* [3]*Grassland Service Station of Chifeng City, Chifeng, 024000, China*

Keywords: *Astragalus adsurgens* cv. Shadawang, degradation, phosphorus, root rot disease

Introduction Diseases partly account for reductions in *Astragalus adsurgens* stand longevity. The effect of some cultural practices on the control of pests and diseases have been reported (Hou, 1986; Nan, 1996), but few reports have detailed the relationship among soil fertiliser status, diseases and premature stand decline. This study was conducted to investigate these relationships in order to extend the longevity of *Astragalus adsurgens* stands.

Materials and methods *Astragalus adsurgens* for this research was established in 1993 at 6-7kg/ha. No irrigation or fertilisers were applied after establishment. Disease incidence was calculated as

$$\text{Disease incidence} = \frac{\text{The number of infected plants} \times 100\%}{\text{The total number of investigated plants}}$$

Plants infected index was divided into 5 classes based on symptoms according to Flood & Isaac (1978) and Latunde-dada & Lucas (1982). Disease index of the stands was calculated according Liu Ruo (1998) as

$$\text{Disease index} = \frac{\sum (\text{Plant infected index} \times \text{plant number}) \times 100\%}{5 \times \text{total number of plants}}$$

Results In the third and fifth year after establishment, phosphorus content in soil, plant leave and stem significantly decreased compared to the first year (Table 1); however, root rot diseases and plant mortality increased. The pathogens accounting for the root rot disease were *Fusarium solani, Fusarium oxysporum* and *Fusarium moniliforme*. The stand showed premature degradation in the 5th to 6th year after establishment.

Table 1 The changes of phosphorus in soil and in *Astragalus adsurgens* plants

Years after establishment	Phosphorus content (P_2O_5 mg/ 100g)			Stand condition (%)	
	Soil	Leaf	Stem	Plant mortality	Disease index
0	1.86 a	—	—	—	—
1	1.54 a	0.5401 a	0.0710 a	66.3	54.8
2	—	—	—	78.9	55.6
3	0.54 b	0.1407 b	0.0291 b	100.0	77.7
4	—	—	—	100.0	86.2
5	0.11 c	0.1066 c	0.0233 c	100.0	92.6
6	—	—	—	100.0	95.0

Note: data in a column followed by different letters are significantly different ($p<0.05$)

Conclusion Premature degradation of *Astragalus adsurgens* were attributed to phosphorus deficiency in soil and root rot diseases caused by the fungi *Fusarium solani, F. oxysporum*, and *F. moniliforme*.

References

Flood J.& I. Issac (1978). Reaction of some cultivars of lucerne to various isolates of *Verticillium albo-atrum*. *Plant Patholology,* 27,166-169.

Hou, T. J. (1986). Diseases of *Astragalus adsurgens* in western Liauning province and Inner Mongolia, *Grasslands of China,* 3, 40-43.

Latunde-dada, A.O. & J.A. Lacas (1982). Variation in resistance to *Verticillium* wilt within seeding populations of some varieties of lucerne. *Plant Patholology,* 31,179-186.

Liu, R. (1998). Science of Grassland Protection. Agriculture Press of China. 259-263.

Nan, Zh. B. (1997) *Astragalus adsurgens* disease and their distribution charactors. *Acta Prataculturae Sinica,* 14, 30-34.

The influence of fertiliser application to strip-sown grasslands on herbage production and quality

A. Kohoutek, P. Komárek, V. Odstrčilová and P. Nerušil
Research Institute of Crop Production Prague. Research Station of Grassland Ecosystems, Jevick, K. H. Borovskeho 461, Czech Republic 569 43, Email: vste@seznam.cz

Keywords: grassland, fertiliser, strip-seeding

Introduction To increase the productivity of dairy cattle in the Czech Republic requires an improvement in herbage quality and an increase in the net energy of herbage. The decrease in cattle numbers by 50 % and the expansion of the grassland area both result in a surplus of feedstuffs of low quality. Introduction of strip-seeding of legumes and grasses into grasslands in interaction with fertilisation provides a possible solution to this problem.

Materials and methods Strip-seeding experiments (Kohoutek *et al*., 2003) were established at a site at Jevicko, Czech Republic in a region that has a mild-warm, mild-humid climate (altitude 330 m, an average annual temperature of 7.5 °C and an annual rainfall of 629 mm). The soil type is a fluvisol. This study evaluates strip-sown grassland (PTP) compared to permanent grassland (TTP). Strip-seeding in 1991 (seeding machine SE 2-024), 1996, 2000 and 2003 (seeding machine for strip-seeding - prototype) was carried out with the same mixture and seed quantity (29 kg/ha). The mixture was of the following composition: Festulolium hybrid (*Lolium multiflorum* Lam. x *Festuca arundinacea* Schreb.), cv. Felina (12 kg/ha), Perennial ryegrass (*L. perenne* L.), cv. Sport (8 kg/ha), Cocksfoot (*Dactylis glomerata* L.), cv. Niva (4 kg/ha), Red clover (*Trifolium pratense* L.), cv. Kvarta (3 kg/ha) and White clover (*T. repens* L.), cv. Huia (2 kg/ha). The strip-sown alternative (PTP) treatments were as follows: (1) zero fertilization; (2) phosphorus (P) and potassium (K) at a rate of 30 kg/ha P as superphosphate and 60 kg/ha K as a potash salt, (3) as treatment (2) plus 90 kg N /ha and (4) as treatment (2) plus 180 kg N/ha. This paper describes dry matter (DM) production and corrected DM production of strip-sown legumes (corrected DM production = DM production x % projective dominance of botanic group / 100) from 1992 to 2004. Concentrations of crude protein and net energy (NEL) were estimated from 1997 to 2004.

Results Strip-seeding into grassland without fertiliser application and with P and K fertiliser (Table 1) had a significantly lower production of DM and had a higher proportion of strip-sown legumes than the alternative treatments that received N fertiliser, where nitrogen supported greater grass DM production and increased grass competitiveness over legumes. The concentration of crude protein and NEL in herbage was higher in the alternatives without N fertiliser application which make them suitable for dairy cow feeding.

Table 1 Dry matter (DM) production and corrected DM production of strip-sown legumes, and concentrations of crude protein and NEL in forage

Fertiliser treatments	DM production (t/ha)	Corrected DM production of strip-sown legumes (t/ha)	Concentration of crude protein (g/kg DM)	NEL concentration (MJ/kg DM)
$N_0P_0K_0$	7.48	1.08	122.4	5.67
$N_0P_{30}K_{60}$	7.48	1.47	125.4	5.61
$N_{90}P_{30}K_{60}$	9.33	0.73	112.3	5.50
$N_{180}P_{30}K_{60}$	10.8	0.52	118.7	5.51
$LSD_{0.05}$	1.35	-	-	-
$LSD_{0.01}$	1.60	-	-	-

Notes: $LSD_{0.05}$ = least significant difference at $P<0.05$; $LSD_{0.01}$ = least significant difference at $P<0.01$

Conclusion Strip seeding of legume-grass mixtures into grasslands without fertiliser application and with P and K fertiliser increased forage quality more than N-fertiliser application, which supported an increase in yield but. decreased the proportion of strip-sown legumes in the herbage.

Acknowledgements The paper results from the support of research project MZe, NAZV, No. QF 3018.

Reference

Kohoutek A., P. Komárek V. Odstrčilová & P. Nerušil (2003). Ecosystem development of permanent strip-seeded and temporary grasslands over an eleven-year period. In: A. Kirilov, N. Todorov and I. Katerov (eds) Optimal Forage Systems for Animal Production and the Environment, Pleven, Bulgaria, pp. 41-44.

The effect of harvest management on forage production and self-reseeding potential of Italian ryegrass (*Lolium multiflorum* L.)

P.W. Bartholomew and R.D. Williams
USDA-Agricultural Research Service, Grazinglands Research Laboratory, PO Box 1730, Langston University, Langston, OK 73050, USA, Email: PBarthol@luresext.edu

Keywords: ryegrass, self-seeding, seed deposition, forage

IntroductionIf Italian ryegrass (*Lolium multiflorum L.*) (IRG) can be managed to produce a seed output sufficient for effective re-establishment, without compromising forage yield, it may provide an alternative to perennial cool-season grasses in the Southern Great Plains of the U.S.A. The reduction in cost of replanting and avoidance of cultivation offered by a self-seeding crop may be particularly useful in low-input production systems. We examined the effect of dates of initial harvest in spring and of partial harvests on forage yield, seed output and re-establishment of Italian ryegrass.

Materials and methods Italian ryegrass was oversown without tilling into dormant unimproved warm-season pasture in the autumn of 2002. In spring of 2003 ("year 1"), initial harvest date treatments, 17 April (H1), 1 May (H2) and 15 May (H3), were combined with forage offtake treatments of 100, 76 and 53% of available forage at each harvest. The different offtake treatments were achieved through use of modified blades on a sickle-bar mower. Following the initial harvest, IRG was allowed to regrow, set seed and to re-establish from seed deposited without further management input. Re-establishment was measured at the end of year 1 and forage production at the end of April 2004 ("year 2"). The effectiveness of reseeding in year 2 was measured by counts of seed deposition in July and of seedling emergence in September.

Results Early harvest and reduced offtake decreased forage yield at first harvest in year 1, but increased the amount of seed deposited and number of seedlings re-established in year 1 (Table 1). There was a residual effect of harvest date and offtake treatments in year 1on forage production of the self-seeded crop that was manifested in a lower mean yield with a late harvest and with the increased offtake treatments. However, in year 2 there was no significant difference (P>0.05) in seed deposition, mean seed weight or emerged seedling numbers among year 1 harvesting treatments. Mean seedling emergence by mid-October of year 2 was 91seedlings/m^2.

Table 1 Effects of first year initial harvest (H1, 17 April; H2, 1 May; H3,15 May) and proportion of offtake on forage yield, seed deposition and re-establishment of Italian ryegrass

	Treatments	Forage yield (MT/ha)		Seed deposition (seeds/m^2)		Re-establishment (seedlings/m^2)	
		Year 1	Year 2	Year 1	Year 2	Year 1	Year 2
	H1	1.03	1.78	7510	2510	4990	92
Initial harvest	H2	1.90	1.93	5410	2020	2800	110
	H3	2.87	1.58	2460	1670	710	71
LSD (P<0.05)		0.385	0.219	2982	NS	1767	NS
	53	1.39	1.82	6530	2120	3540	70
Offtake (%)	76	1.86	1.79	5800	2040	2960	105
	100	2.54	1.69	3050	2050	2010	98
LSD (P<0.05)		0.214	0.134	1387	NS	943	NS

Conclusion Italian ryegrass re-established satisfactorily in its first season of self-seeding and forage yield at the end of April in year 2 was comparable with that obtained from drilled IRG harvested at the same time. However, this satisfactory early-season growth did not translate into an effective second cycle of self-seeding; on average only 4% of seeds deposited by mid-July had emerged as seedlings by mid-October. The seedling population achieved by self-seeding in year 2 was on average 22% of the lowest population achieved in year 1, and only 15% of that produced by drilling IRG at 30kg/ha in the autumn. The poor re-seeding performance in year 2 may have resulted from a combination of high seedling populations, and associated low seed weight, and from premature germination arising from greater than average rainfall in June and July in year 2. The results demonstrate the uncertainty of self-seeding as a means of pasture renewal.

An evaluation of grazing value of maize and companion crops for wintering lactating ewes

E.A. van Zyl[1] and C.S. Dannhauser[2]
[1]*Dundee Research Station, PO Box 626, Dundee, 3000 South Africa, Email: vanzyle@dunrs.kzntl.gov.za,* [2]*School of Agriculture & Environmental Science, University of the North, Private Bag X1106, Sovenga, 0727, South Africa*

Keywords: grazing maize, companion crops, winter-feed, Merino sheep

Introduction Northwestern KwaZulu-Natal (KZN), in South Africa, is well known for its sheep production from natural rangeland in summer (October to May). During winter, however, the nutritional value of the rangeland cannot maintain young growing sheep or pregnant and lactating ewes. With this in mind, Lyle (1991) suggested the use of planted pastures for the winter. Crichton *et al.* (1998) and Esterhuizen & Niemand (1989) suggested the use of maize crop residues for both cattle and sheep during winter, whereas Moore (1997) evaluated grazing (not harvested) maize for this purpose. He found that the crude protein content of the crop was inadequate and for this protein-rich companion crops were evaluated in this study.

Materials and methods Maize, with eight different companion crops, was evaluated as winter feed for lactating Merino ewes and lambs on the Dundee Research Station, KZN, South Africa. One ha plots were planted with 0.5 ha maize and 0.5 ha companion crop adjacent to each other. The following companion crops were used: *Ornithopus sativus (Os), Vicia dasycarpa (Vd), Raphanus sativus (Rs), Ladlab purpureus (Lp), Glycine max (Gm), Vigna unguiculata (Vu), Avena sativa (As), Secale cereale (Sc)* and maize (M) alone. The maize and summer companion crops were planted during late November, each year, and the winter crops during February. Ten ewes and their lambs were allocated to each treatment and they were allowed to select between the maize and the companion crops. Whenever the experimental animals did not ingest sufficient of the available biomass, dry ewes were added. Grazing potential and growth of ewes and lambs were measured. The experiment was conducted over three consecutive years.

Results The rainfall data for the three years (from July-June) were 442.3mm, 792.7mm and 862.6 respectively, with the long-term average being 782.9mm. During year 1 the low rainfall resulted in poor performance of some companion crops and only M+Gm, M+Lp and M+Sc were grazed. During years 2 and 3 all treatments were evaluated. Although the rainfall in the first year was below average, M+Gm managed to carry 24.8 small stock units (SSU)/ha, with a total liveweight gain of 334 kg/ha. In a normal rainfall situation (year 2) M+Lp and M+Rs carried more than 20 SSU/ha and liveweight gains were 318 and 346 kg/ha respectively. During year 3, with a high rainfall, M+Rs and maize alone carried more than 32.5 SSU/ha, with liveweight gains of 376 and 341 kg/ha respectively.

Table 1 Grazing capacity per treatment

Treat-ment	SSU/ha/100days*			
	Season1	Season2	Season3	Mean
M+Gm	24.8	18.3	27.6	23.6
M+Vu	-	16.5	25.7	21.1
M+Lp	19.2	22.2	40.1	27.2
M+As	-	18.8	20.8	19.8
M+Os	-	16.9	27.6	22.2
M+Vd	-	19.6	25.3	22.4
M+Rs	-	20.0	32.5	26.2
M+Sc	20.4	-	-	-
Maize	-	20.0	38.7	29.3

*1 ewe = 1 small stock unit (SSU) and 1 lamb (average 20 kg) = 0.5 SSU

Table 2 Total liveweight gain per treatment (kg/ha)

Treat-ment	Liveweight gain (kg/ha)			
	Season1	Season2	Season3	Mean
M+Gm	334	272	293	300
M+Vu	-	249	361	306
M+Lp	230	318	294	281
M+As	-	247	267	257
M+Os	-	270	238	254
M+Vd	-	233	291	262
M+Rs	-	346	376	360
M+Sc	285	-	-	-
Maize	-	274	341	307

Conclusions When below average rainfall seasons are experienced, maize + *Glycine max* (soybeans) can be expected to be an appropriate winter feed. During normal rainfall seasons maize + *Lablab purpureus* (dolichos), maize + *Raphanus sativus* (Japanese radish) and maize alone are appropriate winter feeds. Maize + *Vigna unguiculata* (cowpeas) may be an appropriate winter feed during a higher rainfall season.

References

Esterhuizen, C.D. & S.D. Niemand (1989). Oesreste van ses kontantgewasse vir die oorwintering van skape. Inligtingsdag, Nooitgedacht-Navorsingstasie, Ermelo, Bl.85-107.

Lyle, A.D. (1991). The use of supplementary licks for sheep on summer and winter Sourveld. Sheep in Natal. Co-ordinated Extension Committee of Natal (5.2.1991). KwaZulu-Natal Department of Agricultural, Private Bag X9059.Pitermaritzburg. 3200.

Yield and mineral concentration changes in maize and Italian ryegrass cropping systems

S. Idota and Y. Ishii
Division of Grassland Science, Faculty of Agriculture, University of Miyazaki 889-2192, Japan Email: sidota@cc.miyazaki-u.ac.jp

Keywords: cropping system, forage crops, minerals

Introduction Mineral balance between plants and soil to which fertilizer has been applied is important in sustainable agriculture. Cropping systems are chosen based on considerations of crop yield, soil physical and chemical properties and climatic conditions. Thus, the sustainability of a forage cropping system should only be assessed after continuous cultivation has been practiced for several years. Forage crop production is employed in the rice paddies of Japan during summer. Thus, the objective of this study was to evaluate the yield and mineral concentration of forage crops cultivated in hard-textured soils for 4 years.

Materials and methods A typical cropping system for maize (sown from early May to mid-July) and Italian ryegrass (sown from October to November) was applied to the clay soils of the experimental field at the University of Miyazaki from 2001 to 2004. Cattle manure at 30 ton/ha, fused magnesium phosphate at 800 kg/ha, magnesium-calcium carbonate at 1 ton/ha, 100 kg/ha of N, 200 kg/ha of P_2O_5 and 200 kg/ha of K_2O were applied as a basal fertiliser. From the initial soil analyses (April 2001), the pH was 5.9 and available phosphoric acid and total nitrogen content were 88 mg/kg and 1.8 g/kg, respectively.

Results Dry matter yield (DMY) for different parts of maize plants is shown in Figure 1. Total DMY for maize was significantly affected by the year due to the occurrence of typhoons and rainfall distribution (Figure 1). Total nitrogen (T-N) and total phosphoric acid (T-P) concentrations of maize were negatively correlated with DMY (r = - 0.966 for T-P), except for the lowest DMY and T-N concentration in 2004. Concentrations of T-N and T-P ranged between 0.25-0.42% and 0.19-0.29%, respectively. Total DMY of Italian ryegrass at the time of every harvest is shown in Figure 2. It was higher in 2001-02 and 2003-04 than in 2002-03. Total nitrogen (T-N) and total phosphoric acid (T-P) concentrations of Italian ryegrass were negatively correlated with DMY (r = - 0.785, and - 0.980 for T-N and T-P, respectively). Concentrations of T-N and T-P ranged between 0.55-0.14% and 0.23-0.41%, respectively.

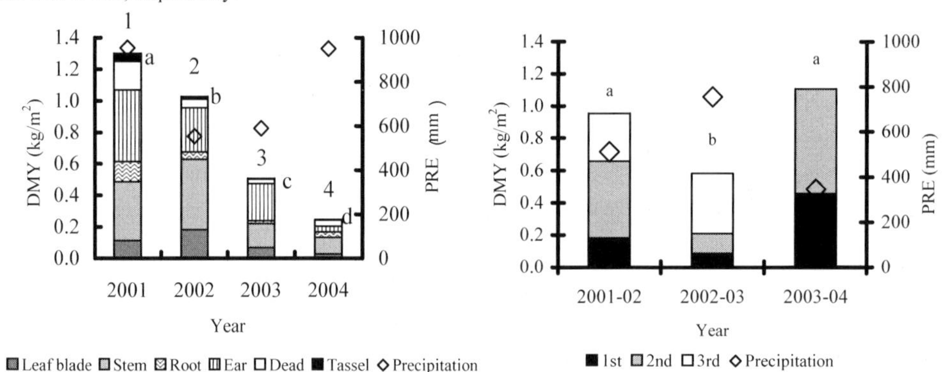

Figure 1 Changes in dry matter yield (DMY) of each part of the maize plant and precipitation (PRE) from 2001 to 2004. Figures above the column indicate typhoon frequency. The values followed by different letters in whole plant were significantly different between years at $P<0.05$

Figure 2 Changes in dry matter yield (DMY) for each harvest of Italian ryegrass and precipitation (PRE) from 2001 to 2004. The values followed by different letters in the total DMY were significantly different between years at $P<0.05$

Conclusions DMY in maize was highly variable between years due to climatic conditions. The lower T-N concentration of maize is attributed to wet soils during the wet summer season. Annual total yields in both maize and Italian ryegrass were relatively stable and ranged between 1.62 to 2.26 kg/m^2.

Constraints on dairy cattle production from locally available forages in Bangladesh

M.A.S. Khan
Department of Dairy Science, Bangladesh Agricultural University, Mymensingh 2202, Bangladesh,
Email: s-khan@royalten.net

Keywords: constraints, protein, parasitism, rice straw, grass

Introduction The productivity of milk producing animals in Bangladesh is low because of low individual yield and poor fertility. The reasons for the low productivity are complex but, in order of priority, appear to be (a) the imbalanced nature of the nutrients that arise from the digestion of the forage resources, (b) the incidence of disease/parasitism, and (c) the often harsh climatic circumstances. Thus, the purpose of this study was to find out the practical constraints on dairy cattle production from locally available forages under small holding village conditions of Bangladesh.

Materials and methods One typical village – Boira, which is about 2 km from Bangladesh Agricultural University, Mymensingh, was chosen for this study. The villagers were mostly resource-poor farmers. The cows of the villagers were used for multipurposes such as draught, dairy and meat. Rice straw was the main roughage source for the animals. Limited seasonal cut and carry grasses were also used. Animals were mainly stall fed. Sixty seven post-partum cows were taken from 65 smallholder farms. Nutrient metabolites were measured by FAO/IAEA Nutritional Metabolite kits according to Kaneko (1989) and concentration of milk progesterone (P4) for calving to first ovulation were measured using the solid-phase Radioimmunoassay (RIA) kits supplied by FAO/IAEA according to Plaizier (1993).

Results Among the blood metabolites studied, a considerable change in plasma urea values were noticed as shown in Table 1. It represents an important nutritional constraint to productivity. The spring value was low enough to suggest a shortage of RDP in the rations at that time. The calving to first service was higher than that of calving to first ovulation as shown in Table 2. It means that farmers were unable to detect heat of their cows at the proper time. The calving to first service interval was lower in autumn than in the other three seasons. It would be tempting to relate this to the urea levels which were highest in the autumn. Average parasitic egg counts were 54/g and were mainly of *Fasciola gigantica.*

Table 1 Group metabolite means within each season

Seasons	BHB (mmol/L)	SEM	Globulin (g/L)	SEM	Albumin (g/L)	SEM	Urea (mmol/L)	SEM	Pi (mmol/L)	SEM
Summer	0.342^a	0.05*	37.00^a	7.40* *	38.00^c	2.34* *	4.52^b	0.98* *	1.33^c	0.12*
Autumn	-------	------	36.67^{ab}	7.80* *	33.67^c	2.78* *	7.20^a	0.09*	1.61^a	0.12*
Winter	0.362^a	0.06*	--------	-----	35.00^b	2.74* *	5.46^b	0.93* *	1.42^b	0.09*
Spring	0.292^b	0.04*	35.00^a	5.00* *	38.18^a	3.85* *	3.25^c	1.08* *	1.46^b	0.13*

* p<0.5; * * p<0.01, ^{abc}Figures with dissimilar superscript in the same column differ significantly (p<0.05)

Table 2 Reproductive intervals by seasons (days)

Seasons	Calving to 1st ovulation	SEM		Calving to 1st service	SEM		Calving to conception	SEM		Calving interval	SEM	
Summer	66	42	NS	272	147	NS	283	148	NS	544	162	NS
Autumn	67	25	NS	120	60	NS	136	67	NS	419	72	NS
Winter	187	105	NS	191	81	NS	197	92	NS	489	84	NS
Spring	51	14	NS	216	38	NS	223	38	NS	501	39	NS

NS p>0.05

Conclusions Feed protein deficiency, improper heat detection and parasitic infestation were the constraints on dairy cattle production from locally-available forages in Bangladesh.

References

Kaneko, J. (1989). Biochemistry of Domestic Animals. Third ed., Academic Press, New York.

Plaizier, J.C.B. (1993). Validation of the FAO/IAEA RIA kits for the measurement of progesterone in skim milk and blood plasma. In: Improving the Productivity of Indigenous African Livestock, IAEA-TCDOC-708, IAEA, Vienna. 151-156.

Pasture management in deer farms in Mauritius

P. Grimaud[1], P. Thomas[2] and J. Sauzier[3]
[1]*Cirad-Elevage, 7 Chemin de l'IRAT, F97410 Saint Pierre, Email: grimaud@cirad.fr,* [2]*UAFP, PK 23, F97418, Plaine des Cafres and* [3]*MDFCSL, Holp Building, Quatre Bornes, Mauritius*

Keywords: Mauritius, *Cervus timorensis russa*, tropical grass, pasture management

Introduction Rusa deer (*Cervus timorensis russa*) were introduced into Mauritius in 1639, and with a population of 65,000 deer now provide the first source of red meat in the island. From June to September, the hunting period, extensive ranches ensure the venison market, while intensive farms provide it for the rest of the year. A pasture survey was carried out over 3 years to advise deer farmers on their sward management.

Materials and methods On 10 farms, soil and grass herbage were collected for 3 years at each season (CS, cold season; HS, hot season) from 3 to 6 paddocks. From each paddock, a soil analysis combined 7-9 samples collected by auger within the 5-15 cm horizon. Values of pH, exchangeable cations and phosphorus (P) content were determined. Grass herbage was harvested to a height of 5 cm from 6-10 quadrats, located in areas of pure grass representative of the studied paddock. Total green weight was recorded in the field and a representative 1 kg sample was taken for DM (80° C for 48 h) and ash content (550° C for 6 h) determination. The indices of mineral element dilution in the forage (IN, Ip and IK) were calculated from analysis of N (Kjeldahl method), P (atomic absorption spectrophotometry) and K (continuous flew colorimetry) according to Blanfort (1998). Values of IN, IP and IK were compared to standard values (100, 80 and 100, respectively).

Results Low IP values confirmed a soil-P deficiency, while very high IK values suggested a luxury consumption of potassium that soil analyses did not reveal (Table 1). Soil pH was non-limiting. The low IN values suggested low availability of N, which inhibited pasture growth potential in the CS.

Table 1 Soil analysis and available biomass of south-western Mauritian pastures according to the dominant grass species

	Soil analysis			Available biomass	
	pH	Phosphorus	Potassium/ Base	DM (%)	t DM / ha
Bothriochloa pertusa (sikin)					
Cold season	5.5 – 6.5	Low / Medium	Very Low/ Low	54	2.9
Hot season	5.5 – 6.5	Low / Medium	Very Low/ Low	39	3.1
Cynodon plechtostachium (star grass)					
Cold season	5.5 – 6.5	Low / Medium	Low / Medium	51	3.6
Hot season	5.5 – 6.5	Low / Medium	Low / Medium	35	3.8
Ischaemum aristatum (silver grass)					
Cold season	6.0 – 6.5	Very Low/Low	Very Low/ Low	34	4.2
Hot season	6.0	Very Low/Low	Very Low/ Low	26	3.5

Star grass (*C. plechtostachyum*) and silver grass (*I. aristatum*) were more productive than sikin (*B. pertusa*). High DM contents provide evidence that consumption of the grass at the end of HS will have a low nutritive value.

Conclusion Results of the present study suggest that a sward management improvement strategy could allow the rusa deer to reach its growth potential: an adapted pasture fertilisation should provide to the grass the requirements in N, P, and K that they need in the CS, and thus allow the pastures to have significant yields throughout the year. Such a management, combined with a relevant rotation in the paddocks corresponding to the grass seasonal rhythms, has been successful implemented in the neighbour Reunion Island for ten years in cattle farms.

References

Blanfort, V.(1998). Agroécologie des pâturages d'altitude à l'île de la Réunion, PhD thesis, Université de Montpellier, France, 259 pp.

Section 2

Research advances in understanding soil/plant/animal relationships

Heterogeneous nutrient distribution across dairy grazing systems in southeastern Australia

C.J.P. Gourley, I. Awty, P. Durling, J. Collins, A. Melland and S.R. Aarons
Ellinbank Research Institute, PIRVic Ellinbank, RMB 2460 Hazeldean Rd, Ellinbank, Victoria, 3821 Australia, Email: Cameron.Gourley@dpi.vic.gov.au

Keywords: dairy farms, soil nutrients, phosphorus, potassium

Introduction The Australian dairy industry is largely based on a grazed pasture system, although most cows also consume substantial amounts of imported feed (Fulkerson & Doyle 2001). This trend is expected to increase as the Australian dairy industry continues to intensify. Fertiliser inputs of nitrogen (N), phosphorus (P), potassium (K) and sulphur (S) are still viewed as necessary to maintain adequate pasture and milk production despite the fact that most dairy farms are in net positive balance for all of these nutrients (Reuter 2001). Nutrient losses from dairy farming regions and eutrophication of waterways has gained strong public and political attention and intensive pasture systems are no longer seen as 'clean and green'. An important aspect of a viable dairy industry in the future will be more refined nutrient management planning.

Materials and methods Soil samples (0-10 cm) were collected from paddocks on 30 commercial dairy farms (Victoria, Australia) and tested for P, K, S, and pH (water). Farm sizes and intensity of inputs varied, with between 14 and 54 paddocks tested per farm. Soil test information was spatial presented and simple nutrient budgets calculated for each paddock or set of paddocks on each dairy farm, using mostly readily available information such as milk production, purchased feed, harvested hay and silage, and feeding strategies.

Results Soil P, K and S levels were unevenly distributed across the farms. For example a soil phosphorus distribution is provided (Figure 1). In general higher nutrient levels were associated with night paddocks, calving paddocks, sacrifice paddocks and dairy effluent application areas. Low nutrient levels were associated with remote farm locations and hay and silage areas. The most spatially variable measure was K, particularly on farms where the dairy unit was at one end; least variable was soil pH. Farmer response to the spatial presentation of their soil test information and the description of nutrient flows within their farm was highly positive. In most instances, farmers recognised that imported nutrients in purchased fodder could offset fertiliser costs. Changes in fertiliser management included deciding to apply no fertiliser to high fertility areas, changing fertiliser blends to better balance nutrient requirements, or increasing rates to identified areas of nutrient deficiency.

Conclusions The improved adoption and application of tools, such as soil testing and nutrient budgeting can make substantial improvements in fertiliser decisions, increasing productivity and profitability, while reducing adverse environmental impacts. Advances in analytical methods and procedures are continuing to refine fertiliser recommendations and reduce costs, while GPS mapping can provide a greater capacity for 'whole-farm' nutrient planning. Nutrient budgets are gaining acceptance as an indicator of sustainable nutrient practices, and provide a useful educational tool to assist dairy farmers to more effectively account for nutrient inputs, redistribution and losses within the farm in Australia.

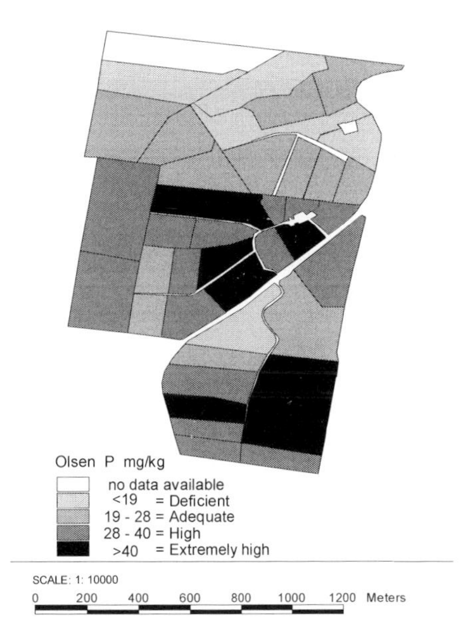

Figure 1 Soil phosphorus distribution across a dairy farm in Victoria

References

Fulkerson, B. & P. Doyle (2001). 'The Australian dairy industry.' (Dept. Natural Resources and Environment: Tatura, Victoria).

Reuter. D.J. (2001). Nutrients – farm gate nutrient balances. Australian national land and water audit 2001. Http://audit.ea.gov.au/ANRA/land/farmgate/Nutrient_Balance.pfd.

Fertiliser responses and soil test calibrations for grazed pastures in Australia

C.J.P. Gourley[1], A.R. Melland[2], K.I. Peverill[2], P. Strickland[1], I. Awty[1] and J.M. Scott[3]

[1]*Primary Industries Research Victoria, Ellinbank, Victoria, Australia, Email: Cameron.Gourley@dpi.vic.gov.au,* [2]*KIP Consultancy Services Pty Ltd, Wheelers Hill, Victoria, Australia,* [3]*University of New England, Armidale, NSW, Australia*

Keywords: fertiliser, nutrients, pasture growth response

Introduction On-farm management of fertiliser is of major economic significance to the Australian grazing industries, based on expenditure on fertiliser and higher farm productivity that fertiliser use supports. However the application of fertiliser has traditionally been an inexact and inefficient process (Peverill *et al.* 1999) and there is increasing pressure for nutrient losses from agriculture to be minimised. The improved adoption and application of tools like soil testing can make substantial improvements in nutrient use efficiency but interpretation needs to be based on the best available information. This paper reports on the collation of current and historical experimental data relating to pasture production - fertiliser response relationships (nitrogen, phosphorus, potassium and sulphur) for various pasture types, climatic zones and soils across Australia.

Materials and methods A national team of scientists and fertiliser agronomists from all states of Australia has contributed to the collation of a comprehensive set of pasture production-fertiliser response data from field studies. This has involved identifying and collating previous reviews, published papers, departmental reports and where available, unpublished material. These data sets have been integrated using a relational database to derive the most appropriate response relationships available for the grazing industries in Australia. The national database has been used to provide regionally specific and scientifically validated soil test calibrations for improved pastures.

Results More than 350 experimental data sets have been collated consisting of circa 2600 sites and >3800 experimental trial years. The number of sites is made up of around 479 N, 662 P, 692 K and 810 S trials, with a total of 615 N, 1313 P, 933 K and 974 S experimental trial year. Not surprisingly, experimental data sets ranged in quality, scope and complexity. Less than 33% had enough statistical rigour to enable nutrient response curves to be generated. Only a few studies involved a number of sites in various regions with different soil types and climatic zones. Most studies provided simpler data sets, mostly at a single field site with a single nutrient applied at only 2 or 3 rates. Such data sets could not be used to establish nutrient response curves. There was a limited capacity to combine experimental data sets as methodologies often differed markedly and site data was inadequate. In some single site studies, soil test levels were strongly related to pasture response to fertiliser applications (variance accounted for (VAF) > 0.9), but when applied to a range of soils and environments, they invariably lack precision (VAF ranging from 0.0-0.5). The addition of other variables such as soil type and climatic zones only marginally improved these relationships.

Conclusions More than 50 years of experimental data relating to the response of pasture to soil nutrient availability has been compiled and analysed across the pasture-based grazing industries of Australia. This extensive exercise has highlighted the lack of precision in many response relationships for N, P, K and S and the difficulty in combining historical data sets to assist in extrapolating across soil types and regions. It is proposed that appropriate soil tests may still require regional calibration and further research should adopt standard experimental methodologies.

References

Peverill K.I., L.A. Sparrow & D.J. Reuter eds. (1999) "Soil Analysis; an interpretation manual" CSIRO Publishing, Australia.

Modelling basal area of perennial grasses in Australian semi-arid wooded grasslands

S.G. Marsden and K.C. Hodgkinson
CSIRO Sustainable Ecosystems, Canberra, ACT 2911, Australia, Email: ken.hodgkinson@csiro.au

Keywords: grasslands, grazing, rainfall, model

Introduction In many semi-arid pastoral systems, landscape processes easily become dysfunctional. Shifts to less functional states may be irreversible, and have long-term consequences for pastoral profitability and social viability of rural communities. Typically, shifts to lower functional states involve a decline in perennial grasses (Hodgkinson, 1994). Here we develop a conceptual basis for modelling the basal area of perennial grasses in a semi-arid grassland and validate the model using data from a 10-year grazing study.

Modelling concepts Change in the basal area of perennial grasses occurs with the growth, decline or death of existing plants or the recruitment of new plants (see Figure 1 c). These processes are primarily dependant on the level of soil water and grazing pressure. The model uses average soil water level over the previous month to determine whether perennial grass basal area increases, decreases or remains unchanged. Average height of the grasses over the previous 12 months (influenced by grazing and growing conditions) is used to determine the magnitude of the change. The basal area derived in the model is an important feedback mechanism for limiting potential perennial grass biomass following drought or overgrazing.

Output and validation The model component, described above, was used to simulate the basal area of palatable perennial grasses at three sheep stocking rates (0, 0.4 and 0.8 sheep/ha). Climate data for Louth, western NSW (average rainfall 270 mm per annum) was used for the simulation. Figures 1 a) and b) display the validation results of the basal area model for the runoff and interception landscape zones respectively.

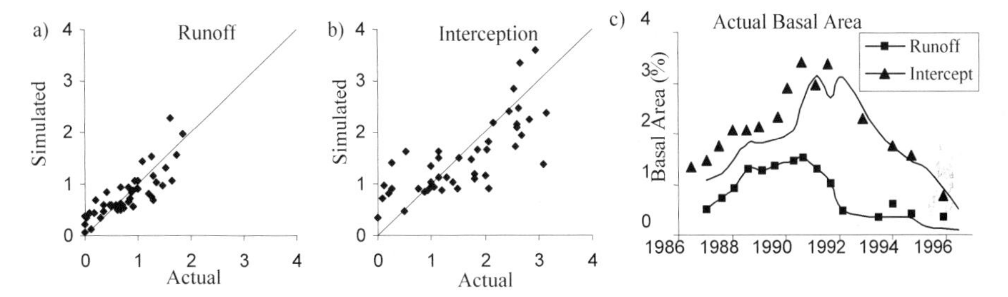

Figure 1 a) and b) Actual versus simulated palatable perennial grass basal area (%) for two landscape zones from 1987-1996 for three sheep stocking rates (0, 0.4 and 0.8 sheep/ha) and c) Actual palatable perennial grass basal area over time in two landscape zones for 0.4 sheep/ha.

Conclusions The basal area of perennial grasses in semi-arid grasslands changes considerably through time for each zone in the landscape, in response to climate variability and grazing levels. A simple modelling approach involving soil water and height of grasses adequately explained these changes. The basal area model is sufficiently robust to simulate basal area over long periods to evaluate the effects of grazing and periodic drought on the basal area and hence biomass production of perennial grasses in semi-arid wooded grasslands.

References

Hodgkinson, K. C. (1994). Tactical grazing can help maintain stability of semi-arid wooded grasslands. *Proceedings of the Seventeenth International Grassland Congress*, pp. 75-76.

Diversity of diet composition decreases with conjoint grazing of cattle with sheep and goats

A.M. Nicol[1], M.B. Soper[1] and A. Stewart[2]
[1]*Agriculture and Life Sciences Division, PO Box 84, Lincoln University, New Zealand, Email: nicol@lincoln.ac.nz,* [2] *PGG Seeds, PO Box 3100, Christchurch, New Zealand*

Keywords: diet selection, cattle, sheep, goats

Introduction Conjoint or mixed grazing can affect the diet selected by each species (Nicol & Collins, 1990). Diet similarity coefficients are often used to compare *pairs* of diets (Krebs, 1999). However this approach is awkward when a number of contrasts are required in a multifactorial comparison. Species diversity is a descriptor of a *particular* environment. Many models provide an estimate of *species* diversity, the most common of these being a log-normal distribution (Tokeshi, 1996). We tested whether this model could be applied to dietary *components* selected from a pasture, and thus provide a coefficient of dietary diversity for the *individual* diets of cattle, sheep and goats when grazed alone or in mixtures, which could then be statistically compared.

Methods Groups of cattle, sheep and goats grazed alone (CA; SA; GA) or as cattle plus sheep (CS; SC) and cattle plus goats (CG; GC), during a progressive defoliation (4400 to 1550 kg DM/ha) of a ryegrass/white clover pasture during 20 days in summer. Oesphageal extrusa (OE) samples were collected from two of each animal species on most days (n = 116 OE samples). The samples were dissected into six botanical components (grass leaf, grass stem, grass seed head, clover leaf and petiole, clover flower and dead material). The dietary diversity coefficient (k) was estimated for each OE by an iterative minimisation of the sum of the squared deviations of the observed proportion of each component and the predicted proportion (P) of each component from $P_R = 100 (1\text{-}k)\, k^{R-1}$, where R = the rank of each component in the observed OE composition. Dietary diversity coefficients were compared by analysis of variance using species (cattle, sheep and goats) and grazing environment (alone or mixed), and their interaction, as fixed effects.

Results and discussion Simple correlations between predicted and observed OE composition ranged from 0.87 to 0.99. The mean k for cattle was significantly greater than that for sheep and goats (Table 1) and the effect of conjoint grazing was to significantly reduce the diversity of the diet.

Table 1 The dietary diversity coefficient (k) and *in vitro* dry matter digestibility of oesophageal extrusa of cattle, sheep and goats grazed alone or conjointly

	Dietary diversity coefficient (k)				*In vitro* DMD (g DMD/kg DM)			
	Alone	Conjoint	Average	sem	Alone	Conjoint	Average	sem
Cattle	0.582	0.522	0.552 c		613	595	604 c	
Sheep	0.505	0.431	0.468 d	0.025	619	641	630 d	6
Goats	0.504	0.401	0.452 d		617	632	624 d	
Average	0.53 a	0.451 b		0.021	616 a	624 b		5

Values followed by a different letter (a or b in rows, c,d in colomns) are significantly different (P<0.01)

Decreased diet diversity was associated with a higher *in vitro* digestibility of OE for sheep and goats, especially when they were conjointly grazed with cattle. This probably reflects their ability (smaller incisor arcade breadth), and opportunity when mixed with cattle (SC and GC), to exploit their dietary preferences. In contrast the reduced diet diversity of CS compared with CA (significant interaction), was associated with a reduction in *in vitro* digestibility, suggesting that the quality of the diet of cattle suffered when they were in competition with sheep.

References

Krebs, C.J. (1999). Ecological methodology. Benjamin/Cummings, Addison Wesley Longman, California, USA, 701 pp.

Nicol A.M. & H.A. Collins (1990). Estimation of the horizons grazed by cattle, sheep and goats during single and mixed grazing. *Proceedings of the New Zealand Society of Animal Production* 50, 49-53.

Scott, D. (1993). Constancy in pasture composition? *Proceedings Seventeenth International Grassland Congress,* 1604-1606.

Tokeshi, M. (1996). Species co-existence and abundance: patterns and processes. In: A. Takuya, S.A. Levin, M. Higashi (eds.) Biodiversity – An ecological perspective. Springer-Verlag New York,, 35-55.

Spatial scale of heterogeneity affects diet choice but not intake in beef cattle

S.M. Rutter, J.E. Cook, K.L. Young and R.A. Champion
Institute of Grassland and Environmental Research, North Wyke, Okehampton, Devon EX20 2SB UK, Email: mark.rutter@bbsrc.ac.uk

Keywords: diet selection, spatial scale, heterogeneity

Introduction Previous research has shown that sheep (Champion *et al.*, 1998) and dairy cattle (Nuthall *et al.*, 2000) have a partial preference for clover of 70%, and achieve higher daily intakes when offered grass and clover as separate but adjacent monocultures compared with animals grazing mixed swards. This intake benefit could be utilised to increase intake and production on farms by grazing from adjacent strips of the two herbages. This study aimed to establish the minimum strip width required to achieve the benefits of monocultures.

Materials and methods Intake and diet preference were measured using n-alkanes in four groups of two yearling Simmental x Holstein beef heifers. They were rotated in a Latin Square around four paddocks, each containing adjacent white clover and ryegrass strips at different spatial scales: 108 cm, 36 cm and 12 cm, and a mixed sward. Heifers spent 5 days on each paddock. They were dosed daily with a cellulose bung impregnated with 420mg of C_{32} n-alkane. Daily dosing started 8 days prior to the start of the Latin Square rotation to ensure steady-state conditions were achieved. Herbage samples were collected on the fourth day that the heifers were on each treatment paddock. Faecal samples were collected by following the animals and collecting a sample of naturally-voided faeces on several occasions over the final 24 h that the heifers were on each treatment paddock. The herbage and faecal samples were freeze-dried an analysed for n-alkane content. These data were processed using EatWhat? (Dove & Moore, 1995) to determine diet composition, and further analysed to estimate daily intake.

Results There was a significant effect of scale on the proportion of clover in the diet ($F_{3,15}$=5.46, P=0.038). Preference showed a step-function response with 59-60% clover in the diet for 108 cm and 36 cm and 36-38% for 12 cm and the mixed sward (Figure 1). There were no significant differences in daily intake between the four treatments (9.2, 9.9, 8.6 and 9.9 kg DM day^{-1} for 108 cm, 36 cm, 12 cm and mixed sward respectively, Figure 2).

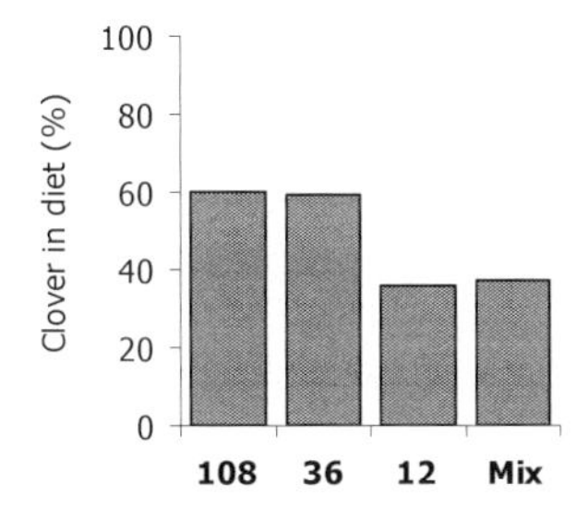

Figure 1 The proportion of clover in the diet and the associated with each of the four treatments

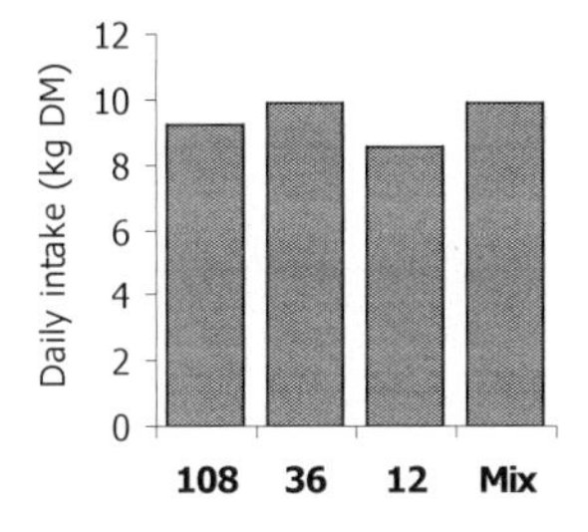

Figure 2 The daily intake (kg DM) associated with each of the four treatments

Conclusions These results indicate that cattle were unable to select the desired proportion of clover at the two smaller scales, but could at the two larger scales, with the critical scale lying between 12 and 36 cm. The fact that the cattle could achieve a higher proportion of clover at the larger scales suggests that, at these scales, they are effectively 'separate monocultures'. However, unlike previous studies with lactating sheep and dairy cattle, intakes from these monocultures were no higher than that from the mixed sward. Further research is needed to establish if this latter result is associated with beef cattle or with using strips.

References

Champion, R.A., S.M. Rutter, R.J. Orr & P.D. Penning (1998). Costs of locomotive and ingestive behaviour by sheep grazing grass or clover monocultures or mixtures of the two species. Proceedings of the 32nd International Congress of the ISAE, Clermont Ferrand, 21-25 July 1998, p 213.

Dove, H. & A.D. Moore (1994). Using a least-squares optimisation procedure to estimate botanical composition based on the alkanes of plant cuticular wax. *Australian Journal of Agricultural Research*, 46, 1535-1544.

Nuthall, R., S.M Rutter & A.J. Rook (2000). Milk production by dairy cows grazing mixed swards or adjacent monocultures of grass and white clover. 6th BGS Research Meeting, Aberdeen, Sept. 2000, pp. 117-118.

A simple vegetation criterion (NDF content) may account for diet choices of cattle between forages varying in maturity stage and physical accessibility

C. Ginane and R. Baumont
INRA, Unité de Recherches sur les Herbivores, Centre de Clermont-Theix-Lyon, F-63122 Saint-Genès Champanelle, France, Email: ginane@clermont.inra.fr

Keywords: diet choice, prediction, cattle, intake rate, neutral detergent fibre

Introduction The management of extensively grazed pastures requires an understanding and prediction of the diet choices of herbivores grazing on vegetation that is qualitatively (maturity stage) and quantitatively (biomass, sward height) heterogeneous. The Optimal Foraging Theory (OFT, Stephens & Krebs, 1986), bases its predictions on the relative energy intake rate (EIR) of forages. However, as EIRs are difficult to assess at pasture and are subject to wide intra- and inter-individual variations, another vegetation criterion was sought (accessibility, quality), by-passing the animal's influence, to predict cattle diet choices quantitatively.

Materials and methods The results of two grazing and two complementary indoor experiments (Ginane *et al.*, 2002; Ginane *et al.*, 2003) were pooled. Eighteen-month old heifers were able to choose, throughout the day for approximately 7 days, between two forages (standing swards or hays), varying in relative maturity stage (vegetative *vs.* reproductive) and physical accessibility (sward height). Forages were characterized by their protein and fibrous chemical composition, their digestibility (measured *in vitro*), and their intake rates (measured *in situ* on the animals used in the choice experiments). These measurements yielded EIR values.

Results Diet choices were significantly and positively linked to forage EIR ratio (EIR of the vegetative forage/EIR of the reproductive forage, Figure 1A), consistent with OFT. Among the different criteria tested the difference in NDF content between forages (reproductive-vegetative) was the one most closely related to diet choices (Figure 1B). The close relation between diet choices and the neutral detergent fibre (NDF) criterion may arise because the NDF content is linked to (i) forage prehensibility, as it takes into account sward resistance to defoliation and mastication (Sauvant *et al.*, 1996), (ii) forage ingestibility, as it partly expresses forage fill effect in the rumen (Mertens, 1994), and (iii) forage digestibility, which varies inversely with NDF content.

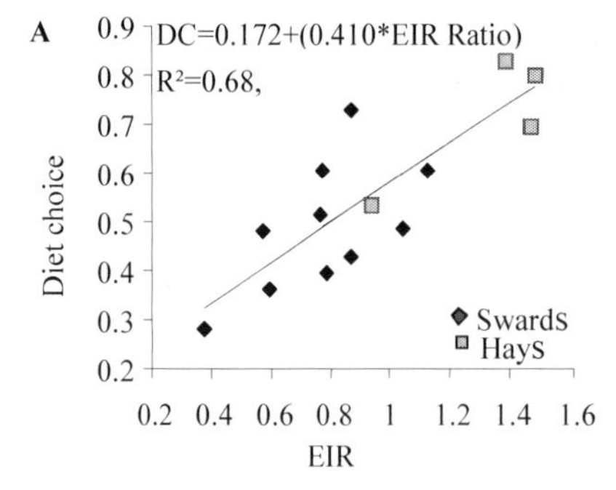

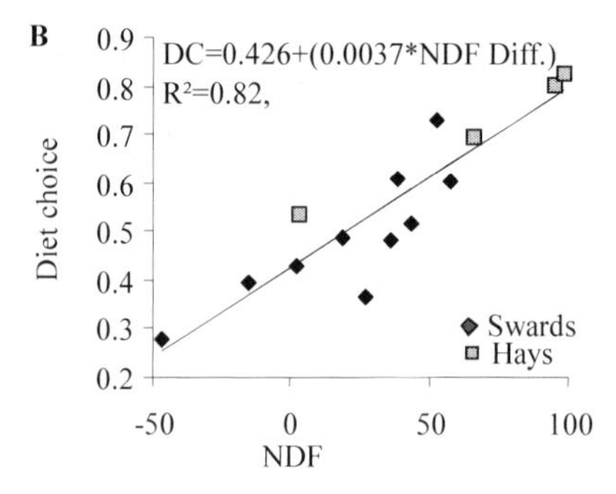

Figure 1 Linear regressions equations of observed diet choices (DC) with (A) EIR ratio of forages (vegetative/reproductive) and (B) difference in NDF content of forages (reproductive-vegetative)

Conclusions This study indicates that forage NDF content, a common and easy-to-measure criterion, may be useful for predicting cattle diet choices in heterogeneous pastures.

References

Ginane, C., R. Baumont, J. Lassalas & M. Petit (2002). Feeding behaviour and intake of heifers fed on hays of various quality, offered alone or in a choice situation. *Animal Research*, 51, 177-188.

Ginane, C., M. Petit & P. D'Hour (2003). How do grazing heifers choose between maturing reproductive and tall or short vegetative swards? *Applied Animal Behaviour Science*, 83, 15-27.

Mertens, D.R. (1994). Regulation of forage intake. In: G.C. Fahey Jr (eds.) Forage quality, evaluation and utilisation. American Society of Agronomy, Madison, Winsconsin, 450-493.

Sauvant, D., R. Baumont & P.Faverdin (1996). Development of a mechanistic model of intake and chewing activitites of sheep. *Journal of Animal Science*, 74, 2785-2802.

Stephens, D.W. & J.R. Krebs (1986). Foraging theory. Princeton University Press, Princeton, NJ, USA.

Do species and functional diversity indices reflect changes in grazing regimes and climatic conditions in northeastern Spain?

F. de Bello[1], J. Leps[2] and M.T. Sebastià[1]
[1]*Laboratory of Plant Ecology and Forest Botany, Forestry and Technology Centre of Catalonia, pujada seminari s/n. E-25280 Solsona, Lleida, Spain, Email: fradebello@ctfc.es,* [2]*Department of Botany, Faculty of Biological Sciences, University of South Bohemia. Na Zlate stoce 1 CZ-370 05 Ceske Budejovice, Czech Republic*

Keywords: land use, disturbance, Mediterranean rangelands, sheep

Introduction Understanding the mechanisms that maintain biodiversity in various ecosystems enables the development of management practices that prevent degradation (Canals & Sebastia, 2000). Each diversity index reflects some compositional properties and could be influenced differently by stress and disturbance factors (Magurran, 2004). In this study, we aim to reveal 1) which management practices and environmental factors affect biodiversity in rangelands of northeastern Spain and 2) the relationship between species diversity and functional diversity (SD and FD).

Methods Species frequencies were measured in natural slope vegetation areas, along a gradient of sheep grazing pressure (high, low, abandonment). Five locations were selected across a steep altitudinal and corresponding climatic gradient from Mediterranean rangelands to natural sub-alpine grasslands. The factorial design was 3 sheep grazing intensities x 5 locations x 2 aspects x 2 replicates = 60 plots. We calculated: 1) species richness (number of species); 2) species diversity according to Shannon (H^1) and Simpson (1-D); 3) species evenness according to Pielou and Camargo (see Canals & Sebastia, 2000); 4) species rarity (number of species classified "rrr") and 5) functional diversity (FD). The average of the pair-wise species differences weighted for their relative frequencies was used as a measure of the functional diversity in such a way that, if difference in all species pairs equals 1, the Simpson index is obtained (Shimatani, 2004). Species differences were calculated on the basis of 8 life history traits for the 467 species found. Three-way ANOVA was used and slope inclination (°) was covariate in the model. Duncan post-hoc test was performed to detect mean differences.

Results The effect of the studied factors on the diversity indices are given in Table 1. Climatic variables were the most important factor affecting species and functional diversity. Generally, diversity indices were lowest in a water-stressed environment (in the driest sites and southern aspects) and increased toward moist areas. FD reached its peak at intermediate elevations but the values in the wettest site resembled those in dry sites. Grazing enhanced species diversity, but no effect was found on functional diversity; species rarity was higher in abandoned areas. There was no clear relationship between species and functional diversity indices.

Table 1 Results of 3-way ANOVA for various diversity indices. Letters indicate different means. (*=P<0.05;**=P<0.01;***=P< 0.001)

	Richness	Shannon	Simpson	E Pielou	E Camargo	Rarity	FD
Slope (covariable)	NS	NS	NS	NS	NS	NS	NS
Aspect (south-north)	*	**	*	**	0.055	NS	***
Location (along climatic gradient)	***	***	***	***	***	***	**
Grazing intensity	0.060	**	*	***	***	*	NS
Aspect x Location	*	0.091	*	NS	NS	0.072	***
Aspect x Grazing	NS	NS	NS	NS	NS	**	NS
Location x Grazing	NS	*	*	*	**	***	***
Aspect x Location x Grazing	NS	NS	NS	*	**	NS	***
R^2 adj	0.60	0.74	0.76	0.73	0.71	0.67	0.86
GRAZING (abban/low/high)	a/b/b (+)	a/b/b (+)	a/b/b (+)	a/a/b (+)	a/a/b (+)	b/a/a (-)	NS

Conclusion Water stress decreases species diversity. Species rich grasslands are not functionally different. Grazing increased species diversity only, having no clear effect on functional diversity. Adequate grazing pressure on natural slope vegetation maintained the diversity of species.

References

Canals, R.M. & M.T. Sebastià (2000). Analyzing mechanisms regulating diversity in rangelands through comparative studies: a case in the south-western Pyrennees. *Biodiversity and Conservation*, 9, 965–984.
Magurran, A.E. (2004). *Measuring Biological Diversity.* Blackwell Publishing, Oxford. 260 pp.
Shimatani, K (2004). On the measurement of species diversity incorporating species differences. *Oikos* 93, 135-147

Species richness affects grassland yield and yield stability across seasons, sites and years

D.J. Barker, R.M. Sulc, M.R. Burgess and T.L. Bultemeier
Department of Horticulture and Crop Science, The Ohio State University, 202 Kottman Hall, 2021 Coffey Rd, Columbus OH 43210, USA, Email: barker.169@osu.edu

Keywords: biodiversity, season, species richness, yield stability, variation

Introduction The benefits of biodiversity (specifically species richness) are proposed to include both greater yield and greater stability of yield in a variable environment (Sanderson *et al.*, 2004). Experimental evidence showing yield benefits is inconsistent (White *et al.*, 2004). There is relatively little experimental data showing the effects of species richness on yield stability. The objective of this study was to measure the yield from mixtures with up to 12 species, and to measure the variability of yield between 2 sites, between spring and summer, and in 2 successive years.

Materials and methods Twelve grassland species were planted in 22 treatments comprising one (12 monocultures), three (5 mixtures), six (3 mixtures), nine (1 mixture) or 12 species (1 mixture). All treatments were planted in 10 m x 2.5 m plots in May 2001 at 1000 viable seed/m^2 at 2 sites (Columbus and Utica, Ohio). Herbage mass was measured for 2 years using a calibrated rising plate meter, pre- and post-grazing, for 4 grazing events in each of 2002 and 2003. For the analysis in this paper, spring yield was the total of the first 2 grazing events, and summer/autumn yield was the total of the third and fourth grazing events.

Results The number of species sown per plot was positively related to mean yield in spring, but was poorly related to yield in summer (Figure 1). In spring, the 12-species mixture never exceeded the best yielding monoculture, however, the identity of that monoculture varied between season, site and year. Plot error (i.e. standard error of the 3 replicates, within site, year and season) was greatest for the monocultures (mean = 566 kg DM/ha) but was similar for the 3-, 6-, 9- & 12-species mixtures (mean = 448 kg DM/ha). Treatment yield was standardized by dividing by the overall mean (proportion of average). The standard deviation of standardized yield between years for the monocultures (0.71) was nearly twice that for the 12-species mixture (0.48), with the other treatments intermediate (0.55). The standard deviation of standardized yield between sites for the monocultures (0.86) was more than twice that for the 12-species mixture (0.34), with the other treatments intermediate (0.58). The standard deviation of standardized yield between seasons for the monocultures (1.17) was almost 5-times that for the 12-species mixture (0.24), with the other treatments intermediate (0.44).

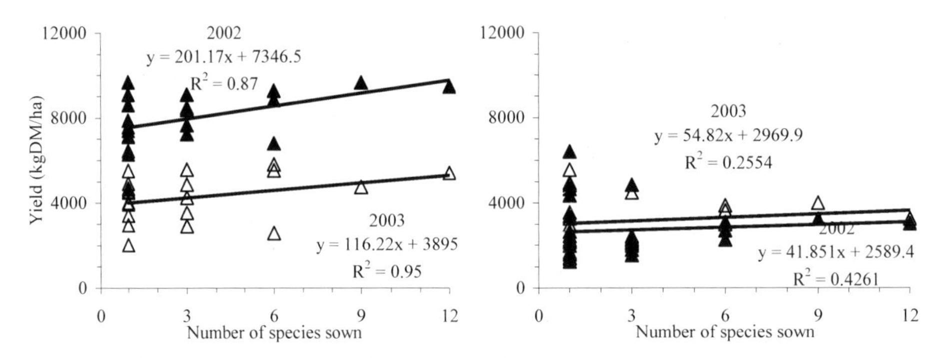

Figure 1 Relationship between species richness and yield in spring (left) and summer/autumn (right) for 2 years. Data are for Columbus only, similar responses were found at Utica. Symbols are the mean of 3 replicates.

Conclusions Yield responses to species richness were only found during spring, and not during summer/autumn. Yield stability across seasons, sites and years increased with species richness for almost every comparison tested.

References

Sanderson, M.A., D.J. Barker, G.R. Edwards, B. Tracy, R. H. Skinner & D. Wedin (2004). Plant species diversity and management of temperate forage and grazing land ecosystems. *Crop Science,* 44, 1132-1144.

White, T.A., D.J. Barker & K.J. Moore (2004). Vegetation diversity, growth, quality and decomposition in managed grasslands. *Agriculture Ecosystems and Environment,* 101, 73-84.

The biodiversity value of 'improved' and 'unimproved' saline agricultural land and adjacent remnant vegetation in South Australia

M.L. Hebart, N.J. Edwards, E.A. Abraham and A.D. Craig

South Australian Research and Development Institute, Struan Agricultural Centre, P.O. Box 618, Naracoorte, SA, 5271, Australia, Email: hebart.michelle@saugov.sa.gov.au

Keywords: dryland salinity, puccinellia, landscape function

Introduction Since European settlement of the Upper South-east of South Australia, the distribution and abundance of much of the native flora and fauna of the region has been affected by clearing of native vegetation and drainage of wetlands to facilitate agricultural production. Only 8.3% of the original vegetation and less than 7% of the original swamps now remain in the region and much of what is left exists as small isolated remnants (Croft & Carpenter, 1996). Furthermore, as a consequence of the demise of large areas of agriculturally productive lucerne in the late 1970's and early 1980's, the rise of saline groundwater has resulted in the deterioration of some of this remnant vegetation. A key question being addressed in a major research project into the productive use of saline farming land is "How do saltland pastures influence ecosystem function?". This paper reports some preliminary observations from this assessment.

Materials and methods Landscape functional analysis (LFA), as described by Tongway & Hindley (1995), was used to determine how ecosystem resources such as water, soil, organic matter and nutrients are being retained, utilised and cycled in time and space. It uses 10 measures of soil surface condition to derive indices of water infiltration, soil surface stability and nutrient cycling. Comparisons were made between approximately 100 ha of remnant vegetation (subset of a 400-ha area), a salt scald and three pasture types on approximately 30 ha of grazing land that was cleared of native vegetation in the early 1950s, sown to lucerne for 30 years and more recently established to a mixed pasture of puccinellia (*Puccinellia ciliata*), tall wheat grass (*Agropyron elongatum*), and balansa clover (*Trifolium michelianum*). Since being established to these improved pastures, areas have either reverted to an 'unimproved' sward of predominantly sea barley grass (*Hordeum marinum*), samphire (*Halosarcia spp.*) and salt scalds (Pasture 1), been maintained as improved puccinellia-based pasture with up to 10% balansa clover (Pasture 2) or been further 'improved' to include up to 50% balansa clover with a strong base of puccinellia (Pasture 3).

Results The remnant vegetation provides a benchmark for the LFA indices in the study region. Clearly, the ability of the soil from the 3 improved pastures to withstand erosive forces and reform following disturbance (stability), to partition rainfall into soil-water available to plants and runoff water (infiltration) and recycle organic matter (nutrient cycling) was lower than the remnant vegetation, but significantly greater than the salt scald. The difference between the 3 pastures in LFA was not significantly different due to the similarity between botanical composition, soil type and condition. However, there was greater species diversity in the remnant vegetation, with 105 species from 76 genera noted in the remnant vegetation and only 13 species from the 3 cleared pastures.

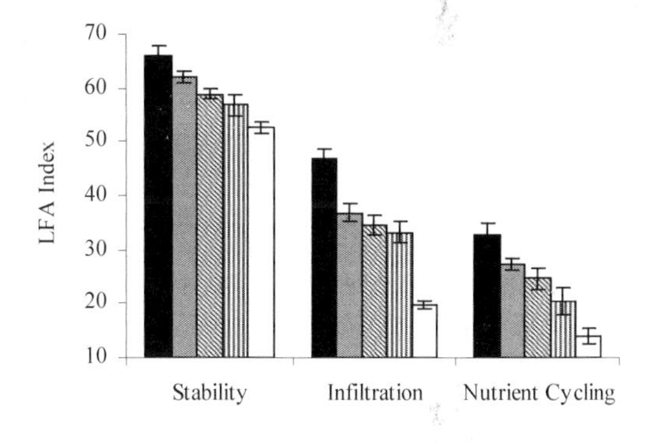

Figure 1 LFA indices for the 5 land types

Conclusions The 3 'improved' pastures, whilst not functioning at the same level as the remnant vegetation, were still significantly better than salt scalds. The critical issue is whether these systems are improving, degrading or remaining stable over time.

Acknowledgements This work is part of the Sustainable Grazing on Saline Land (SGSL) research into profitable use of saline land - an initiative of Australian Wool Innovation Pty Ltd with Land and Water Australia.

References

Croft, T. & G. C. Carpenter (1996). The biological resources of the South East of South Australia. Native Vegetation Conservation Section, Department of Environment & Natural Resources, Adelaide, Australia.

Tongway, D. & N. Hindley (1995). Manual for the Assessment of Soil Condition of Tropical Grasslands, CSIRO Wildlife and Ecology, Canberra.

Grazing impacts on rangeland condition in semi-arid south-western Africa

A. Rothauge[1], G.N. Smit[2] and A.L. Abate[3]

[1]*Neudamm Agricultural College, P/Bag 13188, Windhoek, Namibia, Email: arothauge@unam.na,* [2]*Dept of Animal, Wildlife and Grassland Sciences, University of the Free State, Bloemfontein, South Africa and* [3]*Dept of Animal Science, University of Namibia, P/Bag 13301, Windhoek, Namibia*

Keywords: rangeland condition, degradation, bush encroachment, stocking rate

Introduction The savannah biome, consisting of a dense herbaceous layer and a relatively open woody layer in competitive balance, constitutes 64% of the land surface of Namibia, an arid country in south-western Africa, and is used mainly for extensive cattle and sheep ranching. About half of the savannah area is affected by dense to moderately dense bush-thickening, resulting in a ten-fold decrease in the rangeland's grass-based carrying capacity and a concomitant loss in meat production of about US$115 million per year (De Klerk, 2004). Bush-encroached areas typically have densities > 2 000 bushes/ha with > 90% belonging to a single species. High grazing pressure by specialist grazers, such as domestic cattle, is often blamed for rangeland degradation. There is an urgent need to understand the dynamics of bush encroachment and devise grazing strategies to contain it.

Materials and methods Observations on diet selection of free-ranging beef cattle in a semi-arid savannah of central Namibia quantified dietary preferences and diet composition of cattle exposed to a three-fold increase in the stocking rate (SR) from 1 large stock unit (LSU)/30 ha to 1 LSU/10 ha (Rothauge, 2005). The diet selection study was supported by detailed botanical surveys determining the canopy cover of the soil and botanical composition of the rangeland in the treatment plots, average size, 142 (±28.9) ha, by systematic point sampling, as well as by clipping herbaceous yield in quadrats before each observation of diet selection. Yield and density of grass tufts was determined by counting the clipped tufts per species. Abundance data was subjected to an arcsine transformation. All data were pooled for the SR treatments, cattle type (CT) and season. Observed rangeland condition was the result of 17 consecutive years of imposition of the treatments.

Results and discussion Stocking rate had a much bigger impact on rangeland condition than season or CT. The increase in SR caused the canopy cover of the soil to decrease by 9% ($P < 0.01$), total herbaceous yield to decrease by 17% ($P < 0.05$), density and yield of tufts of grass species preferred by cattle to decrease by 36% and 97%, respectively (both $P < 0.01$), and a concomitant increase in the density of grass species not preferred by cattle by 250%, while their tuft yield declined by 71% although the grass sward was still dominated by perennial grasses at both SRs. Species compositional changes in grasses could not be detected by the traditional method of clipping all grasses irrespective of cattle's dietary preference for different species. Even a decline of 17% in herbaceous yield might not be apparent without a long-term data set for comparison, due to highly variable annual rainfall.
Simultaneously, the increase in SR caused six known invasive species of woody plants to become 85% more abundant ($P < 0.01$), while the abundance of woody species unlikely to become invasive (n=4) increased by only 2% ($P > 0.05$). Overall abundance of woody species increased by 39% ($P < 0.05$), to 17%. In contrast to grasses, woody plants became perceptibly denser although still far from forming a dense stand dominated by only one species. Many woody species were still struggling for supremacy in the rangeland of the treatment plots. Predictably, beef production of cattle in the treatment plots declined by 8% per cow, but increased by 241% per hectare with the three-fold increase in SR. Annual calving rate (calves born/cows mated) decreased by 2% ($P < 0.05$) and the inter-calving period increased by 3% ($P < 0.01$).

Conclusions and recommendations A three-fold increase in the stocking rate of beef cattle had, after 17 years, not caused typical bush encroachment, although rangeland condition had clearly deteriorated in terms of soil cover, abundance of preferred forages and woody plants. Many of the initial changes, especially in species composition, are not readily visible unless monitored intensively and compared to a long-term data set. Ranchers are thus seduced into stocking up their cattle because beef production per unit area is still increasing, albeit at a slower rate, in accordance with the model of Jones & Sandland (1974). The bush encroachment so typical of large parts of Namibia must have been caused by stocking rates in excess of 1 LSU/10 ha.

Acknowledgements The support of the IFS (grant B/3183-1) is gratefully acknowledged.

References

De Klerk, J.N. (2004). Bush Encroachment in Namibia. Report on Phase 1 of the Bush Encroachment Research, Monitoring and Management Project, Ministry of Environment and Tourism, Windhoek, Namibia.

Jones, R.J. & R.L.Sandland (1974). The relation between animal gain and stocking rate. Derivation of the relation from the results of grazing trials. *J. Agricultural Science, Cambridge,* 83, 335-342.

Rothauge A. (2005). The diet of free-ranging beef cattle in a semi-arid savanna of eastern Namibia. *Proceedings of the Twentieth International Grassland Congress,* ??.

Soil, plant and livestock interactions in Australian tropical savannas

L.P. Hunt and T.Z. Dawes-Gromadzki
CSIRO Sustainable Ecosystem and Tropical Savannas CRC, PMB 44 Winnellie, Northern Territory, 0822 Australia, Email: Leigh.Hunt@csiro.au

Keywords: soil processes, plant processes, grazing impacts, soil macrofauna, vegetation patches

Australian savannas and grazing impacts This paper considers the various soil, plant and livestock interactions occurring in Australia's wet-dry savanna rangelands. These regions are relatively intact compared to most of the world's rangelands. However there is increasing pressure for more intensive use of the landscape, especially from pastoralism. This potentially threatens landscape health, function and productivity through reduced soil health and a loss of digestible perennial plants, especially given the low soil fertility and highly variable rainfall characteristic of these regions. There is an obvious need for understanding these impacts to devise sustainable management practices that promote soil health and viable perennial plant communities, and the restoration of soil health where required.

Retaining and optimising the utilisation of nutrients and water in the landscape is critical to maintaining landscape function and productivity. Patches of perennial vegetation play an important functional role in retaining and cycling nutrients and water. Soil macro-faunal assemblages are vital to the maintenance of fertile vegetation patches. For example, through their feeding and nesting activities they create soil macropores that facilitate water infiltration and retention. Grazing has the potential to reduce the capacity of the landscape to capture and retain water and nutrients through changes in patch and vegetation structure and dynamics.

Grazing effects can be manifested at scales ranging from small patches to whole landscapes. At larger scales, grazing can affect the structure, size, function, spatial arrangement and demography of vegetation patches. The tendency for livestock to repeatedly graze preferred locations also results in the development of degraded vegetation patches. At smaller scales, grazing can influence soil and plant processes in a number of ways; for example, the beneficial effects of macrofauna on soil processes can be adversely affected by grazing (T. Z. Dawes-Gromadzki, unpublished data). Demographic processes in perennial plant populations can also be disrupted by grazing and lead to local extinction of desirable species (Hunt, 2001). All these impacts affect the capture, retention and distribution of nutrients and water, with potential negative consequences for landscape function and productivity in the long term. Understanding how these processes interact and knowing which are the key drivers of system health are fundamental requirements for successful management but our current level of knowledge is poor.

Potential pathways of grazing impact It is unlikely that the various soil and plant processes are equally affected by grazing, so there may be different pathways through which grazing effects can lead to declines in range condition, landscape health and productivity. Hypothetical pathways for grazing effects include:

1. Direct changes in soil processes occur more rapidly than plant changes and subsequently limit plant growth and reproduction. For example, are the effects of trampling on soil compaction, surface-sealing and macro-fauna activity in restricting nutrient and water availability to plants the overriding limitation on the persistence of desirable species?
2. Plant changes occur more rapidly than soil changes. For example, are changes in plant population and community processes (e.g. reproduction, growth and species composition) the overriding limitation on the persistence of desirable species, with soil changes being of less importance?
3. Soil and plant changes occur simultaneously and at similar rates. If so, what are the critical processes?
4. Grazing only affects plant processes directly, with consequences that then flow on to soils (e.g. a reduction in transpiration that alters soil moisture status in cracking soils).
5. Grazing only affects soil processes directly with consequences that then flow on to plants.

Possible implications for research and management The potential importance of these interactions demonstrates the need for a systems approach with integrated studies. Such a framework should indicate what the critical issues are for management to focus on. For example, to avoid degradation should management focus on plant processes, whereas for restoration, should soil processes be the focus?

References

Hunt, L.P. (2001). Heterogeneous grazing causes local extinction of edible perennial shrubs: a matrix analysis. *Journal of Applied Ecology,* 38, 238-252.

The effect of different grazing managements on upland grassland

Pavlů[1], M. Hejcman[2], L. Pavlů[3] and J. Gaisler[1]

[1]*Grassland Research Station, Research Institute of Crop Production, Prague, CZ-460 01 Liberec, Czech Republic, Email:pavlu@vurv.cz;* [2]*Department of Forage Crops and Grassland Management, Czech University of Agriculture, Kamýcká 957, CZ-165 21 Prague, Czech Republic and* [3]*Jizerské Mts. Protected Landscape Area Administration, U Jezu, CZ-460 01 Liberec, Czech Republic*

Keywords: pasture, set stocking, sward, plant species composition

Introduction A transformation process in the Czech economy led to a rapid decrease in livestock numbers in the Czech Republic and an enlarged area of grasslands at the beginning of 1990's. The result was extensification of grassland management and also abandonment in marginal areas. The main purpose of this study was to reveal how different managements affect plant species diversity of previously abandoned grassland.

Materials and methods An experiment was carried out from 1998 to 2004 in an experimental pasture in the Jizerské hory mountains, Czech Republic. The treatments applied were: intensive grazing (IG), a first cut in June followed by intensive grazing (ICG), extensive grazing (EG), a first cut in June followed by extensive grazing (ECG), and unmanaged grassland (U) as the control were arranged in two complete randomized blocks. The sward was continuously grazed by young heifers to target heights of 5 cm and 10 cm under IG, ICG and EG, ECG treatments, respectively. Percentage cover of all vascular species was measured in permanent 1 m x 1 m plots in four replications in each paddock. Redundancy analysis (RDA) followed by a Monte Carlo permutation test was used to analyze the multivariate data.

Results and conclusion Effect of interaction of year and treatment (Figure 1) explained 11.7 % and 17.2 % (P<0.001), whereas successional development independent of experimental treatments explained 13.3 % and 19.4 % (P<0.001) of the variability by the first and all canonical axis, respectively. Tall forbs and tall grasses had higher abundance on treatment U. Species associated with managed treatments were *Agrostis capillaris, Taraxacum spp., Trifolium repens* and *Ranunculus repens.* This result indicates a replacing of tall dominants by short grasses. and prostrate forb species.

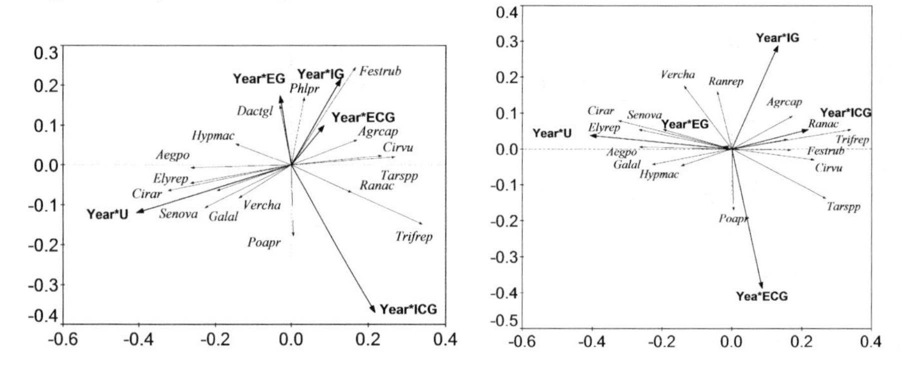

Figure 1 Ordination diagram showing the result of RDA analysis a) first and second axis b) first and third axis of plant species composition data. Abbreviations: *indicates interaction of environmental variables, Aegpo-*Aegopodium podagraria*, Agrcap- *Agrostis capillaris*, Cirar-*Cirsium arvense*, Cirvu-*Cirsium vulgare*, Dactgl-*Dactylis glomerata*, Elyrep-*Elytrigia repens*, Festrub-*Festuca rubra*, Galal-*Galium album*, Hypmac-H*ypericum maculatum,* Phlpr-*Phleum pratense*, Poapr-*Poa pratensis*, Ranac-*Ranunculus acris*, Ranrep-*Ranunculus repens*, Senova-*Senecio ovatus*, Tarspp-*Taraxacum* spp., Trifrep-*Trifolium repens* and Vecha-V*eronica chamaedrys*.

Acknowledgement The study was supported by the Grant Agency of the Czech Republic (no. 526/03/0528).

The effect of fertiliser treatment on the development of rangelands in Argentina

E.F. Latorre and M.B. Sacido

The Faculty of Agronomy. Azul National University of the Centre of the Buenos Aires Province. Argentine. CP: 7300. Commission of Investigation Scientific, Buenos Aires Province. Argentine. Email: msacido@faa.unicen.edu.ar

Keywords: rangelands, fertiliser application, productivity

Introduction In Argentina grazing of rangelands may result in a decrease in winter gramineous species with an increase in summer weeds such as *Cynodon dactylon*. *Lolium multiflorum* is an important forage resource for grazing in the autumn, winter and spring. A delay in its emergence may occur because of summer weeds, which reduces the germination rate. The proportion of the seed bank as ryegrass allows the recovery of natural grassland and facilitates an increase in the productivity of livestock. The objective of this study was the evaluation of the impact of application of fertiliser in the short term on the relationship with botanical composition at different herbage availabilities.

Materials and methods The experimental site was a typical natural pasture in the Flooding Pampas, with three level of availability of herbage mass (high, >5000 kg dry matter (DM)/ha; medium, between 5000 and 3000 kg DM/ha; and low <3000 kg DM/ha). The treatments were with and without fertilization. On the fertilized treatment the technique of promotion of rangeland ("promoción de pasturas") involved the annual application of ammonium phosphate (80 kg/ha) and glyphospate (5 l/ha). Measurements were made at 45 and 75 days after fertiliser application of floristic composition, herbage availability and species abundance. Correlation (software STADISTIC 5.0) and similarity coefficients by the Czekanowski, Dice, Sørensen Index were determined.

Results The floristic composition, at the three different availabilities of herbage, in both treatments (-F without fertilized and F fertilized) is shown in Figure 1. It shows the variation in floristic composition, especially for *Lolium multiflorium* Lam, between the fertiliser and no fertiliser treatments ($P<0.01$). Within a fertiliser treatment there were no significant differences in floristic composition between the three levels of availability. In relation to differences between fertiliser treatments in floristic composition, these were significant ($P<0.05$) using a Dependent Samples test.

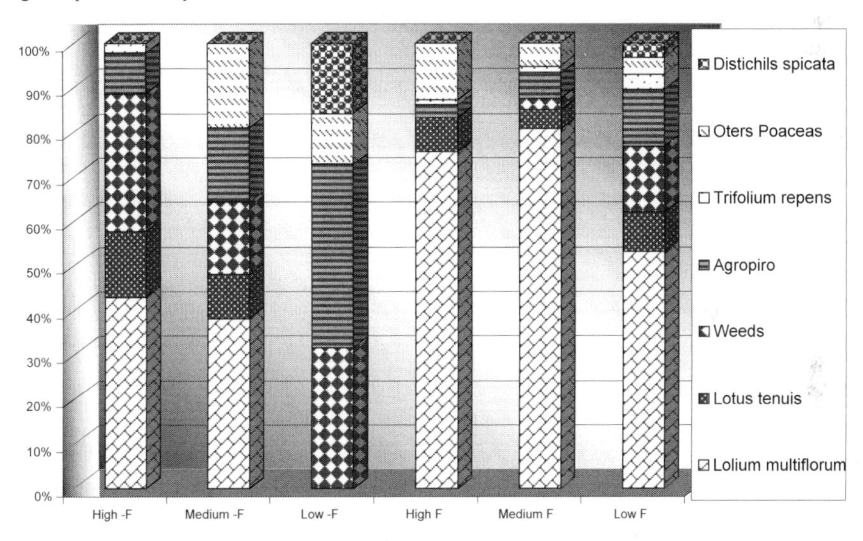

Figure 1 The floristic composition of the pastures as affected by three levels of availability of herbage and fertilizer treatment (-F, without fertilizer; and F, fertilised)

Conclusion The composition in the forage offered differed according to the fertiliser application regime. The high fertilizer treatment increased the composition of forage species, especially *Lolium multiflorum,* and decreased the proportion of weeds.

Effect of pre-planting seed treatment options on dormancy breaking and germination of *Ziziphus mucronata*

A. Hassen[1]*, N.F.G. Rethman[1] and W.A. van Niekerk[2]
[1]*Department of Plant Production and Soil Science, University of Pretoria, Pretoria 0002, Republic of South Africa, Email: hassenabubeker@yahoo.com,* [2]*Department of Animal and Wildlife Science, University of Pretoria, Pretoria 002, Republic of South Africa *Permanent Address: Adami Tulu Research Centre, Zeway, Ethiopia*

Keywords: *Ziziphus mucronata*, dormancy, hard seed coat, germination, seed mortality

Introduction *Ziziphus mucronata* (Buffalo thorn) is a multipurpose tree, widely adapted to a range of ecological conditions and tolerant of extreme climatic conditions, including frost and drought (Venter & Venter, 1996). It is a valuable fodder tree for livestock and game animals, especially in the drier parts of Africa (Rothauge *et al.* 2003). Similar to many other leguminous species, establishment is constrained by low and erratic germination of the seed, which has been attributed mainly to the physical barrier of the stony endocarp and dormancy associated with seed coat impermeability . This experiment aimed to compare the suitability of various seed treatment options as practical methods to break seed dormancy and enhance germination.

Material and methods *Z. mucronata* seeds were isolated from the dried fruit by light hammering on a concrete floor covered with a cloth sheet. Seeds were subjected to the following pre-planting treatment options: untreated seed (control); scarification using sandpaper; immersion in boiling water for 1 or 5 minutes; and immersion in 94% sulphuric acid for 10, 20, 30 or 45 minutes. Treated seeds were placed in petri dishes on moist filter paper and kept in a germination compartment adjusted to day/night temperature of 30/20°C with 12h of light. Germination of seeds was counted every day while non- germinating seeds were categorized into hard and dead after 15 days. The data were subjected, after arcsine transformation, to analysis of variance and means were compared using Tukey's test at the threshold of $P<0.05$. Arcsine transformed means were back transformed for presentation.

Results Hard seed percentage was significantly ($P<0.05$) lower than the control in seeds subjected to sandpaper scarification or treated with sulphuric acid for >/=10 minutes (Table 1). However, sulphuric acid treatment for >/=20 minutes was the most effective way to break hard seed dormancy. Regardless of its duration, boiling water treatment was not effective to break hard seed dormancy for *Z. mucronata.* Scarification with sandpaper or sulphuric acid treatment for 20 minutes significantly ($P<0.05$) increased the percentage germination compared to the control but the latter also resulted in significantly ($P<0.05$) higher seed death (Table 1).

Table 1 Effect of pre-planting seed treatment options on germination rate and proportions of hard and dead seed remaining of *Z. mucronata* after 2 weeks' incubation

Treatment options	Seed (%)		
	Hard	Germinated	Dead
Control	77.7a	10.7b	11.6d
Sandpaper scarification	23.9c	65.4a	10.7d
Immersion in boiling water for 1min	52.2abc	26.8ab	21.1cd
Immersion in boiling water for 5min	57.4ab	11.0b	31.7cd
Immersion in 94% H_2SO_4 for 10min	49.5bc	12.0b	38.5bcd
Immersion in 94% H_2SO_4 for 20min	2.2d	57.5a	40.4abc
Immersion in 94% H_2SO_4 for 30min	0d	37.2ab	62.9ab
Immersion in 94% H_2SO_4 for 45min	0d	31.5ab	68.5a

Means followed by the same letter within a column differ significantly at $P<0.05$

Conclusions Scarification using sandpaper was the best way to maximize germination without increasing the risk of seed death significantly. The duration of immersion was critical in the case of sulphuric acid treatment; 20 minutes was the optimum time of immersion.

References

Rothauge, A., G. Kaendji & M. L. Nghikembua (2003). Forage preference of Boer goats in the highland savanna during the rainy season II: Nutritive value of the diet. Agricola, pp 43-48
Venter, F. & J. A. Venter (eds.) (1996). Making the most of indigenous trees. Briza publications, Arcadia 0007. Pretoria, South Africa. pp. 304

A genecological study of the widespread Australian native grass *Austrodanthonia caespitosa* (Gaudich.) H.P. Linder.

C.M. Waters[1,2], J. Virgona[1] and G.J. Melville[2]
[1] *CRC for Plant Based Management of Dryland Salinity, Charles Sturt University, PMB 588, Wagga Wagga, NSW 2678 Australia, Email: cathy.waters@agric.nsw.gov.au, and [2]NSW Department of Primary Industries, PMB 19, Trangie, NSW 2823, Australia*

Keywords: native grass, provenance, ecotype

Introduction The lack of commercial quantities of seed is preventing the use of native grasses in large-scale revegetation programmes. Sourcing wild-land non-local provenance seed from distant locations brings with it risks associated with maladaptation and potential genetic pollution. Understanding of intra-specific ecotypic variation and its adaptive consequences is required to both increase seed supply and retain adaptive characteristics in native plant revegetation programmes. A recently commenced genecological study on the widespread Australian native grass, *Austrodanthonia caespitose,* aims to examine quantitative traits in a common garden study and genetic structure (using DNA analysis) of 35 populations collected from a large geographic range. Examination of the adaptive significance of these traits using reciprocal transplant experiments will aid in the development of provenance guidelines for Australian native grasses. In this paper we report the initial findings for one of many characteristics being measured in a common garden study, namely plant transpiration efficiency.

Materials and methods Leaf, stem and inflorescence samples were harvested from a subset of the main plant collection. This subset included ten populations represented by 8 individuals (4 pairs) covering a range of bio-geographical characteristics. Ground and dried plant samples were analysed for $\partial^{13}C$ according to standard methods. A higher value of $\partial^{13}C$ indicates higher discrimination and therefore less carbon fixed per unit of water lost (low water-use efficiency).

Results No significant difference in $\partial^{13}C$ among sites was detected ($F_{(9, 63)} = 1.451$, $P > 0.18$). Calculation of the variance components showed that most of the variation was between plant pairs within a site (1.62, $P < 0.001$), indicating genetic differences were occurring over small distances, perhaps <1 m. Exposure to solar radiation from a northerly aspect had a significant effect on $\partial^{13}C$ values (Table 1) which became increasingly more negative with decreasing sunlight. Whilst the relationship between altitude and $\partial^{13}C$ was significant, it was also weak. Here, $\partial^{13}C$ increased with increasing altitude but this may have been confounded by soil type as higher altitude sites were associated with shallower soils than those at lower altitudes. The relationship with altitude was strongest for those micro-sites that were heavily shaded; in these situations the close proximity to shrubs and trees would provide competition for soil moisture with the *Austrodanthonia caespitosa* plant.

Table 1 ANVOA for stepwise regression for response variable $\partial^{13}C$. $R^2 = 0.24$

Term	d.f.	Sums of Squares	Mean Square	F-value	P
Horizon (North)	1	2.98	2.98	6.47	0.01
Altitude	1	1.94	1.94	4.22	0.04
Logging/clearing	1	1.81	1.81	3.93	0.05 ns
Horizon (South)	1	1.46	1.46	3.16	0.07 ns
Exchangeable Aluminium	1	1.63	1.63	3.55	0.07 ns
Residual	67	30.89	0.46		

All $\partial^{13}C$ values were within the range of -25 to -29 reported previously for C_3 plants. It was expected that $\partial^{13}C$ would be highest (less negative) in the drier environments compared to wetter ones but the opposite was found. This suggests that larger scale landscape characteristics may be less important than micro-site characteristics in driving genetic differentiation at least for this trait.

Conclusions These preliminary results suggest that there is more within-site than between-site variation in $\partial^{13}C$ reflecting little ecotypic variation within this species. Distance appeared to be a poor indicator of variation, rather variation was more pronounced at a much finer scale. In the case of $\partial^{13}C$, differences in micro-site characteristics and altitude appear to have the greatest influence on $\partial^{13}C$ suggesting that it is perhaps more important to match restoration site conditions with those of the seed collection site. These results suggest that a broad provenance range for *A. caespitosa* and that seed could be sourced widely with little apparent consequence for performance in plant transpiration efficiency. However, this is not the only variable that should be considered in delineating a provenance boundary.

An agronomic evaluation of grazing maize combined with companion crops for sheep in northwestern KwaZulu-Natal, South Africa

C.S. Dannhauser[1] and E.A. van Zyl[2]

[1]*School of Agriculture & Environmental Science, University of the North, Private Bag X1106, Sovenga, 0727, South Africa, Email: chrisd@unorth.ac.za,* [2]*Dundee Research Station, PO Box 626, Dundee, 3000 South Africa*

Keywords: grazing maize, companion crops, winter feed

Introduction Northwestern KwaZulu-Natal (KZN), in South Africa, is well known for its sheep production from natural rangeland in summer (October to May). During winter however, the nutritional value of the rangeland cannot maintain young growing sheep or pregnant and lactating ewes. With this in mind Lyle (1991) suggested the use of planted pastures for the winter. Crichton, Gertenbach & Henning (1998) and Esterhuizen & Niemand (1989) suggested maize crop residues for both cattle and sheep during winter, whereas Moore (1997) evaluated grazing maize (not harvested) for this purpose. He found that the protein content of the crop was inadequate and for this reason, protein rich companion crops were evaluated in this study.

Materials and methods Maize was intercropped with 14 different crops for three consecutive seasons, on the Dundee Research Station, Dundee, KZN, South Africa. Ten of the crops are given in Table 1 (see the conclusions for the others). Yellow maize was planted in blocks of 60m x 40m in a tramline layout. Rows alternated with a spacing of 90cm and 270cm. The companion crops were planted by hand between the maize rows on two different planting dates: PD1 during late January and PD2 during late February. A randomised plot design with two replications was used. The first season was used for technique evaluation and in the next two seasons maize grain yield, total dry matter (DM) production of companion crops and nutritional values were measured.

Results The rainfall for the three different seasons (July–June) was 575, 620 and 964 mm respectively (long-term average is 782.9mm). In the second season the maize grain production amounted to 3.9t/ha and in the third season to 4.6t/ha. The DM production of the companion crops and their nutritional value are given in Table 1.

Table 1 The average DM production (t/ha) and nutritional value of the companion crops

Companion crop	DM Production (t/ha)				Nutritional value
	Season 2		Season 3		
	PD 1	PD2	PD 1	PD2	CP (%)
Raphanus sativus	4.40a	1.74 bc	2.18a	1.20 b	20.16
Avena sativa	3.03 b	3.27a	0.75 b	1.14 b	4.23
Pennisetum glaucum	2.35 bc	2.49ab	0.41 b	0.63 bc	10.00
Ornithopus sativus	2.39 bc	1.71 bc	1.94a	2.04a	12.23
Vicia dasycarpa	1.65 cde	2.04 bc	1.92a	1.21 b	18.05
Secale cereale	1.95 c	1.11 bc	0.18 b	1.18 b	13.43
Triticale hexaploide	1.48 cde	0.65 bc	0.00 b	0.21 c	15.04
Lablab purpureus	0.72 ef	1.46 bc	0.66 b	0.14 c	14.65
Glycine max	0.17 f	0 c	0.13 b	0.00 c	15.06

Figures with the same roman letters do not differ significantly ($P \leq 0.05$) for DM production

Conclusions From a DM production point of view *Ornithopus sativus* and *Raphanus sativus* can be recommended for their high production and high nutritional value (CP). *Vicia dasycarpa* had a high nutritional value. The DM production of *Pennisetum glaucum* and *Avena sativa* was relatively high, but the nutritional value was marginal. *Glycine max* and *Lablab purpureus* showed a high nutritional value, but DM production was low. [*Lolium multiflorum, Sorghum, Trifolium vesiculosum, Eragrostis teff* and *Bromus wildenowii* (not mentioned in Table 1) cannot be recommended].

References

Crichton, J. S., W. D. Gertenbach & P. W. van H. Henning (1998). The utilization of maize-crop residues for over wintering livestock (1). Performance of pregnant beef cows as affected by stocking rate. *South African Journal of Animal Science,* 28 (1), 9-15.

Esterhuizen, C.D. & S.D. Niemand (1989). Oesreste van ses kontantgewasse vir die oorwintering van skape. Inligtingsdag, Nooitgedacht-Navorsingstasie, Ermelo. Bl.85-107.

Lyle, A.D. (1991). The use of supplementary licks for sheep on summer and winter Sourveld. Sheep in Natal. Co-ordinated Extension Committee of Natal (5.2.1991). KwaZulu-Natal Department of Agricultural, Private Bag X9059.Pitermaritzburg. 3200.

Moore, A. (1997). Proewe wys lam-afronding op mielies werk. *Landbouweekblad,* 7 Maart 1997:10 -12.

Growth, nitrogen and phosphorus economy in two *Lotus glaber* Mill. cytotypes grown under contrasting P-availability

D.H. Cogliatti[1,2], L.A. Lett[1], M.S. Barufaldi[1], P. Segura[1] and J.A. Cardozo[2]

[1]*Facultad de Agronomía de Azul, Universidad del Centro de la Provincia de Buenos Aires. Av. República de Italia 780, CP 7300 Azul, Provincia de Buenos Aires, Argentina, Email: dhc@faa.unicen.edu.a, and* [2]*Consejo Nacional de Investigaciones Científicas y Técnicas, Argentina*

Keywords: *Lotus glaber*, forage legume, induced autotetraploid, P-economy, N-economy

Introduction *Lotus glaber* Mill. (lotus) is a forage legume with its origin in Europe which has shown an excellent adaptation to the Depressed Pampas of the Province of Buenos Aires, Argentina. The soils colonized by lotus usually have poor drainage, moderate sodium and low extractable P concentrations. An experiment was performed with the aim of comparing the early growth and economy of phosphorus (P) and nitrogen (N) within two *L. glaber* cytotypes differing in their ploidy level, a commercial diploid versus an induced autotetraploid population (Barufaldi *et al.*, 2001).

Materials and methods Plants were grown outdoors under two contrasting P-availabilities from January to April 2003 in soil-filled pots kept at field capacity. The experimental design was a 2x2 factorial consisting of two *L. glaber* cytotypes (tetraploid population-denominated Leonel and diploid cv. Chaja) and two P-fertilisation levels (0 and 100 ppm of P as triple superphosphate). The original extractable P concentration was 4 ppm. At transplanting, all germinated seeds were inoculated with 10^{10} cells per plant of a commercial inoculant (strain LL32). Twenty plants were harvested at days 46, 74 and 102 after germination and in each plant component, P and nitrogen concentrations were determined.

Results In addition to the greater growth of all fertilised plants, both ploidy types showed a similar growth rate (RG) and relative growth rate (RGR) at each P availability. As shown in Figure 1, variability of biomass of dry matter was greater for tetraploid plants than for diploid ones. Tetraploid plants showed the heaviest individual plants.

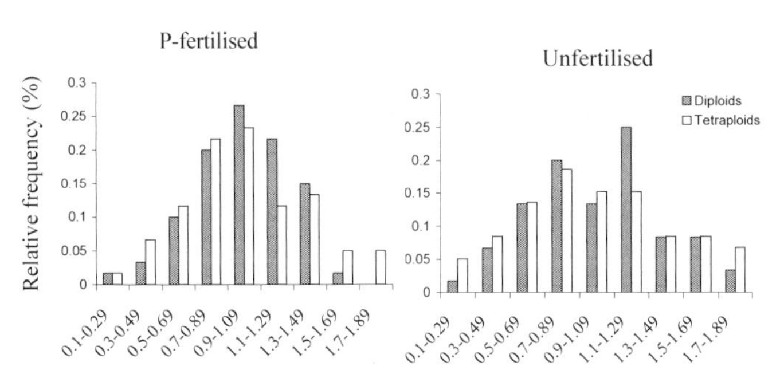

Figure 1 Distribution of relative frequencies of standardised plant biomass in two *L.glaber* cytotypes. At each harvest, plant biomass was standardised by dividing the actual biomass by the corresponding mean value.

P concentration was similar between ploidy types at both P levels. No differences were found either for P-absorption and P-utilisation efficiencies or P-partitioning between shoots and roots. As was expected, P-fertiliser application increased P-uptake and reduced P-utilisation efficiencies, but partitioning was not affected with either ploidy type. Neither N concentration nor N-use efficiency were different between ploidy types. Differences in N content observed between P availabilities were partially attributed to the higher number of nodules observed in high P plants.

Conclusion This preliminary study demonstrates that no differences were found in the analysed variables when comparing an induced polyploidy of *L. glaber* with a commercial diploid cultivar. In the tetraploid, superior plants for biomass were observed which could lead to improvements in this character by means of a recurrent selection programme.

References

Barufaldi M. S., H. N. Crosta, M. F. Eseiza, R. H. Rodríguez & E. Sánchez (2001). Proceedings of the XIX International Grassland Congress, Sao Pedro, Sao Paulo, Brazil. pp. 488-489

Belowground meristem populations as regulators of grassland dynamics

H.J. Dalgleish and D.C. Hartnett
Division of Biology, Kansas State University, Manhattan, KS 66506, USA. Email: hjdal@ksu.edu

Keywords: bud, meristem limitation

Introduction Studies of plant populations are critical for linking organism to ecosystem-level phenomena and for understanding mechanisms driving responses to global change. In perennial grasslands, the below-ground population of meristems (the bud bank) plays a fundamental role in local plant population recruitment, persistence and dynamics. We explore two aspects of the bud bank in North American grasslands. It has been hypothesized that low variability in arid biomes is explained by meristem limitation, which constrains responses to pulses of high resource availability. Our research tests this hypothesis by comparing bud-bank populations across six sites in the United States that vary 3-fold in precipitation and 4.5-fold in productivity. In addition, we are examining the effects of management practices, such as fire and grazing, on bud-bank populations using replicated long-term treatments at Konza Prairie LTER site located in north-central Kansas.

Materials and methods At each of six study sites, stem densities, above-ground biomass and bud densities of grasses and forbs were measured within a 25 cm x 25 cm x 15 cm depth sample. We also sampled replicate watersheds in a factorial design at Konza Prairie that have been subjected to annual or four-year burning cycles and are either grazed by Bison (*Bos bison*) or are ungrazed.

Results Data from November 2003 and March 2004 support the hypothesis that bud densities increase from east to west along a precipitation/productivity gradient (Figure 1). Bud densities increased significantly at all sites between sampling dates (P=0.001). Letters indicate differences among sites in March at P<0.05 (LSD). Grass bud density increased from November 2003 to March 2004, but there were no significant treatment effects (grazing, P=0.215 and fire, P=0.339). There was a trend of higher grass bud density in the annually burned, ungrazed treatment compared to the four-year burning cycle. However, this trend disappears with the introduction of grazers. Forb bud density did not change from November 2003 to March 2004 and there was a significant treatment interaction (P=0.001, Figure 2). Annually burned and grazed tallgrass prairie had the highest bud density, but this difference disappeared in the absence of grazers.

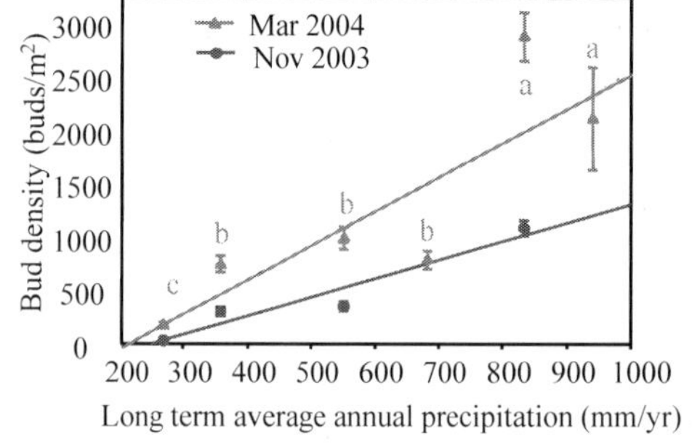

Figure 1 Bud density along a precipitation gradient in central North America

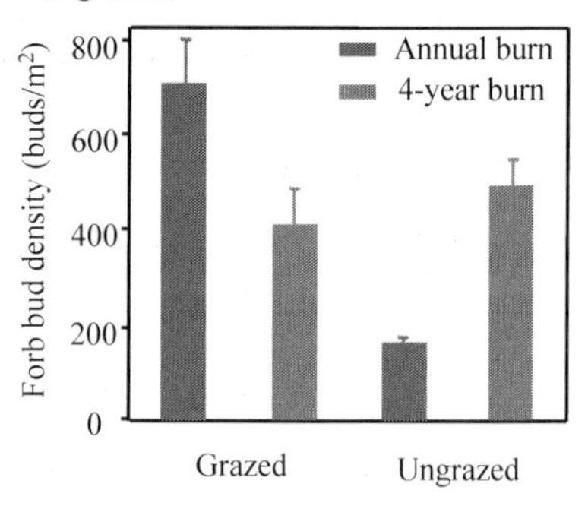

Figure 2 Effects of fire and grazing on forb bud populations

Conclusions Below-ground meristem (bud) density increased with increasing precipitation and productivity. Overall lower bud density in arid grasslands supported the hypothesis that meristem limitation constrains production variability in these grasslands. Grass and forb bud populations had different responses to fire and grazing in tallgrass prairie, with forb bud density increasing in grazed areas while grass bud density remained unchanged. Therefore, demographic mechanisms may be important for forbs in these grasslands.

The impact of vegetation structure and spatial heterogeneity on invertebrate biodiversity within upland landscapes

L. Cole, M.L. Pollock, D. Robertson, J.P. Holland and D.I. McCraken
Scottish Agricultural College, Auchincruive, Ayr, KA6 5HW, UK, Email: lorna.cole@sac.ac.uk

Keywords: invertebrates, vegetation structure, grazing intensity, agricultural management

Introduction Livestock grazing influences vegetation structure and composition at both the patch and wider landscape scale (Milne *et al.,* 1998), and this may have effects on upland invertebrate communities, which in turn influence bird abundance and distribution (Fuller & Gough, 1999; Cole *et al*., 2002). Of particular importance are open grasslands and wet flushes where invertebrates are abundant and more accessible to birds. However, there have been few studies of invertebrates associated with upland habitats, and most of these have focused on heather moorland, blanket bog, or very fine-scaled structure within grasslands (Dennis *et al.* 1997; 1998; 2001). This study addresses the relationship between upland invertebrate biodiversity and the spatial and structural diversity of vegetation.

Materials and methods Four large (>40 ha) plots were established at two sites: Kirkton in Perthshire, Scotland and Sourhope, The Borders, Scotland. At each site, two grazing treatments (light summer-only grazing and moderate year-round grazing) were applied as part of a larger research project. Forty areas of vegetation (i.e. 10 within each plot) were chosen to represent the range of vegetation types and structures within each plot, with emphasis on areas containing *Nardus stricta*. Surface-active invertebrates were sampled within each area using a line of nine pitfall traps. Vegetation height and species composition data were collected from a transect around the pitfall traps. The vegetation patches within a 30 m diameter circle (centred on the line of traps) was mapped, and sward heights measured within the patches.

Results Two-way ANOVA indicated that neither site nor grazing treatment influenced the number of ground beetle species recorded in the pitfall traps (Site: $F=0.02$, $P>0.1$, df = 1,39; Treatment: $F=1.98$, $P>0.1$, df = 1,39). Differences in ground beetle assemblage structure were evident between sites, with altitude (and hence potentially exposure) appearing to have greater influence on invertebrates at Kirkton, while at Sourhope there was a relationship between dominant vegetation type and invertebrates. Across both sites, there was no relationship between invertebrate occurrence and vegetation height or the number of different vegetation patches within the 30 m circle. Further analyses will, however, consider the potential influence of spatial location of these different vegetation types on ground beetles.

Conclusions The results of this study demonstrate that livestock management, in the short term, has less impact on ground beetle assemblages than site factors such as geographical location, climate, altitude and vegetation type. The influence of different variables was site-specific thus indicating that management prescriptions should be tailored to particular sites rather than the adoption of a prescription of one-fits-all.

References

Cole, L.J., D.I McCracken, S. Blake, I.S Downie, G.N Foster, A. Waterhouse, P. Dennis, J.A Milne, K.J. Murphy, M.P. Kennedy, R.W. Furness & A. Milligan (2002). The influence of grassland management and vegetation structure on invertebrates and birds. In: I. Hulbert & E. McEwan (eds). Biodiversity, plant structure and vegetation heterogeneity: interactions with the grazing herbivore, Proceedings of Meeting 9-10 October 2001,. SAC Hill & Mountain Research Centre, Crianlarich, UK, pp 6-10. (CD-Rom)

Dennis, P., M.R. Young & C. Bentley (2001). The effects of varied grazing management on epigeal spiders, harvestmen and pseudoscorpions of *Nardus stricta* grasslands in upland Scotland. *Agriculture, Ecosystems & Environment*, 86, 39-57

Dennis, P., M.R Young & I.J. Gordon (1998). Distribution and abundance of small insects and arachnids in relation to structural heterogeneity of grazed, indigenous grasslands. *Ecological Entomology*, 22, 253-264.

Dennis, P., M.R. Young, C.L. Howard & I.J. Gordon (1997). The response of epigeal beetles (Col: Carabidae, Staphylinidae) to varied grazing regimes on upland *Nardus stricta* grasslands. *Journal of Applied Ecology*, 34, 433-443.

Fuller, R.J. & S. J. Gough (1999). Changes in sheep numbers in Britain: implications for bird populations. *Biological Conservation*, 91, 73-89.

Milne, J. A., C. P. D. Birch., A. J. Hester, H. M. Armstrong & A. Robertson (1998). The impact of vertebrate herbivores on the natural heritage of the Scottish uplands – a review. Scottish Natural Heritage Review No 95. Scottish Natural Heritage, Edinburgh.

The potential for summer-dormant perennial grasses in Mediterranean and semi-arid pastures

F. Lelièvre[1], F. Volaire[1], P. Chapon[1] and M. Norton[2]
[1]*INRA, UMR SYSTEM, 2 place Viala, 34060 Montpellier, France, Email: lelievre@ensam.inra.fr, and* [2] *NSW Dept. Primary Industries, PO Box 408, Queanbeyan, NSW 2620,Australia*

Keywords: *Dactylis glomerata*, cocksfoot, orchard grass, dormancy, drought survival, water uptake

Introduction In rain-fed Mediterranean and semi-arid areas, herbage production of perennial grasses depends on their ability to grow efficiently during the rainy seasons and to persist over the dry summer. A key survival strategy in these harsh conditions is summer dormancy (Volaire, 2002). Within the species *Dactylis glomerata* L., two cultivars (cvs.), contrasting in this trait, were compared in order to analyse their suitability in terms of yield and survival in these environments.

Materials and methods The drought-resistant, non-dormant cv. Medly (Southern France origin) and the drought-resistant, summer-dormant cv. Kasbah (Moroccan origin) were compared at Montpellier, France. In two field experiments (A & B), plots were fully irrigated except from June to October (summer water deficits of 400-600 mm). Aerial biomass was measured regularly and plant survival was counted after autumn rehydration. In Experiment B, soil cores were taken by auger to 160 cm depth at the beginning, middle and end of the drought. A similar experiment (Experiment T) was conducted in 2 m-long soil columns within transparent plastic tubes (3 plants/tube). Soil water use was assessed by weighing the tubes and root depth was measured.

Results In Experiment A, in both years (Table 1), cv. Medly exhibited a higher spring yield potential. Due to its endo-dormancy, cv. Kasbah ceased growing at the end of spring and senesced irrespective of the environmental conditions and its summer yield under irrigation was negligible. In contrast, under a severe drought (year 2003), its survival and recovery after autumn rehydration were significantly greater than that of cv. Medly. In Experiment B, tiller density of cvs Medly and Kasbah recovered to 90-100% of initial density in the autumn. In Experiments B and T, the overall soil water use over droughts was significantly different between cultivars (Table 2) due to greater uptake of cv. Medly while it was still growing during the first part of the drought. In Experiments B and T, at the end of the drought, a substantially higher amount of transpirable soil water remained under cv. Kasbah than under cv. Medly (+34 mm, +40 mm) even though cv. Kasbah`s rooting depth (135 cm) was similar to that of cv. Medly (Experiment T).

Table 1 Seasonal biomass production (g m^{-2}) of 2 cvs. of cocksfoot in field experiment A. Year 2002: moderate summer drought; year 2003: intense summer drought. L.s.d. (5%) given when significant differences at $P<0.05$

Cultivar	Spring 2002	Summer 2002		Autumn 2002	Spring 2003	Summer 2003		Autumn 2003
	Irrigated	Irrigated	drought	after drought	Irrigated	Irrigated.	drought	after drought
Kasbah	340	39	0	160	428	7	0	265
Medly	455	437	0	188	586	246	0	149
L.s.d.	-	127	-	25	72	110	-	71

Table 2 Soil water use under drought (mm water per period) for 2 cvs of cocksfoot (Kasbah and Medly) in a tube experiment (T) . and a field experiment (B). L.s.d. (5%) given when significant differences at $P<0.05$

Expt T Period	Kasbah	Medly	l.s.d.	Expt B Period	Kasbah	Medly	l.s.d
10/07-30/07	37	83	13	30/06-30/07	63	82	-
31/07-10/10	42	38	-	30/07-28/09	34	49	-
10/07-10/10	79	121	25	30/06-28/09	97	131	31

Conclusions The summer-dormant cocksfoot exhibited a lower yield potential, particularly under summer irrigation. Nevertheless, this adaptative strategy conferred superior survival under intense drought, by reducing water consumption during the usually critically dry summer. To improve perennial forage plants by combining higher water use efficiency (production) and summer dormancy (persistence) is now an objective of breeding programs in the Mediterranean (Euro-project PERMED '2004-2008').

Acknowledgements The financial support of Meat and Livestock Australia Ltd. is appreciated.

References

Volaire, F. (2002). Drought survival, summer dormancy and dehydrin accumulation in contrasting cultivars of *Dactylis glomerata*. *Physiologia Plantarum*, 116, 42-51

Improvement of native perennial forage plants for sustainability of Mediterranean farming systems

F. Lelièvre and F. Volaire
Institut National de Recherche Agronomique, UMR SYSTEM, 2 place Viala, 34060 Montpellier, France, Email: volaire@ensam.inra.fr

Keywords: perennial grasses, lucerne, drought survival, water use efficiency

Introduction The amount of water available to agriculture in the Mediterranean is declining because of increasing population pressure and greater incidence of drought. Therefore, the efficiency of the use of water for agricultural production must be maximized and, in this context, perennial forage species have a number of advantages in comparison to the predominantly-used annuals. They can utilize water throughout the whole year besides being able to halt rangeland degradation, restore soil fertility and enhance forage production, thereby contributing to greater sustainability of rain-fed agricultural systems in the southern European Union and North Africa. Despite these advantages, the small size of individual national markets has so far worked against the development of a viable forage industry based on perennials. By adopting a multi-national approach and targeting the key breeding objectives of superior drought-resistance and water-use efficiency (WUE), an European Commission-funded project aims to produce commercially cultivars of a number of species of broad regional interest and adaptation.

Methods The project PERMED (1 October 2004 – 31 September 2008) includes ten research groups from southern Europe (INRA France-F. Lelièvre, CNR Italy- C. Porqueddu, INIAP Portugal-M. Tavares de Sousa, University of Barcelone-S. Nogues, University of Balears- J. Cifre, ISCF Italy – P. Annicchiarico) and North Africa (INRA Algeria- A. Abdelguerfi, INRA Morocco-C. Al Faïz, Institut des Régions Arides, Tunisia-A. Ferchichi, Pôle de Recherches Agronomiques Tunisie, M. Ben Younes). They combine work on species, including lucerne, cocksfoot, tall fescue and sulla, to enhance cultivar development across environments ranging from the sub-humid to arid. Complementary work packages (Figure 1): (i) completing North African forage germplasm collection and evaluation, (ii) assessing the use of molecular genetics in breeding of drought-resistant lucerne, (iii) evaluating elite forage populations across the region for high WUE and adaptation to drought as bases for new cultivars, (iv) enhancing knowledge of physiological traits for drought survival and WUE, and (v) determining optimal use of perennial forages in four representative farming systems. The results will contribute to the development of technical packages for easy on-farm adoption across the western Mediterranean, thereby ensuring a long-term interest of the seed industry.

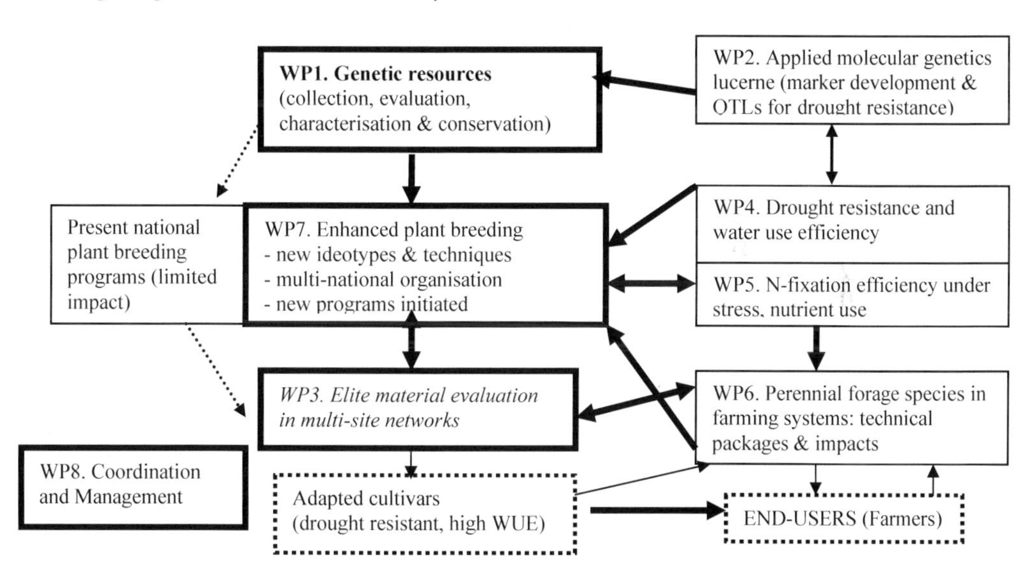

Figure 1 Relationships between work packages in the European-INCO project 'PERMED'

The influence of management on health status of *Festuca rubra* in mountain meadows

B. Voženílková, F. Klimeš, J. Květ, Z. Mašková, B. Čermák and K. Suchý
University of South Bohemia, Studentská 13, 370 05 České Budějovice, Czech Republic, Email: vozenil@zf.jcu.cz

Keywords: *Festuca rubra, Holcus mollis, Fusarium* spp.

Introduction Snijders & Winkelhorst (1996) investigated swards in West Europe and showed that it was not the snow mould (*Microdochium nivale*) but other species of the genus *Fusarium* (*F. cerealis* (Cooke) Sacc., *F. graminearum* Schwabe, *F. culmorum* (Wm. G. Sm.) Sacc. and *F. acuminatum* Ellis & Everh.) that caused serious damage to grasslands where *Lolium perenne* L. and *Festuca rubra* L. were dominant components. In this study the spread and harmfulness of pathogeneous fungi involved in damage to and death of some species (*Festuca rubra* L., *Holcus mollis* L.) in grass swards was examined.

Material and methods The spread and harmfulness of pathogenic fungi involved in the senescence of two dominant species of grasses (*Festuca rubra* and *Holcus mollis*) was observed mainly from the viewpoint of phytopathology. The investigation was carried out in 1999 – 2003 in controlled experiments. The experimental site (Zhůří in the Šumava Mountains) was situated at an altitude of 1150–1180 m. In the experiments, three treatments (Mo – mowing, Mu -mulching, F – fallow) were imposed with three replicates (30 m^2 per plot) per treatment.
Phytopathological analysis of plants with symptoms indicating an attack of *Fusarium* fungi was carried out in May of each year. The method for evaluation of symptoms in plants, according to Dixon & Doodson (1971), was used. The projective dominance of the grasses was estimated in mid-July of each year (prior to mowing or mulching of the Mo or Mu treatments).

Results During the observation of *Fusarium* fungi in selected grass species, there were significant (p<0.001) differences in the extent of attack on *F. rubra* from that on *H. mollis*. The least extent of *Fusariium* symptoms was recorded in *H. mollis* whereas the most serious damage was found in *F. rubra,* especially in the unharvested fallow treatment.
In this treatment, *F. rubra* also showed the greatest decrease of coverage from the original value of 30-40 % D to only 3-6 % D (% D = projective dominance). In the mulched treatment, a trend in a reduction in *F. rubra* to 20-30 % D was also recorded owing to the fungal attack. The increased attack by *Fusarium* fungi found in the mulched treatment and an even more severe attack on unharvested *F. rubra* created worse conditions for subsequent fodder crop production. The infestation of *F. rubra* with the *Fusarium* fungi was twice as intense in the mulched treatment (28.9%) and almost 4.5 times as intense in the unharvested treatment (65.2%) in comparison with the mown treatment (14.0%).
A high correlation was found between the attack of *Fusarium* fungi (in May) and subsequent reduction in *F. rubra* along with a simultaneous increase (parallel with *F. rubra* retreat) in *H. mollis* (Figure1).

Conclusions The least extent of damage caused by *Fusarium* fungi was recorded in *H. mollis* whereas the most serious damage was found in *F. rubra,* especially in unharvested herbage.

Acknowledgement The study was supported by GAČR: 206/99/1410, NAZV: QF 3018, MSM 6007665806

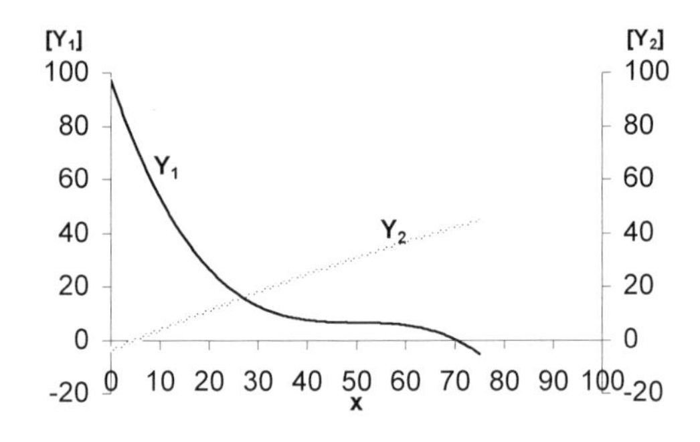

Figure 1 Correlation between *Fusarium* fungi attack of *Festuca rubra* in May (x in %) and consequent cover change of *F. rubra* (Y_1 in %D) and *Holcus mollis* (Y_2 in %D) in July: $Y_1=97.678-5.362x+0.105998x^2-0.000704x^3$ ($R^2=0.816$; $P<0.01$); $Y_2=-6.873+0.750x-0.001742x^2$ ($R^2=0.744$; $P<0.01$)

References

Dixon, G. R. & J. K. Doodson (1971). Assessment keys for some diseases of vegetable, fodder and herbage crops. *Journal of National. Institute of Agricultural Botany, 12,299 – 307.*

Snijders, C.H. A. & G.D. Winkelhorst (1996). An artificial inoculation method to screen for resistance to *Fusarium* rot in grasses. *The Second International Conference on Harmful and Beneficial Microorganisms in Grassland, Pastures and Turf,* pp. 265-271.

Nutritive value of Alopecurus pratensis, Festuca rubra, Arrhenatherum elatius and Lolium perenne grown in the South of Belgium

A. Nivyobizi[1], A.G. Deswysen[2]†, D. Dehareng[2], A. Peeters[1] and Y. Larondelle[3]

[1]*Laboratory of Grassland Ecology, Catholic University of Louvain (UCL), Place Croix du Sud, 5 Box 1 1348 Louvain la Neuve Belgium, Email: nivyobizi@ecop.ucl.ac.be,* [2,3]*Catholic University of Louvain (UCL), Faculty of Bio-engineering, Agronomy and Environment, Units of applied Genetics (GENA[2]) and Biochemistry of Nutrition (BNUT[3]), Place Croix du Sud, 2 Box 14 1348 Louvain la Neuve Belgium*

Keywords: nutritive value, secondary grasses, *in vivo*, sheep

Introduction In Europe, recent strategies have aimed at encouraging farmers to use production techniques more efficient in preserving the environment and maintaining natural areas. Those strategies have encouraged the use of secondary grass species in forage production systems. However, the nutritive value of those grasses is not well known. Therefore, the aim of the present study was to evaluate the energy and nitrogen values of *Alopecurus pratensis* (ALPR), *Festuca rubra* (FERU) and *Arrhenatherum elatius* (AREL) under moderate rates of nitrogen (N) application (60 kg N/ha per cut) and a hay-cutting regime (2 cuts/year: 25 May and 9 July). *Lolium perenne* cv. Bastion (LOPE) was used as a control. The first cut of ALPR was a mixture of 18 April and 25 May cuts.

Materials and methods For each cut, organic matter digestibility (OMD) was measured with 4 Texel crossbred male sheep (35 ± 5 kg), according to a Latin square (4 x 4) experimental design. Maintained in digestion crates, the sheep were fed each grass hay in 2 daily equal meals (restricted feeding level: 40 g DM/kg $LW^{.75}$/d) for 21 days (a 13-d adaptation period followed by an 8-d period of total faeces collection). Effective degradability of crude protein (Deg CP) was also evaluated (outflow rate k = 0.02/h) on sheep: an *in sacco* procedure was performed on 3 other Texel crossbred male sheep (91 ± 7 kg) fitted with a rumen canula. Observed degradability data were then fitted to the lagged exponential model. Finally, nutritive value was expressed in terms of net energy (UF) and intestinally digestible protein (PDI) according to the French feeding system. Only digestibility data could be submitted to a variance analysis. Dry matter, organic matter, crude protein (CP) and neutral detergent fibre (NDF) contents were determined following classical procedures.

Results With the exception of ALPR (1st cut only), OMD (P<0.05) and UF values of secondary grasses were lower than those of LOPE, reflecting differences in CP and NDF contents (Table 1). The Deg CP was similar for AREL and LOPE, while it was either higher (1st cut) or lower (2nd cut) for ALPR and FERU than for LOPE (P<0.05). Nevertheless, reflecting variations in CP content, PDI values of secondary grasses were similar to or even higher than those of LOPE. Overall, based on DM yields, the calculation of total UF and PDI production (2 cuts) showed that the studied secondary grasses provided lower (ALPR) or higher (FERU, AREL) UF yield and higher PDI yield than LOPE.

Table 1 Nutritive value of secondary grasses

Grasses	DM Yield (t/ha)*	CP	NDF	OMD	Deg CP	UFL	UFV	PDIE	PDIN
		(g/kg DM)		(g/g)		(/kg DM)		(g/kg DM)	
First cut									
ALPR	4.05	177	601	.680^{a}	.573^{a}	0.77	0.70	103	110
FERU	6.13	93	697	.595^{c}	.438^{b}	0.66	0.57	91	66
AREL	7.86	72	714	.599^{c}	.391^{c}	0.66	0.57	90	58
LOPE	5.36	80	635	.648^{b}	.404^{c}	0.73	0.65	91	58
Second cut									
ALPR	1.56	217	745	.659^{b}	.445^{c}	0.76	0.67	127	133
FERU	2.72	171	715	.644^{b}	.522^{b}	0.72	0.64	98	101
AREL	2.74	154	721	.645^{b}	.563^{ab}	0.73	0.64	91	91
LOPE	0.83	156	656	.701^{a}	.594^{a}	0.81	0.74	91	89

*Average data previously reported by Peeters A. (2004)

Conclusions Because of their relatively valuable nutritive value as compared to that of *Lolium perenne*, the studied secondary grasses, particularly *F. rubra*, could be recommended for forage production systems related to preserving the environment.

Reference

Peeters, A. (2004). Wild and sown grasses. Profiles of temperate species selection: ecology, biodiversity and use. FAO and Blackwell Publishing, Rome, 311p.

Increasing the productive potential of permanent grasslands from the forest steppe area of Romania

V. Vintu, C. Samuil, T. Iacob and St. Postolache
University of Agricultural Sciences and Veterinary Medicine from Iasi, Romania, Street Mihail Sadoveanu 3, 700490 Romania, Email: vvintu@univagro-iasi.ro

Keywords: improved grassland, fertilisation, floristic structure

Introduction In Romania permanent grasslands represent 32 % of the total agricultural area, stretching over 4,872 million hectares, out of which 340,000 ha are located in the forest steppe area, on less productive soils, which explains their inadequate botanical composition and low quality and yields (Vintu, 2003). One of the main measures taken to increase the productivity of grasslands is through fertilisation (Birch, 1999). This paper presents the results obtained during 2000-2004 on the effect of organic fertiliser on degraded grasslands made up of *Festuca valesiaca L.*

Material and methods The unreplicated experiment comprised the following treatments: V1 – control (unfertilised); V2 – 10 t/ha cattle manure annually + $N_{50}P_{36}$ annually; V3 - 10 t/ha cattle manure every 2 years + $N_{100}P_{72}$ annually; V4 – 20 t/ha cattle manure annually + $N_{50}P_{36}$ annually; V5 - 20 t/ha cattle manure every 2 years + $N_{100}P_{72}$ annually; V6 – 10 t/ha sheep manure annually + $N_{50}P_{36}$ annually; V7 - 10 t/ha sheep manure every 2 years + $N_{100}P_{72}$ annually; V8 – 20 t/ha sheep manure annually + $N_{50}P_{36}$ annually; V9 - 20 t/ha sheep manure every 2 years + $N_{100}P_{72}$ annually. Harvesting was carried out within a hay field pattern when the dominant grasses were at the start of flowering.

Results The data, presented in Table 1, show the positive influence of fertilisation upon production depending on the fertiliser rate, manure application period and manure type. The dry matter (DM) production varied between 4.46-4.90 t /ha on those treatments fertilized with cattle manure and between 5.24-5.67 t/ha on those fertilised with sheep manure, compared to a 3.21 t/ha obtained on the unfertilised control treatment. Fertiliser application changed the botanical composition (Table 2) by increasing the cover of graminaceae from 64 % up to 67 %, that of leguminous species from 13 % up to 17 % and by decreasing the cover other species from 23 % to 16 %.

Table 1 Production of herbage and difference between the control and other treatments (Dif.) (t DM/ha)

Treatment	Mean 2000-2004		Significance
	t/ha	Dif.	
V_1	3.21	-	-
V_2	4.46	1.25	*
V_3	4.76	1.55	*
V_4	4.54	1.33	*
V_5	4.90	1.69	**
V_6	5.24	2.03	**
V_7	5.64	2.43	***
V_8	5.27	2.06	**
V_9	5.67	2.46	***

*, $P<0.05$; **, $P<0,01$; ***, $P<0.001$

Table 2 Vegetation cover (%) in 2000 and 2004

Treatment	2000			2004		
	G	L	OS	G	L	OS
V_1	60	12	28	71	11	18
V_2	60	15	25	68	16	16
V_3	60	14	26	64	20	16
V_4	68	10	22	68	15	17
V_5	60	16	24	63	18	19
V_6	70	11	19	72	14	14
V_7	64	13	23	68	17	15
V_8	71	10	19	64	17	19
V_9	61	14	25	62	20	18
Average	64	13	23	67	17	16

G - grasses; L – legumes; OS – other species

Conclusions Permanent grasslands of *Festuca valesiaca L.*, from Romania present a very good answer to the organic-mineral fertilisation; thus, yield outputs of 39 up to 53% are obtained by the variants fertilised with cattle muck in comparison with the variants fertilized with sheep muck, which presented an output of 63 up to 77%. The use of fermented farmyard manure is an important measure of rehabilitating the permanent grasslands from Romania with the framework of promoting the concept of organic agriculture.

References

Birch N.V., A.M. Avis & A.R. Palmer (1999). Changes to the vegetation communities of natural rangelands in response to land use in the mid-Fish River valley, South Africa. Proceeding of VIth International Rangeland Congress. Vol. I, Townsville, Australia, pp.319-320.

Vintu V. I. Avarvarei, T. Iacob, N. Dumitrescu. & C. Samuil (2003). Improvement of the degraded rangeland of the Romania forest steppe by organic and mineral fertilisation. Proceedings of VIIth International Rangeland Congress. 26 July-1 August 2003, Durban, South Africa, pp. 1267-1269.

The influence of harvest period and fertilisation on the yield of some mixed grass and leguminous species under the forest steppe conditions of North-east Romania

V. Vintu, C. Samuil, T. Iacob and St. Postolache

University of Agricultural Sciences and Veterinary Medicine from Iasi, Romania, Street Mihail Sadoveanu 3, 700490 Romania, Email: vvintu@univagro-iasi.ro

Keywords: harvest period, mixtures of perennial grasses, floristic structure

Introduction In the forest steppe area of North-east Romania, temporary grasslands represent an important source of high quality fodder but they have a short period of exploitation, associated with some changes in the floristic composition (Vintu, 2003). Fertiliser application and harvest period have an important role in maintaining high productivity (Hopkins, 1991). The aim of this paper is to determine the influence of harvest period and fertilization on the yield of some grass and leguminous species in the forest steppe conditions of North-east Romania.

Material and methods The experiment was a randomised block design with the following factors: factor A – harvest period: a_1 – repeated mowing; a_2 – hay cut; factor B: mixture of grasses and perennial leguminous species: b_1 –*Dactylis glomerata* 20% + *Lolium perenne,* 30% + *Medicago sativa* 50%; b_2 –*Festuca pratensi,s* 25% + *Lolium perenne,* 25% + *Agropyron pectiniforme,* 10% + *Lotus corniculatus,* 10% + *Medicago sativa,* 30%; factor C: fertiliser: c_1 – unfertilized; c_2 – manure 20 t/ha annually; c_3 – N_{100} fertiliser annually; c_4 – manure 20t/ha + N_{50} fertiliser annually.

Results The average yield of dry matter (DM) varied between 4.50–7.45 t/ha for repeated mowing and between 5.23 – 7.77 t/ha for the hay cut (Table 1). The DM yield was higher for the b_2 mixture.. The highest yield was obtained from the fertilised treatment with manure at 20 t/ha + N_{50} used annually, regardless of the harvest period or the type of mixture of perennial grasses. After 4 years of treatments, species other than grasses and legumes have increased to 4–10%, while the cover of grasses and legumes has declined (Table 2).

Table 1 Yield of dry matter (t/ha)

Factor A	Factor B	Factor C	Yield	%	Significance
a1	b1	c1	4.50	100	
		c2	5.64	125	*
		c3	5.92	131	**
		c4	7.50	166	***
	b2	c1	5.36	119	*
		c2	6.43	143	***
		c3	6.31	140	***
		c4	7.45	165	***
a2	b1	c1	5.23	116	
		c2	7.22	160	***
		c3	6.80	151	***
		c4	7.56	168	***
	b2	c1	6.09	135	**
		c2	7.48	166	***
		c3	7.26	161	***
		c4	7.77	173	***

*, *P*<0.05; **, *P<0.01;* ***, P<0.001

Table 2 Vegetation cover (%)

Factor A	Factor B	Factor C	G	L	OS
a1	b1	c1	60	36	4
		c2	53	40	7
		c3	57	38	5
		c4	52	40	8
	b2	c1	53	31	6
		c2	58	35	7
		c3	65	30	5
		c4	61	32	7
a2	b1	c1	51	44	5
		c2	44	48	8
		c3	53	40	7
		c4	51	42	7
	b2	c1	59	34	7
		c2	55	35	10
		c3	58	33	9
		c4	54	36	10

G - grasses; L – legumes; OS – other species

Conclusions Temporary grasslands from the North–east forest steppe of Romania reach the highest productivity level used as hayfields, regardless of the type of mixture or the fertiliser application rate. After four years of use, the floristic composition shows important changes from the original sward; grasses, legumes and other species represented 44-65%, 30-48% and 4-10% respectively..

References

Hopkins A. (1991). Grassland improvement by the use of fertilisers compared with reseeding experience of multi-site trials in Great Britain. Proceedings of Conference held at Graz, Austria, 18-21 September, 1991, pp.161-162.

Vintu V., T. Iacob, N.Dumitrescu, C. Samuil & A. Trofin (2003). The reabilitation of the degraded rangelands from Romania's hill zone by the radical recovery of the vegetation cover. Proceedings of VIIth International Rangeland Congress, 26 July-1 August, Durban, South Africa, 2003, pp. 1270-1272.

Productivity of Sahiwal and Friesian –Sahiwal crossbreds in marginal grasslands of Kenya

W.B. Muhuyi, F.B. Lukibisi and S.N. ole Sinkeet
National Animal Husbandry Research Centre, P.O. Box 25 Naivasha, Kenya, Email: karinaiv@kenyaweb.com

Keywords: Sahiwal, Friesian-Sahiwal crossbreds, grasslands, productivity

Introduction Dual-purpose cattle can be used to exploit the production potential of semi-arid grasslands of Kenya for milk and meat production. Although the Sahiwal is adapted to these grasslands, its productivity is low. In order to increase milk and meat productivity, the Sahiwal has been crossed with the Friesian to produce Friesian-Sahiwal crossbreds (McDowell *et al.*, 1996) adapted to the tropical environment. The objective of this study was to evaluate the productivity of Sahiwal and Friesian-Sahiwal crossbreds.

Materials and methods Data were obtained from Sahiwal and Friesian-Sahiwal crossbred cattle at Naivasha Research Centre, situated in agro-ecological zone IV. The average rainfall is 680mm per annum. Cattle grazed on natural pastures. The herd consisted of young and breeding cattle. Data on performance traits were analysed using a fixed effect model (Harvey, 1990) to obtain mean estimates. Measures of productivity were production efficiency (milk yield in kg/d of calving interval) and the Pry productivity index (total output value in Kenya Shillings per kg dry matter intake), derived using the Pry productivity model (Baptist, 1988).

Results Age at first calving was significantly different ($P < 0.01$) between the two genotypes, with mean values (± standard deviation) of 35.0±4.05 and 41.2±5.42 months for the Friesian-Sahiwal crossbreds and Sahiwal, respectively. The decrease in this parameter is similar to that reported in India (Bhat *et al.,* 1978). Similarly the calving interval was significantly lower ($P < 0.01$) among the crossbreds (400±61 days) compared to the Sahiwal cattle (446±105 days. Mean milk yield was 2,210±808.9 kg and 1,500±551.7 kg for the crossbred and Sahiwal cattle, respectively ($P < 0.01$) and the values for the crossbred are similar to those reported by McDowell *et al.* (1996). Productivity indices were different with the Friesian-Sahiwal crossbreds superior to the Sahiwal (Table 1).

Table 1 Productivity indices of the Sahiwal and Friesian-Sahiwal crossbreds

Productivity Index	Sahiwal	Friesian-Sahiwal
Production efficiency (kg/dci[1])	3.46	5.54
Pry Productivity index (Ksh*./kg DMI[2])	5.71	6.47

[1]Milk yield in kg/d of calving interval, [2] Total output value in Kenya shillings per kg of dry matter intake,
*1 US$ = 75 Kenya Shillings.

Conclusion The crossbred genotype gave improved productivity for all the measured parameters. The Friesian - Sahiwal crossbreds, which combine the high production of the Friesian and hardiness of the Sahiwal, have high overall productivity. The utilisation of crossbreds is of economic importance for smallholder farmers, since the Friesian - Sahiwal is well suited to multiple uses in marginal areas.

Acknowledgement The senior author is grateful to the Kenya Agricultural Research Institute for sponsorship.

References

Baptist, R. (1988). Herd and flock productivity assessment using the standard offtake and the demogram. *Agricultural Systems*, 28, 67-78.

Bhat, P.N., V.K. Taneja & R.C. Garg (1978). Effects of crossbreeding on reproduction and production traits. *Indian Journal of Animal Science*, 48, 71-78.

Harvey, W.R. (1990). Mixed model least squares and maximum likelihood computer program. Ohio State University, Columbus.

McDowell, R.E., J.C. Wilk & C.W Talbott (1996). Economic viability of crosses of *Bos taurus* and *Bos indicus* for dairying in warm climates. *Journal of Dairy Science*, 79, 1292-1303.

Rainfall and grazing impacts on the population dynamics of *Bothriochloa ewartiana* in tropical Australia

D.M. Orr and P.J. O'Reagain
Dept. Primary Industries and Fisheries, PO Box 6014, Rockhampton Mail Centre, Queensland 4702, Australia, Email: david.orr@dpi.qld.gov.au

Keywords: plant survival, plant size, *Bothriochloa ewartiana*

Introduction *Bothriochloa ewartiana* (desert bluegrass) is a palatable, native perennial (C4) grass of considerable importance to the northern Australian grazing industry. However, little is known of the interaction between grazing pressure and the highly variable rainfall found in this area, on its population dynamics. This paper reports interim results (1998-2004) from a long-term study, in which its population dynamics were examined under 3 grazing strategies.

Materials and methods An extensive grazing study was established in December 1997, at "Wambiana" ($20^0$34' S $146^0$07'E) near Charters Towers, to assess the relative ability of five grazing strategies, replicated twice, to cope with rainfall variability in terms of their effects on animal production, economics and resource condition (O'Reagain and Bushell 1999). Paddock sizes are 93-117 ha and are arranged so that each paddock contains similar portions of each of 3 soil types. Long-term mean annual rainfall is 630 (range 109-1653) mm, with most falling between November and March. The vegetation is open *Eucalyptus* savanna overlying an herbaceous layer of C_4 grasses. Permanent quadrats (n=20; each 50 x 50 cm) delineating 40 *B. ewartiana* plants, were established in each replicate of 3 grazing treatments to examine the persistence of *B. ewartiana* under grazing. These treatments were light stocking (8 ha/steer), heavy stocking (4 ha/steer) and rotational rest (6 ha/steer with 33% of the pasture rested annually during the wet season). The dynamics of these *B. ewartiana* plants were charted annually between 1998 and 2004 using the methodology of Orr *et al.* (2004).

Results Plant size increased between 1999 and 2001 in response to above average summer rainfall but fell dramatically after 2002 in response to severe drought. There were no differences (P>0.05) in plant size between the 3 treatments although there was a clear trend for plant size to be greater under light grazing (Figure 1a). Plant survival did not differ (P>0.05) between treatments with high survival between 1998 and 2002 but there was evidence of accelerated plant death between 2002 and 2004 in response to the severe drought (Figure 1b).

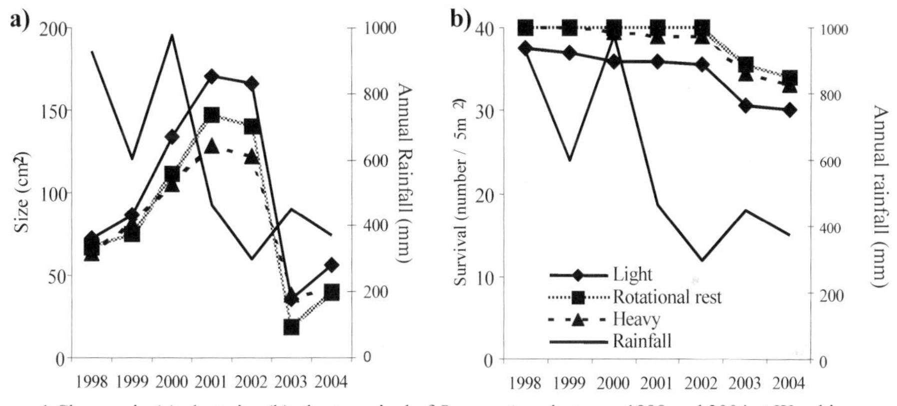

Figure 1 Changes in (a) plant size (b) plant survival of *B. ewartiana* between 1998 and 2004 at Wambiana

Conclusions Rainfall variability rather than grazing pressure appeared to have the greater impact on the dynamics of *B. ewartiana* between 1998 and 2004. This result indicates some resilience under grazing by this grass but we expect a more pronounced grazing impact in future years of this study.

References

O'Reagain, P. J. & J. J. Bushell (1999). Testing grazing strategies for the seasonally variable tropical savannas. *Proceedings 6th International Rangelands Congress*, Durban, 485-486.

Orr, D.M., Paton, C.J. and Reid, D.J. (2004). Dynamics of plant populations in *Heteropogon contortus* (black speargrass) pastures on a granite landscape in southern Queensland. 1. Dynamics of *H. contortus* populations. *Tropical Grasslands,* 38, 17-30.

Herbage quality of dwarf Napier grass under a rotational cattle grazing system two years after establishment

Y. Ishii[1], A.A. Sunusi[2], M. Mukhtar[1], S. Idota[1] and K. Fukuyama[1]
[1]*Division of Grassland Science, Faculty of Agriculture, University of Miyazaki, Miyazaki, 889-2192, Japan, Email: yishii@cc.miyazaki-u.ac.jp and* [2]*Faculty of Animal Science, Hasanuddin Univ., Makassar, 90245, Indonesia*

Keywords: herbage quality, dwarf Napier grass, rotational grazing, beef cattle

Introduction Dwarf Napier grass (*Pennisetum purpureum* Schumach) of a late-heading type (dwarf-late, DL), introduced by the Dairy Promotion Organization, Thailand, has a high over-wintering ability and is suitable for grazing. The objective of this study was to examine the digestibility and crude protein (CP) concentration of DL Napier grass both before and after rotational grazing in relation to the daily liveweight gain of cattle 2 years after establishment in the lowland area of Kyushu, Japan.

Materials and methods Four paddocks of DL Napier grass pasture (20 m × 25 m/paddock) were established by rooted tillers on 6 May, 2002. The DL pasture was rotationally grazed for one week with a 3-week rest period by 3 breeding cattle for 3 cycles from August to December in 2002 and by 2-3 beef cattle (Japanese-Black) for 6 cycles from June to December in 2003. Live weight (LW) of cattle was measured at 1100 h when cattle switched paddocks, and no concentrate or roughage was fed during this rotational grazing. Initial LWs were 451 and 359 kg/head in 2002 and 2003, respectively. *In vitro* dry matter digestibility (IVDMD) and CP concentration were measured in the herbage cut at 10 cm above the ground surface at every cycle in 2002 and at cycles 2, 4 and 6 in 2003, which are designated as period I, II and III, respectively.

Results Concentration of CP and IVDMD are shown for 2003 in Figure 1(A) and (B), respectively. Annual mean CP concentration and IVDMD were 99 g/kg DM and 0.623-0.652, respectively, and they were higher than the averages of tropical grasses (about 93 g/kg DM and 0.54, respectively). Both IVDMD and CP concentration tended to be higher in period II than in periods I and III in 2002, and decreased from period I to period III in 2003. Differences in IVDMD between plant parts were small and CP concentration tended to be lower in the stem than in the leaf blade in both years. Live weight of grazing cattle was at least maintained in 2002 and increased at 0.30-0.48 kg/head/day during periods I and II in 2003.

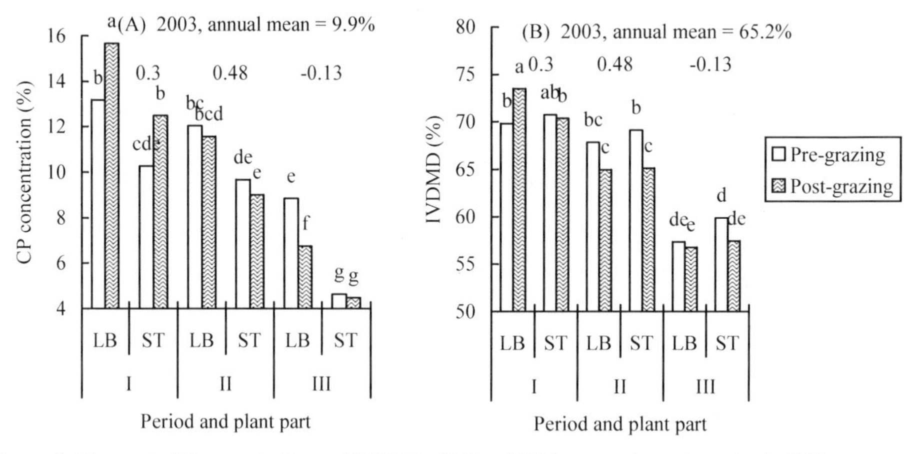

Figure 1 Changes in CP concentration and IVDMD of LB and ST from pre- to post-grazing in 2003
Figures with different letters denote significant difference between periods and plant parts at 5% level
Figures on the column denote the daily gain of grazing cattle (kg/head/day)

Conclusion The results of this study demonstrate that the herbage quality of DL Napier grass can be higher than the average of tropical grasses, and the change in herbage quality almost corresponded with the daily LW gains of grazing beef cattle in Kyushu, Japan.

Grazing suitability of various Napier grass varieties in paddocks of different ages

Y. Ishii[1], M. Mukhtar[1], S. Tudsri[2], S. Idota[1], Y. Nakamura[1] and K. Fukuyama[1]
[1]*Division of Grassland Science, Faculty of Agriculture, University of Miyazaki, Miyazaki 889-2192, Japan, Email: yishii@cc.miyazaki-u.ac.jp, and [2]Faculty of Agriculture, Kasetsart University, Bangkok, 10900, Thailand*

Keywords: dwarf Napier grass, normal Napier grass, rotational grazing, dairy cows

Introduction Previous studies have demonstrated that late-heading type dwarf (DL) Napier grass (*Pennisetum purpureum* Schumach) introduced to Japan from Thailand by the Dairy Promotion Organization of Thailand was able to overwinter in the lowland areas of southern Kyushu (Mukhtar *et al.*, 2003). The species has a higher proportion of leaf blade than other normal and dwarf varieties. These studies were conducted to assess the suitability of the various Napier grass varieties for grazing (Mukhtar *et al.*, 2004). The objective of this study was to examine the grazing suitability, herbage quality and wintering ability of three Napier grass varieties for dairy cows on newly-established and four-year-old pastures in 2003 in Kyushu, Japan.

Materials and methods Three varieties of Napier grass were used: the normal variety of Wruk wona (Wr) and dwarf varieties of late-heading (DL) and early-heading (DE) types obtained from the Department of Livestock Development, Thailand. One paddock (9 m × 36 m), divided into 6 plots with 2 replicates, was prepared using rooted tillers of the 3 varieties in mid-September, 2000 and again on May 3, 2003. Rotational grazing was practiced in the paddock employing rest periods of 30-77 and 21-42 days in 2000-2002 and 2003, respectively. Length of day-grazing was 7.5 hours and 15-16 head of dairy cows were used. Wr was cut twice to reduce stem elongation after grazing on 1 July and 2 September, 2003.

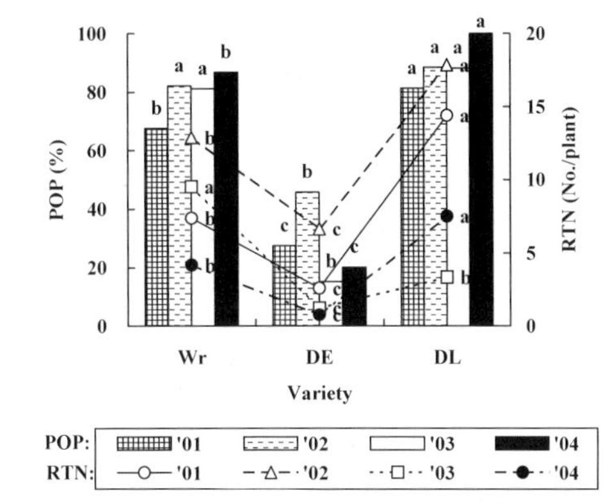

Figure 1 Changes in POP and RTN among 3 varieties in 2001-4
Values with different letters denote significant difference among varieties in the same year at 5% level

Results The percentage of overwintered plants (POP) and number of tillers that resprouted (RTN) over the 4-year period are shown in Figure 1. Both POP and RTN were highest in DL, followed by Wr and lowest in DE over the 4-year period. Changes in *in vitro* dry matter digestibility (IVDMD) and crude protein (CP) concentration of leaf blades (LB) from pre-grazing herbage are shown in Figure 2. Both IVDMD and CP concentration tended to be higher in dwarf varieties than in Wr. Percentage utilization was higher in dwarf varieties than it was in Wr except where grazing followed cutting.

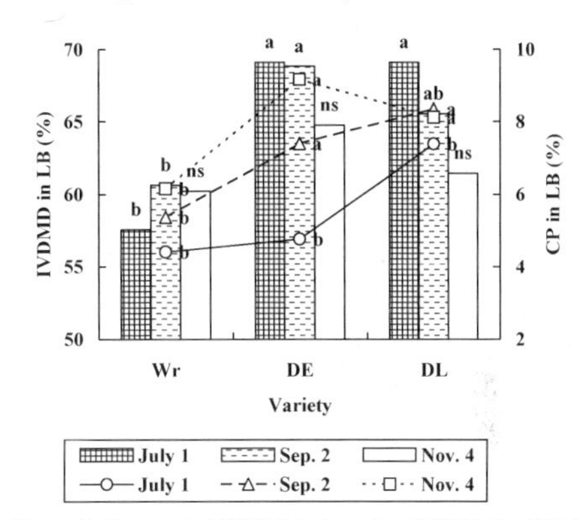

Figure 2 Changes in IVDMD (column) and CP (dot) of LB in 2003
Values with different letters denote significant difference among varieties in the same date at 5% level

Conclusion This study demonstrated that under conditions of grazing by dairy cows at low altitudes in Kyushu, Japan, DL Napier grass exhibited superior overwintering characteristics and maintained higher quality of herbage when compared to other Napier grass varieties.

References

Mukhtar, M., Y. Ishii, S. Tudsri, S. Idota & T. Sonoda (2003). Dry matter productivity and overwintering ability of the dwarf and normal napiergrasses as affected by the planting density and cutting frequency. *Plant Production Science*, 6, 65-73.

Mukhtar, M., Y. Ishii, S. Tudsri, S. Idota, K. Fukuyama & T. Sonoda (2004). Grazing suitability of normal and dwarf Napier grasses transplanted on a Bahia grass pasture. *Grassland Science*, 50, 15-23.

Detecting fauna habitat in semi-arid grasslands using satellite imagery

N.A. Bruce[1], I.D. Lunt[1], M. Abuzar[2] and M. Mitchell[3]

[1]*The Johnstone Centre, Charles Sturt University, PO Box 789, Albury NSW 2640, Australia, Email: nbruce@csu.edu.au,* [2]*Dept Primary Industries, Institute of Sustainable Irrigated Agriculture, Tatura Vic 3616, Australia and* [3] *Dept of Primary Industries, Rutherglen Research Institute, Rutherglen Vic 3685, Australia*

Keywords: grassland habitat, remote sensing, south-eastern Australia

Introduction Managing grasslands for biodiversity conservation is a relatively recent phenomenon and there is uncertainty over the most effective strategy. Past research has found that intermediate levels of disturbance (e.g. burning or grazing) may be required to maintain the natural mosaic of small-scale patterning required for a diverse range of flora and fauna species. For sustainable grassland management, appropriate methods of spatial assessment and temporal monitoring are required, to facilitate understanding of how past and present climate, land management and landscape features influence vegetation structure. Due to the expense and time-consuming nature of conventional ground-based monitoring, satellite remote-sensing techniques offer a feasible approach.

Plains-wanderers (PWs, *Pedionomus torquatus*), a small ground dwelling bird, are endangered due to the loss of native grasslands in south-eastern Australia, through cropping and inappropriate grazing. Plains-wanderers require an open habitat structure, and cannot survive in grasslands that are too dense or too open. Terrick Terrick National Park (TTNP) is one of the most significant PW habitats in Victoria, Australia. A major role for the park is to provide a drought refuge for PWs when heavy grazing reduces habitat in the surrounding landscape. This study explores the use of satellite imagery in detecting grassland habitat for PWs at TTNP during the drought in 2002.

Methods Visual estimates of PW habitat classes (vascular plant cover) were collected at TTNP during mid-spring 2002. One hundred and forty-five sites were sampled, using a stratified random sampling regime, to ensure optimal coverage of variation in vegetation patterns. A cloud-free Landsat 7 ETM+ image (path 93/row 85), acquired on 11 October 2002, was used in this study. Relationships between Landsat TM spectral values and ground-truthed cover data from October 2002 were examined using raw TM bands and various vegetation indices and ratios.

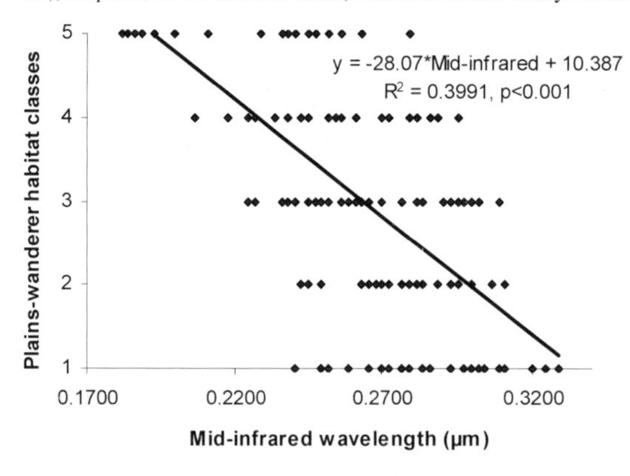

Figure 1 Relationship between the mid-infrared band and Plains-wanderer habitat classes (1-Much too sparse, 2-Slightly too sparse, 3-Ideal, 4-Slightly too dense, 5-Much too dense)

Results Linear regression analyses highlight the ability of the mid-infrared band (TM 7) to predict PW habitat classes (Figure 1). Results indicate that approximately 34.6% of TTNP supported suitable PW habitat during the 2002 drought. Spatial patterning of habitat classes suggests that PW habitat suitability was affected by inter-paddock differences in stocking levels and with paddock differences in factors including stock movement patterns, soils and vegetation type. Comparison of results against soil patterns (results not shown) suggests that grazing patterns may be the major factor affecting habitat availability.

Conclusion Results show that satellite imagery can be used to monitor spatial patterns of PW habitat and to identify areas where grazing management can be refined. The strong contrast between the abundance of PW habitat in the east and west of the reserve, for instance, is primarily related to differences in stocking levels during the 2002 drought, rather than to underlying soil or vegetation patterns. In this study, the mid-infrared band (TM 7) showed the strongest correlation to ground-truthed PW habitat values. This strong correlation probably reflects the fact that PWs require a specific amount of open ground cover. Other vegetation attributes (such as greenness which can be detected by the near-infrared band) appear to be less important to PWs. This study demonstrates that remote sensing can be used to identify large-scale patterns of habitat suitability for endangered ground animals. This approach is currently being extended to explore temporal changes in PW habitat in TTNP over the last decade.

The effect of manipulated conservation margins in intensively grazed dairy paddocks on the biodiversity of Pteromalidae and Braconidae (Hymenoptera: Parasitica)

A. Anderson, G. Purvis, A. Helden and H. Sheridan
Department of Environmental Resource Management, Faculty of Agri-Food and Environment, University College Dublin, Belfield, Dublin 4, Ireland, Email:annette.anderson@ucd.ie

Keywords: biodiversity, Hymenoptera, field margins

Introduction Conserving field margins provides an opportunity to enhance biodiversity in agricultural landscapes. The parasitoid Hymenoptera represent one of the most diverse and biologically specialised of all insect groups and play an important role in insect pest control (LaSalle & Gauld, 1993). The diversity of parasitiods in any habitat is theoretically likely to reflect the diversity of host taxa. The aim of this study was to assess the effects of field margin manipulations on the diversity of parasitoids as a wider indication of effects on general arthropod diversity.

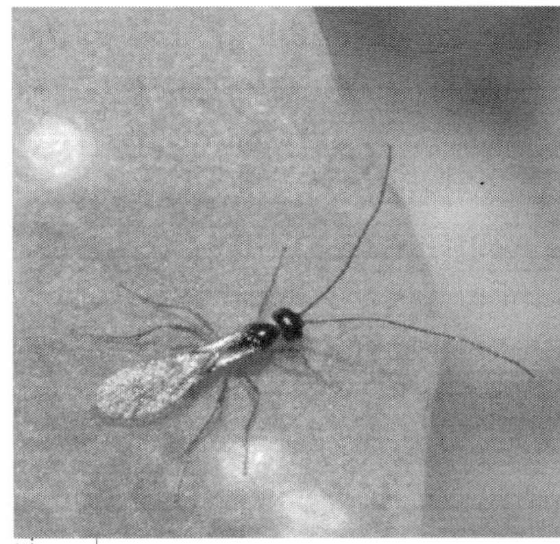

Materials and methods The experiment was arranged in 9 m x 90m paddock margin strips (blocks), within each of which were 3 m x 30m manipulated field margin treatments: i) Fenced Control - existing paddock sward, excluding grazing and external inputs; ii) Rotovated and Fenced - sprayed with a herbicide, rotovated, rolled and allowed to naturally revegetate; and iii) Reseeded and Fenced - rotovated and reseeded with a grass and wild flower seed mixture. A 30 m Control paddock margin adjacent to each block was grazed, fertilised and cut for silage in a similar manner to the rest of the paddock. Field vegetation arthropods were sampled using a Vortis suction sampler in June 2004. One composite sample was collected from each plot, comprising 10 areas individually sampled for 10 s duration. Parasitoid Hymenoptera were initially identified to family and subsequently to genus. As few significant differences were found between marginal manipulations, the most abundant parasitiod groups in the Control strips were compared directly to the mean of the marginal treatments using ANOVA.

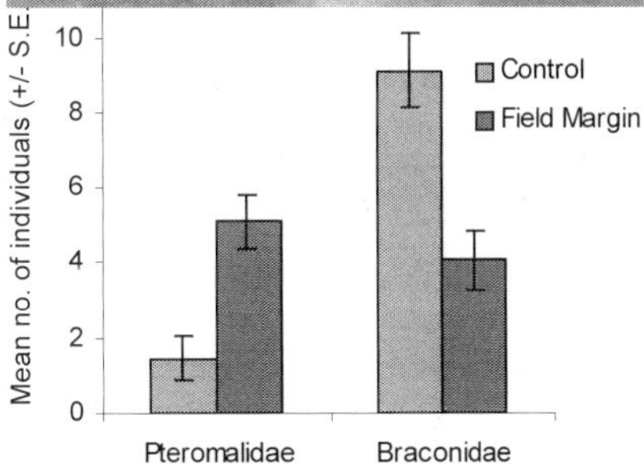

Figure 1 Effects of field margin treatments on the abundance of Pteromalidae and Braconidae

Results Significant treatment effects were found for the abundance (number of individuals) of both Pteromalidae ($F_{1,18}$=16.50, P<0.01) and Braconidae ($F_{1,18}$=8.73, P<0.01) (Figure 1). Treatment effects were also found for the number (diversity) of pteromalid ($F_{1,18}$=13.00, P<0.01) and braconid genera ($F_{1,18}$=6.75, P=0.03) (Figure 2).

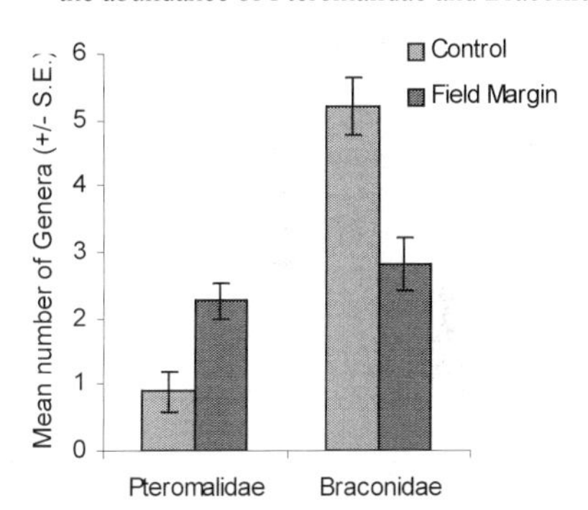

Figure 2 Effects of field margin management on the diversity of Pteromalid and Braconid genera

Conclusions Simply fencing the existing paddock margins resulted in higher diversity and abundance of Pteromalidae. However, braconid diversity and abundance were greater under conventional sward management. This may reflect differences in the host ranges of these two families - the braconid genera collected being mainly parasitoids of aphids or dung feeding Diptera, while the pteromalid genera were predominantly parasitoids of plant-mining Diptera.

References

LaSalle, J. & I.D.Gauld (1993). Hymenoptera: diversity, and their impact on the diversity of other organisms. In: J. LaSalle and I. D. Gauld (eds) Hymenotpera and Biodiversity. CAB International, London, pp 1-26.

Effect on sward botanical composition of mixed and sequential grazing by cattle and sheep of upland permanent pasture in the UK

J.E. Vale, M.D. Fraser and J.G. Evans
Institute of Grassland and Environmental Research, Bronydd Mawr, Trecastle, Brecon, Powys LD3 8RD, UK, Email: jim.vale@bbsrc.ac.uk

Keywords: sward composition, grazing, cattle, sheep, clover

Introduction Previous work has shown benefits of sequential grazing by cattle and sheep, with superior liveweight gains being recorded for lambs grazing swards previously grazed by cattle. A preliminary study of the effects of mixed sheep and cattle grazing also suggested that more extensively grazed swards offer scope for complementary grazing between sheep and cattle. The aim of this experiment was to directly compare these two approaches of integrating the grazing of cattle and sheep.

Material and methods The experiment was conducted on a ryegrass/white clover-dominated permanent pasture. Four treatments were compared: 1) sheep only grazing during the growing season from May to October (S/S); 2) cattle only grazing during May to July, followed by sheep only grazing from August to October (C/S); 3) cattle and sheep grazing during May to July, sheep only grazing for the rest of the growing season (C+S/S); and 4) cattle and sheep grazing for the whole grazing season from May to October (C+S/C+S). Each treatment was replicated three times. On cattle-grazed plots Charolais-cross steers were used throughout the season. On sheep-grazed plots, Beulah Speckle Face ewes and their Suffolk cross lambs were used from the start of the experiment until weaning at the end of July, and then from weaning until termination of grazing the plots were grazed by weaned lambs. Sward height was maintained at 6 cm using a 'put and take' stocking system. Other experimental details were as given in Fraser *et al.* (2005). To determine the effect of grazing treatment on sward composition quadrat cuts were taken at turn-out of livestock at weaning in late July. On each occasion, six 145 cm x 14 cm quadrats were cut to ground level within each experimental plot and a botanical separation carried out on a representative sub-sample of the material collected.

Results There were no statistically significant effects of grazing treatment on botanical composition at weaning in the first year of the experiment (2001). However, over the subsequent two years significant between-treatment differences in the proportions of key sward components likely to influence nutritive value of herbage, such as white clover, flower stem and dead material, were identified.

Table 1 Effect of grazing treatment on sward composition (proportion x 100)

		Treatment					
Year	Sward category	S/S	C/S	C+S/S	C+S	s.e.d.	Significance
2002	Grass lamina (live)	58	53	57	61	2.5	*
	Grass pseudostem (live)	8	9	8	9	1.5	ns
	Grass flower stem (live)	15	5	9	11	3.1	**
	Clover (live)	6	11	7	9	1.9	*
	Other dicots (live)	10	10	6	6	2.2	*
	Dead material	17	26	24	19	3.0	*
2003	Grass lamina (live)	52	50	54	54	2.2	ns
	Grass pseudostem (live)	4	3	4	2	1.0	ns
	Grass flower stem (live)	12	9	8	8	3.0	ns
	Clover (live)	4	12	9	15	2.0	***
	Other dicots (live)	6	10	5	9	2.6	ns
	Dead material	29	28	30	24	2.4	*

Conclusions The results demonstrate that, even under relatively controlled conditions, choice of grazing system can influence sward composition, which in turn are likely to be linked to effects on animal performance (Fraser *et al.*, 2005).

Acknowledgements This work was funded by Department of Environment, Food and Rural Affairs in the UK and Scottish Executive Environment and Rural Affairs Department.

References

Fraser, M.D., J.E Vale & G.E Evans (2005) Effect of mixed and sequential grazing by cattle and sheep of upland permanent pasture on liveweight gain. *This meeting.*

Effect of mixed and sequential grazing by cattle and sheep of upland permanent pasture on liveweight gain

M.D. Fraser, J.E. Vale and J.G. Evans
Institute of Grassland and Environmental Research, Bronydd Mawr, Trecastle, Brecon, Powys LD3 8RD, U.K, Email: mariecia.fraser@bbsrc.ac.uk

Keywords: liveweight gain, grazing, cattle, sheep

Introduction Previous work has shown benefits of sequential grazing by cattle and sheep, with superior liveweight gains being recorded for lambs grazing swards previously grazed by cattle. A preliminary study of the effects of mixed sheep and cattle grazing also suggested that more extensively grazed swards offer scope for complementary grazing between sheep and cattle. The aim of this experiment was to directly compare these two approaches of integrating the grazing of cattle and sheep.

Material and methods The experiment was conducted on a ryegrass/white clover-dominated permanent pasture. Four treatments were compared: 1) sheep only grazing during the growing season from May to October (S/S); 2) cattle only grazing during May to July, followed by sheep only grazing from August to October (C/S); 3) cattle and sheep grazing during May to July, sheep only grazing for the rest of the growing season (C+S/S); and 4) cattle and sheep grazing for the whole grazing season of May to October (C+S/C+S). Individual plot sizes were 2.0, 1.0 and 0.5 ha for treatments 3, 4, and 2 and 1 respectively, and each treatment was replicated three times. On cattle-grazed plots Charolais-cross steers with a mean turnout weight of approximately 350 kg were used throughout the season. On sheep-grazed plots, Beulah Speckle Face ewes and their Suffolk cross lambs were used as core group animals from the start of the experiment until weaning at the end of July, and then from weaning until termination of grazing the plots were grazed by weaned lambs. The number of "core" animals on individual plots of the four treatments were: 1) 8 ewes + 11 lambs until end of July; 8 lambs thereafter, 2) 4 steers until end of July; 16 lambs thereafter, 3) 4 steers and 16 ewes + 22 lambs until end of July; 32 lambs thereafter, and 4) 4 steers and 16 ewes + 22 lambs until end of July; 4 steers and 16 lambs thereafter. Sward heights were maintained at a target height of 6 cm using a "put and take" stocking system. Results are presented for years 2 and 3 of an experiment which began in 2001.

Results Liveweight gains for the different classes of livestock before and after weaning during each of the grazing seasons are summarised in Table 1. During the post-weaning phase in 2002, growth rates of lambs were higher on the C/S and C+S/C+S treatments than on the S/S treatment, while in 2003 the growth rates recorded for the lambs grazing the C+S/C+S plots post-weaning were higher than those for lambs on any of the other three treatments.

Table 1 Effect of sequential and mixed grazing of sheep and cattle on liveweight gain (g/d)

Year	Treatment	Pre-weaning			Post-weaning	
		Ewes	Lambs	Cattle	Lambs	Cattle
2002	S/S	37	240	-	40	-
	C/S	-	-	1166	67	-
	C+S/S	73	274	1362	57	-
	C+S/C+S	80	262	1107	79	662
	s.e.d.	16.4	10.6	93.5	11.5	
	Significance	*	*	*	*	
2003	S/S	86	238	-	95	-
	C/S	-	-	1407	97	-
	C+S/S	100	262	1338	97	-
	C+S/C+S	95	253	1295	125	1037
	s.e.d.	21.2	10.0	89.7	11.5	
	Significance	ns	ns	ns	*	

Conclusions The results demonstrate the potential production benefits of mixed grazing. Although systems based on sequential grazing were found to have the potential to improve performance, the treatment which gave the most consistent benefits in terms of post-weaning lamb growth rates was mixed grazing by cattle and sheep throughout the season.

Effects of breed and stage of growing season on the metabolic profile of sheep grazing moorland

V.J. Theobald, M.D. Fraser and J.M. Moorby
Institute of Grassland and Environmental Research, Plas Gogerddan, Aberystwyth, SY23 3EB, UK, Email: vince.theobald@bbsrc.ac.uk

Keywords: energy metabolism, metabolic profile, sheep, breed, heather

Introduction Previous studies have shown that the diet of sheep grazing heather moorland is affected by season (Grant *et al*., 1987) and the proportion of *Calluna vulgaris* cover (Osoro *et al*., 2000). In order to investigate the impact of these factors on the associated nutrient supply of animals grazing heathland, blood samples were taken to monitor the metabolic status of different breeds of sheep at different stages of the growing season when grazing sites with different proportions of heather cover.

Material and methods Experimental sites with low (0.08) and medium (0.36) proportions of heather (*Calluna vulgaris*) cover were grazed over two successive years by two breeds of sheep, Scottish Blackface (SB) and Welsh Mountain (WM). Each year blood samples were collected from six non-productive ewes of each breed with similar body condition scores during two experimental periods, at the middle (July) and end (October) of the growing season, and analysed to determine plasma concentrations of β-hydroxybutyrate (BHB) and glucose, and serum concentrations of non-esterified fatty acid (NEFA), total protein, albumin and urea.

Results The metabolic profiles of the two sheep breeds when grazing the different swards is summarised in Table 1. There were more significant effects of season on metabolite profiles in animals grazing the low heather content sward than on the medium heather sward, in particular for the non-glucose energy metabolites, which may be caused by a greater effect of season on grass nutritional composition than on heather composition. Plasma total protein concentrations were significantly higher in October than in July for sheep on both swards. Plasma urea concentrations were consistently higher in WM than SB sheep on both sward types, which could indicate differences in diet selection between the two breeds, leading to lower dietary protein intakes by SB sheep, higher utilisation of endogenous nitrogen sources by the WM, or a combination of the two.

Table 1 Effects of breed and stage of growing season on metabolic profiles of sheep (WM, Welsh Mountain and SB , Scottish Blackface) grazing heathland with low and medium heather cover. There were no significant season × breed interaction effects

Heather cover	Metabolite	July		October		s.e.d.	Significance	
		SB	WM	SB	WM		Season	Breed
Low	Glucose (mmol/l)	3.85	3.52	3.65	3.72	0.228	ns	ns
	BHB (mmol/l)	0.262	0.307	0.383	0.328	0.0379	*	ns
	NEFA (mmol/l)	648	861	552	520	114.2	*	ns
	Total protein (g/l)	72.4	70.2	78.3	77.8	1.91	***	ns
	Albumin (g/l)	29.9	30.0	31.4	29.4	1.39	ns	ns
	Urea (mmol/l)	3.87	4.58	4.53	6.92	0.674	**	**
Medium	Glucose (mmol/l)	4.13	4.13	3.52	3.57	0.288	**	ns
	BHB (mmol/l)	0.378	0.415	0.405	0.333	0.0399	ns	ns
	NEFA(mmol/l)	317	436	204	404	78.5	ns	**
	Total protein (g/l)	69.3	68.5	73.6	70.7	1.76	*	ns
	Albumin (g/l)	34.5	33.4	33.4	33.0	1.05	ns	ns
	Urea (mmol/l)	2.27	3.07	2.3	3.23	0.362	ns	**

Conclusions Differences in plasma energy and protein metabolites indicate that the diet consumed influenced nutrient absorption. These results, together with corresponding diet selection and intake data, will be used to quantify the consequences of different foraging strategies.

References

Grant, S. A., L. Torvell, H.K. Smith. D.E Suckling. & J. Hodgson (1987). Comparative studies of diet selection by sheep and cattle: blanket bog and heather moor. *Journal of Ecology*, 75, 947-960

Osoro, K., M. Oliván, R. Celaya. & A. Mertinez. (2000). The effect of Calluna vulgaris cover on the performance and intake of ewes grazing hill pastures in northern Spain. *Grass and Forage Science*, 55, 300-308

Characterising the fermentation capabilities of gut microbial populations from different breeds of cattle and sheep grazing heathland

D.R. Davies, M.D. Fraser, V.J. Theobald and E.L. Bakewell
Institute of Grassland and Environmental Research, Plas Gogerddan, Aberystwyth, SY23 3EB, UK Email: david.davies@bbsrc.ac.uk

Keywords: gas production, digestibility, rumen fermentation, *in vitro*, cattle, sheep

Introduction Previous studies have demonstrated differences in the diet composition of sheep and cattle when grazing heather moorland, and such differences may in turn lead to differences in rumen fermentation characteristics and associated adaptation to diet. To investigate this further an *in vitro* gas production experiment was conducted using inocula from different breeds of cattle and sheep grazing heathland.

Material and methods Representative sub-samples of *Calluna vulgaris* (heather)- and *Nardus stricta*-dominated semi-natural rough grazing (grass) were cut in early August, freeze-dried and ground and used as the substrate for the gas production technique. Two breeds of sheep (Welsh Mountain and Scottish Blackface) and two breeds of cattle (Welsh Black and Simmental/Belgian Blue cross (Continental)) had grazed the area for over 4 weeks from which the substrate samples were cut. Fresh faecal material was collected from six mature, non-productive females of each animal type. Individual faecal samples were prepared by diluting (1:1 w/v) in anaerobic medium and homogenising, with the resultant faecal liquor used as inoculum for the *in vitro* procedure. Gas pressure and volume was measured at time points over the initial 6-d period of growth of the faecally-derived microbial populations, and the data fitted to a model of gas production.

Results The fermentation data showed differences between the digestion of grass compared with heather, with the former having a significantly ($P<0.001$) shorter lag time and greater total gas pool size. In addition the grass substrate had a significantly greater ($P<0.001$) final dry matter (DM) loss than the heather. With respect to animal type, there were significant differences between animal species but few differences between animal breeds within a species (Table 1). For heather substrate, only DM loss was significantly ($P<0.001$) different between animal types, with greater digestion with the sheep inocula. In contrast, all fermentation parameters were significantly different between animal species for the grass substrate, with the initial and secondary rates and final DM losses significantly ($P<0.001$) higher for sheep inocula.

Table 1 Effect of substrate and inocula source on rate of gas production and dry matter loss

Substrate		Scottish Blackface	Welsh Mountain	Simmental/ Belgian Blue	Welsh Black	s.e.d.	Significance
Heather	P1	0.988	0.984	0.986	0.979	0.0040	ns
	P4	201	187	179	163	16.6	ns
	Lag (h)	3.93	3.96	4.37	4.05	0.303	ns
	DM loss (g/g)	0.403^{a}	0.416^{a}	0.303^{b}	0.269^{b}	0.0236	***
Grass	P1	0.982^{a}	0.975^{a}	0.996^{b}	0.994^{b}	0.0039	***
	P4	262^{a}	245^{a}	308^{b}	279^{ab}	17.6	*
	Lag (h)	1.95^{a}	2.57^{ab}	3.33^{b}	3.25^{ab}	0.364	**
	DM loss (g/g)	0.521^{a}	0.523^{a}	0.417^{b}	0.399^{b}	0.0246	***

where P1 = initial rate constant; P4 = Predicted asymptote

Conclusions The data confirm that heath grasses are more digestible than heather. The data also indicate that the sheep and cattle possessed gut microflora showing different activities, despite being exposed to the same forage for more than one month. The results from this study, together with corresponding diet selection and intake data, will be used to explore the consequences of different foraging strategies.

Acknowledgements This work was funded by the Department of the Environment, Food and Rural Affairsof England and Wales, the Countryside Council for Wales and English Nature.

The performance of cattle on lowland species-rich neutral grassland at three contrasting grazing pressures

B.A. Griffith and J.R.B. Tallowin
IGER North Wyke, Okehampton, Devon EX20 2SB UK, Email: jerry.tallowin@bbsrc.ac.uk

Keywords: species rich grassland, cattle, animal performance, botanical composition

Introduction Grazing is an essential management practice for maintaining the nature conservation value of lowland semi-natural neutral grassland to control succession and create different faunal habitats via structural heterogeneity within the pasture (Duffey *et al*., 1974). However, there is a paucity of information on what would constitute a sustainable grazing intensity that will deliver the wildlife objectives and what the consequences of this management would be on growth rate of livestock and overall pasture output. An experiment was designed to quantify the ecological and agronomic consequences of imposing different grazing intensities on species-rich neutral grassland. The results will provide sward-based criteria for the integration of such species-rich grassland into commercial livestock systems.

Methods A randomised block experiment with three continuous grazing treatments, based on maintaining sward surface heights of 6-8cm (Severe), 8-10cm (Moderate) or 10-12cm (Lenient), using continental-cross beef heifers was imposed each year from 2000-2004 on species-rich grassland at a site in Somerset, England. The grassland comprised a mixture of *Lolium perenne-Cynosurus cristatus* (MG6) and *Centaurea nigra – Cynosurus cristatus* (MG5) grassland (Rodwell, 1992). A fourth treatment grazed to a surface sward height of 6-8 cm and receiving 130 kg N/ha per year was imposed on agriculturally improved land at the same site. This acted as a control and provided a means of comparison with a conventionally-managed beef system. Treatment targets were maintained by weekly adjustment of animals using a 'put and take' grazing system. Three 'Core' animals remained on each of the plots at all times. The following measurements were made: daily live-weight gain of the core animals, cumulative grazing day totals, weekly surface sward heights, herbage quality (by pluck sampling), sward canopy structure and botanical composition in 10 1m-2 fixed quadrats per paddock.

Results The species-rich grassland supported similar individual growth rates in the cattle to the fertilized control pasture (Table 1). Lower growth rates occurred on the severe compared with the lenient treatment on the species-rich grassland. The severe treatment carried significantly more live weight both per day and over the grazing season than the lenient and moderate treatments. The live weight carried was similar between the moderate and lenient treatments. There was no difference in live weight produced between the three grazing severities on the species-rich grassland. The fertilized pasture produced on average 91% more live weight over the grazing season than the species rich grassland.

Table 3 Animal performance: values represent means of the four experimental years

	Severe	Moderate	Lenient	Improved	s.e.d. (df = 3)	significance
LWG (kg/day)	0.73	0.81	0.92	0.86	0.044	$P < 0.001$
LW carried per day (t/ha)	1.09	0.94	0.85	1.82	0.044	$P < 0.001$
LW carried per season (t/ha)	184.8	147	129.3	315.1	10.64	$P < 0.001$
Output (kg/ha per year)	350.6	319.8	305.5	620.2	47.76	$P < 0.001$

Although canopy structure was affected by grazing pressure on the species-rich grassland, there was no significant overall change in botanical diversity, which remained at 16-18 plant species m^2, across treatments between 2000 and 2004.

Conclusion Under moderate to lenient grazing pressure predictable agronomic output and high individual animal growth rates can be achieved on lowland species-rich neutral grassland whilst maintaining the biodiversity interests – at least in the short term.

References

Duffey, E., M.G. Morris, J. Sheail, L.K Ward, D.A Wells & T.C.E. Wells (1974) Grassland Ecology and Wildlife Management. Chapman & Hall, London.

Rodwell J.S. (1992) British Plant Communities. Volume 3. Grasslands and montane communities. Cambridge University Press, Cambridge.

Nutritional value of pasture forage for sheep in Krkonoše National Park

P. Homolka
Research Institute of Animal Production, 104 00 Prague–Uhříněves, Czech Republic,
Email: Homolka.Petr@vuzv.cz

Keywords: grazing, production of biomass, digestibility

Introduction This study describes the nutritional value of pastures in extreme mountain conditions in Krkonoše Mountains National Park. The Park performs important ecological and environmental functions. Extensive sheep grazing serves in the preservation of rare, protected and endangered species of plants (including endemics) and in the restoration of the biodiversity of meadows. The objective of this experiment was to estimate the production of plant biomass in these pastures and its digestibility by sheep.

Material and methods The experimental pasture is situated in the territory of the Krkonoše Mountains National Park (Czech Republic), at an average altitude of 1240 m (the locality of Zadní Rennerovky). The samples of biomass were taken from twelve fenced small areas (0.5 m x 0.5 m) situated in the pasture. The samples were taken twice, in the middle of the grazing period (June) and at the end of grazing (August) in the years 2001-2003. Twice during the grazing period (9 June – herbage sample A and 16 August – herbage sample B), two samples of herbage were collected in order to determine the digestibility of dry matter (DM), crude protein, ether extract, nitrogen-free extract and crude fibre in *in vivo* digestibility trials. Four mature Merino rams, weighing approximately 85 kg, were used.

Results The above-ground biomass was 5230 kg DM/ha in 2001, 5040 kg DM/ha in 2002 and 4380 kg DM/ha in 2003. The nutrient contents (g/kg DM) of herbages A and B are shown in Table 1. The digestibility coefficients of nutrients in herbages A and B are given in Table 2.

Table 1 Nutrient content of herbages A and B (g/kg DM, except for DM content, g/kg)

	Dry matter	Crude protein	Ether extract	Nitrogen-free extract	Crude fibre	Ash
Herbage A	180	217	28	439	244	72
Herbage B	425	132	23	435	346	64

Table 2 Digestibility coefficients of nutrients of herbages A and B

	Organic matter	Crude protein	Ether extract	Nitrogen-free extract	Crude fibre
Herbage A	0.627	0.680	0.495	0.638	0.566
Herbage B	0.526	0.554	0.479	0.532	0.502
Level of significance	*	**	NS	*	NS

*$P<0.05$; **$P<0.001$; NS: not significant

Conclusions The results of the experiment show a moderate nutritional value for the mountain grassland, which require to be grazed by sheep to preserve its biodiversity.

Acknowledgements This project was supported by the Ministry of Agriculture of the Czech Republic (project MZE 0002701403).

Vegetation dynamics of campos under grazing/fire regimes in southern Brazil

F.L.F. De Quadros, J.P.P. Trindade, D.G. Bandinelli and L. Pötter
Dep. Zootecnia, Universidade Federal de Santa Maria, Santa Maria–RS, Brazil, Email: fquadros@ccr.ufsm.br

Keywords: *Andropogon lateralis*, continuous quadrats, relief, resilience

Introduction Natural grassland vegetation in Southern Brazil, known as campos, has most likely evolved under a disturbance regime that included fire and grazing (Pillar *et al.*, 1997). Nowadays, the composition of the vegetation of campos is grazing- and fire-dependent (Boldrini *et al.*, 1997). Its importance can be evaluated by the fact that it represents 37 % of the state's area and provides 77 % of the slaughtered cattle at Rio Grande do Sul (Barcellos *et al.*, 2002). The objective of this experiment was to evaluate the vegetation dynamics of campos under grazing/fire regimes in order to explore the resilience of the vegetation under the regimes studied.

Materials and methods The effect of fire and grazing regimes on vegetation dynamics was evaluated during eight years on campos vegetation (Pillar *et al.* 1997). The experimental area is located in Santa Maria, in the southern-most Brazilian state. The experiment consisted of eight plots subjected to combinations of two grazing and fire regimes (presence/absence) on two relief positions (convex, concave slope). The annual evaluation started in August 1995 and finished in January 2002. Data on visual estimates of species above-ground biomass, using field procedures of BOTANAL (Tothill *et al.*, 1992), were subjected to ordination analysis.

Results Treatments ungrazed (Figure 1) showed a resilient behaviour tending to a dominance of *Andropogon lateralis*, except on the burning treatment in the concave position. After 8 years under grazing management, typical species were *A. lateralis*, *Desmodium incanum*, *Paspalum notatum* and *Eragrostis augustifoilius.* During this time interval, campos vegetation changed from a dominance of the first species to include the others cited above. This could be due to the plasticity observed in plants of *A. lateralis*, which changed from a tall tussock grass to a sward of decumbent plants.

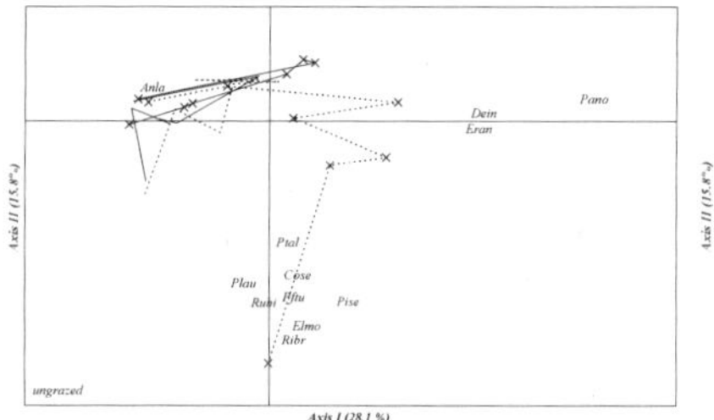

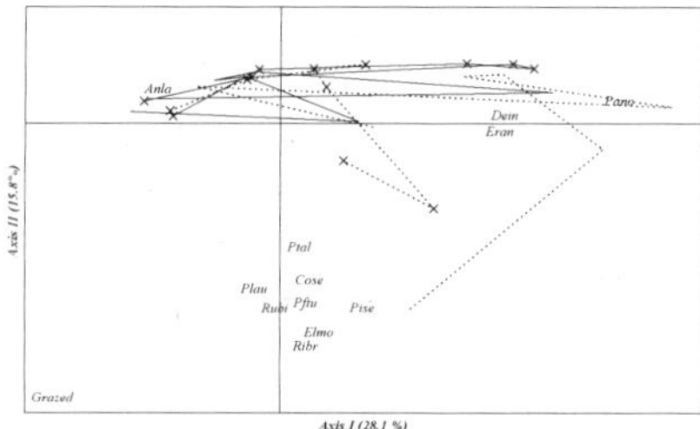

Figure 1 Principal Coordinate Analysis of 64 experimental plots under grazing/fire regimes in eight evaluation years. The lines indicate the campos dynamic and the grazing/fire regimes at two relief positions (lines with X for burned treatments and continuous line for convex slope) for grazed and ungrazed treatments. Characters indicate species most correlated with the axes (r>0.50): *Anla* (*Andropogon lateralis*), *Pano* (*Paspalum notatum*), *Dein* (*Desmodium incanum*), *Ptal* (*Pterocaulon alopecuroides*), *Cose* (*Coelorhachis selloana*), *Pftu* (*Pfaffia tuberosa*), *Eran* (*Eragrostis angustifolis*), *Plau* (*Plantago australis*), *Elmo* (*Elephanthopus mollis*), *Ribr* (*Richardia brasiliensis*), *Rubi* (Undetermined (Rubiacea)) and *Pise* (*Piriqueta selloi*).

Conclusion Disturbance factors like grazing and fire are important in maintaining and enhancing diversity on campos vegetation. Both maintain open spaces allowing the development of species dependent on light and soil resources.

References

Barcellos, J.O.J., E.R. Prates & M.D. Silva (2002). Sistemas pecuários no Sul do Brasil-"Zona Campos": Tecnologias e perspectivas. In: *Reunion del Grupo Técnico en Forrageras del Cono Sur. Zona Campos*, pp. 10-15. Mercedes.

Boldrini, I.I. & L. Eggers (1997). Directionality of succession after grazing exclusion in grassland in the south of Brazil. *Coenoses,* 12, 63-66.

Pillar, V.D. & F.L.F.De Quadros. (1997). Grassland-forest boundaries in southern Brazil. *Coenoses,* 12, 119-126.

Tothill, J.C., J.N.G. Hargreave & R.M. Jones (1992). BOTANAL - a comprehensive sampling and computing procedure for estimating pasture yield and composition. 1. Field sampling. *Tropical Agronomy Technical Memorandum,* 78, 1-24.

Modelling grazing animal distributional patterns using multi-criteria decision analysis techniques

M.R. George[1], N.R. Harris[2], N.K. McDougald[1], M. Louhaichi[3], M.D. Johnson[4], D.E. Johnson[3] and K.R. Smith[3]
[1]Department of Agronomy and Range Science, University of California, Davis CA, 95616, USA, Email: mrgeorge@ucdavis.edu, [2]Palmer Research Center, University of Alaska, Fairbanks, 533 E. Fireweed Ave., Palmer, AK 99645-662, USA, [3]Department of Rangeland Resources, Oregon State University, Corvallis, Oregon 97331-22180, USA and [4]Department of Mathematics, University of Southern California, Los Angeles, CA 90089-253, USA

Keywords: livestock distribution, global positioning, geographic information systems

Introduction Predicting livestock distribution is crucial to reducing livestock impacts on environmentally critical areas. Attempts to model livestock distribution on rangelands have met with varying levels of success. Most of these models described conditions at specific sites and did not work well when they were applied to other sites. In part, the weakness of these models arises from a lack of connection to the spatial arrangement of the study area and the pattern shown by animal distributions. To model the influence of the factors on livestock distribution we developed the Kinetic Resource and Environmental Spatial Systems (KRESS) Modeller. The KRESS Modeler is a multi-criteria decision analysis program that can use GIS layers to predict the suitability of positions in a pasture for animal use.

Methods The KRESS Modeller builds GIS layers from abiotic and biotic landscape data (e.g. slope, distance from water, solar radiation and forage quantity) that can influence cattle distribution. KRESS scales the landscape data so that it can be weighted proportionally, determines the spatial and temporal relationships of the landscape data, builds a spatial model describing landscape suitability for cattle grazing, evaluates the model using GPS-generated cattle positions and applies the model to new landscapes. Spatial data that described abiotic and biotic characteristics of California foothill pastures were entered into the KRESS Modeler. Ground resolution of the information was 10 m by 10 m. Data included digital elevation models, forage standing crop, and distance from water. A series of decision rules defined suitability of positions in the pasture for grazing and resting activities for beef cows based upon our knowledge of animal food requirements, water requirements, ability to travel, and thermal requirements. Factors were weighted and applied using the weighted sum multi-criteria algorithm in the KRESS Modeler. The models were saved so that they could be applied to other pastures in this vegetation type. Models were tested by using data collected by either 24-hour visual observations or with data from cows fitted with GPS collars. The frequency of occurrence of cattle in each cell on the landscape was calculated and the distribution of animals was compared to the patterns predicted by the multi-criteria model using ROC analysis.

Results and Discussion The KRESS Modeller integrated abiotic and biotic landscape characteristics into a cattle suitability map for California foothill pastures containing steep and gentle terrain (Figure 1). The validity of the landscape suitability maps was tested by comparing with beef cow positions from visual mapping and GPS collars using ROC analysis. The results of this study show that the frequency of cow positions located in landscapes mapped as highly suitable for cattle exceeded that expected if the cows were distributed randomly. These results indicate that the KRESS Modeler is a useful tool for predicting and understanding livestock distribution on rangeland.

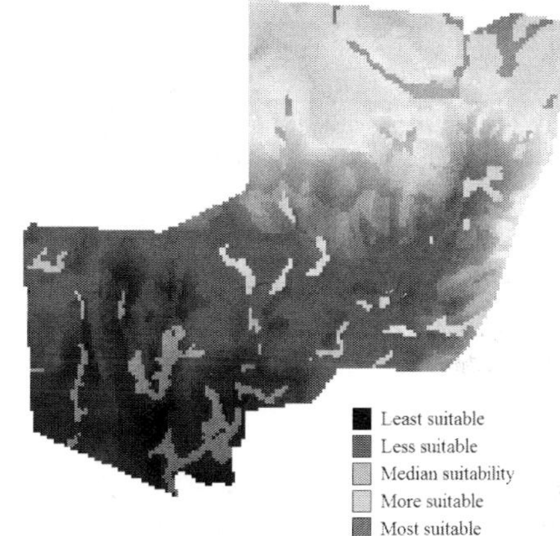

Figure 1 Cattle suitability map

Species richness, species identity and ecosystem function in managed temperate grasslands

S.C. Goslee, M.A. Sanderson and K. Soder
USDA-ARS, Pasture Systems and Watershed Management Research Unit, Building 3702, Curtin Road, University Park, PA, 16802-3702, USA, Email: Sarah.Goslee@ars.usda.gov

Keywords: community dynamics, diversity, invasibility, pasture

Introduction Manipulation of plant species diversity may provide a way to improve the ecosystem functioning of managed systems by increasing productivity and suppressing weedy species. As yet, the functional role of species richness is not well-enough understood to enable practical application. We investigated the effects of differing species richness on community stability and invasion resistance in a grazed temperate grassland.

Materials and Methods Nine plant species from three different functional groups (grass, legume, forb) were planted in the autumn of 2001 at University Park, PA, USA, in mixtures of 2, 3, 6 and 9 species, as follows:
2 species: *Dactylis glomerata, Trifolium repens:*
3 species: *D. glomerata, T. repens, Cichorium intybus:*
6 species: *D. glomerata, Festuca arundinacea, Lolium perenne, Trifolium pratense, Lotus corniculatus, C. intybus:*
9 species: six species mix plus *T. repens, Medicago sativa, Poa pratensis.*
Two replicates of each treatment were planted in 1 ha plots, and rotationally grazed by dairy cattle during 2002 and 2003. Species composition was assessed during April and October of each year using the multiscale modified Whittaker plot. Percentage canopy cover was visually estimated in 10 quadrats of 1 m^2 for each plot, and species lists were compiled in areas of 10, 100 and 1000 m^2. Canopy cover can sum to more than 100%.

Results Higher richness conferred an establishment advantage; planted species richness was positively correlated with total cover in autumn 2001 and spring 2002 (Fig. 1a). Although cover of unplanted (weedy) species declined with increasing richness on these dates, there was no difference in the number of unplanted species. After the first year there were no significant trends in cover or richness between the treatments (Figure 1). Cover in the 2-species treatment was most affected by drought in the summer of 2002, but all treatments had recovered by autumn 2003, and there was no significant difference in total plant cover thereafter. The higher-species treatments lost planted species over time, even when those species persisted in the lower-species mixtures.

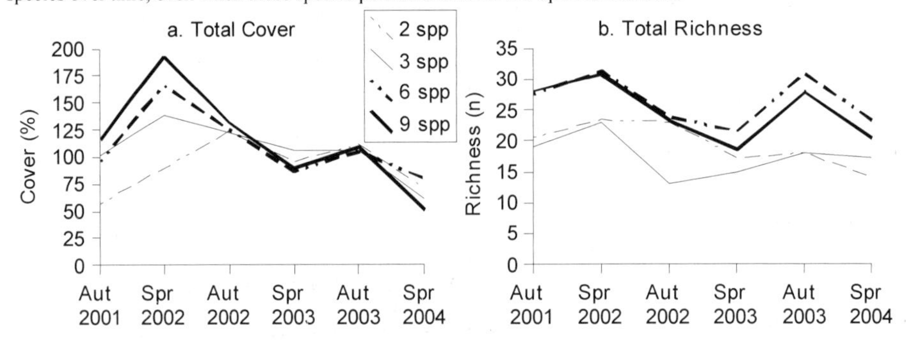

Figure 1 Total plant cover and species richness in grazed pastures of four different initial richness treatments

The forb *C. intybus*, part of the 3, 6 and 9 species mixtures, established rapidly and contributed greatly to the early success of these treatments. Several other grasses and legumes increased in abundance over the first year, then declined. By the end of the study, the same species (*D. glomerata* in all treatments and *T. repens* in 2, 3 and 9-species mixes) were dominant regardless of initial seeding composition. Cover and richness within the 1 m^2 quadrats were only related in the 6- and 9-species mixtures. Richness was significantly correlated between scales for all species treatments.

Conclusions Species identity was more important to cover and invasion resistance than was species richness. Management recommendations based on particular combinations of species may be more effective at improving ecosystem function than those based solely on species richness. If high plant diversity is desired, the management regime must support the maintenance of that diversity or, as in this study, species will tend to disappear from the higher-richness treatments.

Preference of goats for cool-season annual clovers in the southern United States

T.H. Terrill[1], W.F. Whitehead[1], G. Durham[2], C.S. Hoveland[2], B.P. Singh[1] and S. Gelaye[1]
[1]*Fort Valley State University, Fort Valley, GA 31030, USA, Email: terrillt@fvsu.edu, and* [2]*The University of Georgia, Athens, GA 30602, USA*

Keywords: goats, clover, preference

Introduction In the southern U.S.A., annual clovers provide high-quality winter and spring grazing for beef cattle and sheep. New Zealand data on white clover (*Trifolium repens* L.) suggests that goats do not prefer this plant as much as sheep (Clark *et al.*, 1982) but little data are available on willingness of goats to consume different clover types in the USA.

Materials and methods Two cafeteria-style grazing experiments were completed in Fort Valley, GA., USA. For both experiments, 10 replicates of 6 clover cultivars (Experiment 1: Dixie, AU Robin and AU Sunrise crimson clover, Yuchi arrowleaf, Segrest ball and R18 rose clover; Experiment 2: 'BYMV' and Yuchi arrowleaf, Yuchi arrowleaf coated with Apron fungicide, Dixie and AU Sunrise crimson clovers, plus Americus hairy vetch) were planted into 3.05 m x 3.05 m plots on 10 November, 1999 and 11 November, 2000, respectively. After establishment, each block of 6 plots was individually fenced and grazed for two 48-h periods in spring of 2000 and 2001 using 32 Spanish does (3-4 yr old, 4/block) and 40 yearling Spanish-Boer male kids (5/block), respectively. Before each period, does or kids were allowed to graze 2 blocks of plots as a single group to familiarize them with the forages. After the initial grazing, plots were mowed and then allowed to regrow. In Experiment 1, a 0.76 m x 3.05 m strip of forage was cut out of each plot pre- and post-grazing, weighed fresh and subsampled for determination of total forage DM consumed from each plot. For Experiment 2, all plots were visually evaluated to determine extent of pasture use after 4 h, 24 h, and 48 h of grazing. Two observers assigned each plot an ocular preference score (Shewmaker *et al.*, 1997) from 1 to 10 (1=no grazing; 2=<2%; 3=2-5%; 4=5-10%, 5=10-25%; 6=25-40%; 7=40-60%; 8=60-75%; 9=75-90%; 10=completely grazed). Preference data for Experiment 1 (DM consumed) were analyzed as a randomized block design, and for Experiment 2 (ocular preference score after 4, 24, and 48 h of grazing) as a randomized block design with repeated measures analysis.

Results There was no effect of cutting date on DM consumed for the different clover cultivars, so data from the two grazing periods in Experiment 1 were pooled. Total DM consumed by the goats averaged 372, 368, 322, 218, 145, and 100 g for Au Robin crimson, Dixie crimson, AU Sunrise crimson, arrowleaf, rose, and ball clovers, respectively. Consumption of Dixie and AU Robin crimson clover was significantly higher than for ball ($P < 0.05$) or rose ($P < 0.07$) clover. In the initial grazing period in Experiment 2, yearling kids also preferred ($P < 0.05$) the crimson clover cultivars over arrowleaf clovers and hairy vetch, with rose clover least preferred. Differences were most pronounced after 24 h of grazing. With no crimson clover available during the second grazing period, the kids preferred Yuchi arrowleaf clover without fungicide coating significantly ($P < 0.05$) more than the other forages, with rose clover preferred significantly least ($P < 0.05$). The coefficient of variation (CV) for the ocular preference scoring technique used in Experiment 2 averaged 34%, while the CV for cutting before and after grazing to establish preference (Experiment 1) was 49%.

Conclusions Crimson clover was preferred by goats over other clover types and may be suitable as high-quality winter-spring pasture for goat production in the south-eastern USA. Although less preferred than crimson clover, arrowleaf clover has a longer grazing season in this region and may also have potential for goat grazing. Further research is needed with these species to determine performance of goats grazing annual clover as a component of the diet. There was higher repeatability and time savings for the ocular scoring technique compared with cutting before and after grazing and the ocular preference scoring technique appears to be an effective means of establishing grazing preference of goats.

References

Clark, D.A., M.G. Lambert, M.P. Rolston & N. Dymock (1982). Diet selection by goats and sheep on hill country. *Proceedings of New Zealand Society of Animal Production*, 42, 155-157.

Shewmaker, G.E., H.F. Mayland & S.B. Hansen. (1997). Cattle grazing preference among eight endophyte-free tall fescue cultivars. *Agronomy Journal*, 89, 695-701.

Stevens, D.R., M.J. Casey, G.S. Baxter & K.B. Miller (1993). A response of angora-type goats to increases of legume and chicory content in mixed pastures. *Proceedings of XVII International Grassland Congress*, pp. 1300-1301.

Production per animal and use of intake estimatives to predicted animal productivity in *Pennisetum purpureum* cv. Mott and *Cynodon* spp cv. Tifton 85 pastures

F.L.F. de Quadros, A.R. Maixner, G.V. Kozloski, D.P. Montardo, A. Noronha, D.G. Bandinelli, M. da S. Brum and N.D. Aurélio
Departamento de Zootecnia, Universidade Federal de Santa Maria, Santa Maria – RS. CEP: 97105-900. Brazil, E-mail: fquadros@ccr.ufsm.br

Keywords: production per cow, chromium oxide, leaf lamina mass

Introduction Dairy production is a very important activity in southern Brazil, being an essential source of income to small household farms. Milk production from pastures is an alternative to reduce costs in dairy systems. Some C4 grasses, such as dwarf elephant grass (DEG) and Tifton 85, have presented high animal production per animal and per area. Although studies evaluating milk production from these pastures are rare in south Brazil, i*n vitro* studies have demonstrated that the nutritional value of these forages is higher than production registered in grazing. So, it is possible that, in spite of a high intrinsic nutritional value, limitation on cows' productivity is linked to the food's capacity of conversion to milk and/or management conditions that limit forage intake. Leaf mass in pastures is a factor that determinates forage intake, as cows prefer leaf to other parts of plants. In this context, adequate animal performance may be possible if offered enough leaf biomass at pasture. The aim of this experiment was to evaluate the potential of milk production with these two forage species.

Materials and methods The experiment was conducted at Palmeira das Missões/RS, from 11 July to 12 December 2003. Experimental animals were Holstein dairy cows (multiparous) weighing approximately 600 kg and 120 days into lactation. Continuous grazing was used with variable stocking rates to maintain a constant herbage mass on offer. Available leaf mass (kg DM/ha) was evaluated by visual assesment with double sampling every 14 days, herbage being split into leaf lamina and steam+sheath. In the last 12 days of the experimental period, each cow received a chromium oxide capsule for estimating intake by faecal indicators. Values of total digestible nutrients (TDN) was estimated according to NRC (2001). For statistical analysis of treatment effects in the evaluated variables MULTIV software (Pillar, 1997) was used.

Results Individual production of dairy cows was similar between the species evaluated (17.7 kg of milk/cow per day for DEG and 21 kg of milk/cow perday for Tifton 85). Forage species did not affect dry matter intake with values of 2.77 and 3.18 % of live weight for DEG and Tifton 85, respectively. The contents of crude protein (150-160 g/kg DM) were similar and an average TDN content of 0.61 was estimated.

Conclusions The results demonstrated that adequate levels of leaf lamina on offer allowed around 19 kg milk/cow per day in perennial tropical grasses. Observed energy and protein offered by Tifton 85 was similar to predicted by NRC (2001), but for DEG these values were underestimated by approximately 18 %.

References

Pillar, V. de P. (1997). Multivariate exploratory analysis and randomization testing with MULTIV. *Coenoses*, 12, 145-148.

NRC (2001). Nutrient requirements of dairy cattle. 7th edition. National Research Councill, Washington, DC, National Academy Press, 381 pp.

Ingestive behaviour of steers in native pastures in southern Brazil

C.E. Pinto[1], P.C.F. Carvalho[2], A. Frizzo[2], J.A.S.F. Júnior[2], T.M.S. Freitas[2] and C. Nabinger[2]
[1]*Empresa de Pesquisa Agropecuária e Extensão Rural de Santa Catarina, Lages SC, PO BOX 181 Brazil, Email: cassiano@epagri.rct-sc.br, and* [2]*Universidade Federal do Rio Grande do Sul, Porto Alegre, RS, Brazil*

Keywords: native pasture, steers, ingestive behaviour

Introduction The Campos biome, particularly its native pastures, is the main resource for livestock production in southern Brazil (Boldrini, 1997). It has a huge floristic diversity in which more than 400 grass and 150 legume species are found. These pastures are very heterogeneous with horizontal as well as vertical structure. Hodgson (1985) stressed the importance of sward structure upon diet selection, emphasizing the difficulties animals could have to access all pasture layers in temperate pastures. This experiment investigates how variation in herbage allowance along the growing season influences the ingestive behaviour of steers.

Material and methods The treatments consisted of different herbage allowances (kg DM/100 kg liveweight per day, expressed as %) and different combinations of herbage allowance in sequence: 4%, 8%, 12% and 16% allowance during the entire year; 8% forage allowance in spring and 12% in summer/autumn; 12% forage allowance in spring and 8% in summer/autumn; 16% forage allowance in spring and 12% in summer/autumn. The grazing method was continuous with variable stocking, and the experimental animals were crossbred steers. Treatments were in 14 paddocks (64 ha) in a completely randomized block design, with two replicates. Pasture accumulation was measured by four exclosure cages per paddock, herbage mass by the comparative yield method and grazing behaviour variables obtained by visual assessment. Data from summer/autumn 2002 are reported here.

Results The forage allowances observed were different from those imposed (Table 1) although the objective of creating different grazing allowances was attained. There was no significant difference between the herbage allowances in grazing time (P>0.05). In such complex vegetation, classical relationships established with cultivated pastures were not observed. However, when a variable which takes into account and controls for pasture heterogeneity was used, then pasture abundance was associated with animal performance and behaviour (Figure 1). As allowance increased, the frequency of tussocks increased, pastures becoming more heterogeneous. Nevertheless, between-tussock vegetation was less variable, and models fit better when between-tussock sward height was used as the independent variable, and using only that data associated with fixed forage allowances. Treatments with low forage allowance restricted animal intake, and they increased grazing time as a strategy to compensate. Each 1.0 cm increase in between-tussock sward height decreased grazing time by 66 minutes. Animal performance decreased with decreasing between-tussock sward height, indicating animals cannot compensate for the reduction in intake at low forage allowances.

Table 1 Grazing time of steers at different levels of forage allowance in a native pasture in southern Brazil

Forage allowance (kg DM/ 100 kg LW per day)						
4.0	8.0	12.0	16.0	8→ 12	12→ 8	16→12
Observed forage allowance (kg DM/ 100 kg LW per day)						
4.7	12.4	17.7	23.5	13.7	15.4	17.8
Grazing time (minutes/day)						
564.4	498.3	571.7	453.3	512.8	559.2	542.2

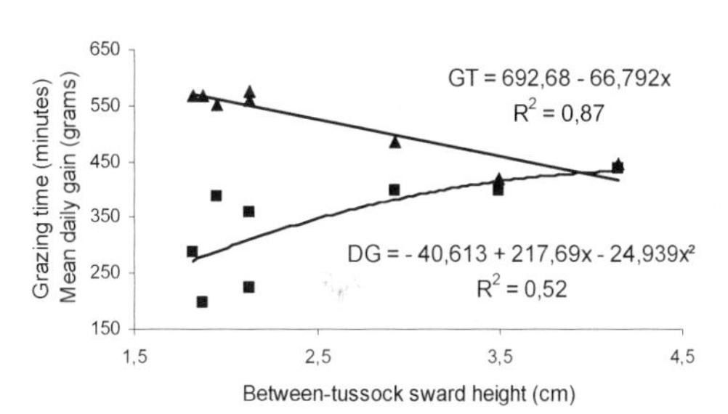

Figure 1 Grazing time (GT, minutes) and mean daily livewight gain (DG, g) of steers in relation to between-tussock sward height (cm.)

Conclusions Contrary to classical relationships based on homogenous cultivated pastures, it seems variables such as forage allowance or herbage mass do not explain sufficiently animal performance and behaviour in more complex vegetation. The characterization of the between-tussock vegetation is essential to construct response functions to obtain advances in knowledge in order to manage effectively native pastures in southern Brazil.

References

Pinto, C. E. (2003). Effect of herbage allowance upon animal production and behaviour in a native pasture in Deprassao Central of Rio Grande do sol. Master of Science dissertation in Forage Science, Faculdade de Agronomia, Universidade Federal do Rio Grande do Sul, Porto Alegre, RS, Brazil (73p.).

Hodgson, J. (1985). The control of herbage intake in the grazing ruminant. *Proceedings of the Nutrition Society*, 44, 339-348.

Challenges in modelling live-weight change in grazed pastures in the Australian sub-tropics

C.K. McDonald and A.J. Ash
CSIRO Sustainable Ecosystems, 306 Carmody Road, St Lucia, Queensland, 4067, Australia, Email: cam.mcdonald@csiro.au

Keywords: liveweight change, pasture, diet, modelling

Introduction In sub-tropical regions there is enormous seasonal, annual and spatial variation in pasture quality and considerable variation in quality between pasture species. The heterogeneous structure of sub-tropical pasture swards means that process based modelling of liveweight change (LWC) is particularly difficult. In response to this complexity LWC has been expressed as a function of the length of the growing season and/or pasture utilization (McKeon *et al.* 2000), green leaf availability, or pasture availability and climate (Hirata *et al.* 1993). However, these relationships vary from year to year, often fail when species composition changes, and generally explain <70% of the variation in LWC. This study used a large set of pasture and animal production data from a native pasture in south-east Queensland to investigate these relationships.

Materials and methods The data consisted of daily temperature, evaporation and rainfall, 6-weekly pasture yield, composition and proportion of green leaf, and monthly animal LWC at 3 stocking rates over 5 years. The LWC was related to a range of pasture and climate variables by means of linear and curvilinear relationships, multiple regression and optimisation routines. Finally, LWC and pasture variables were analysed using cluster analysis and ordination.

Results No pasture or climatic variables could be found that related well to LWC and discriminated between stocking rates and years. Pasture quality, as measured by green leaf percentage or the amount of green leaf, explained up to 70% of the variation in LWC in any one year, but <30% over all years. This suggests that climate-soil-pasture interactions are affecting the quality of green leaf from year to year, or the contribution of forbs to the animal's diet may be confounding these relationships. The progressive number of days from the start of the growing season explained 70-90% of the variation in LWC in the designated period, and was robust across years. However, this relationship was independent of stocking rate, despite the fact that there were observed LWC differences between stocking rates. Cluster analysis of LWC data gave 4 reasonably distinct groups consisting of high and low live weight (LW) gain and loss. Ordination of these groups against pasture vectors (Fig. 1) shows high LW gain (>0.5 kg/hd per day) and loss (>0.25 kg/hd per day) is associated with moderate to large quantities of green and dead material, respectively. Low LW gain (0-0.5 kg/hd per day) is intermediate between these. However, low LW loss (0-0.25 kg/hd per day) can be associated with the whole range of pasture conditions. Hence, given any particular pasture status, it is almost impossible to reliably and accurately predict LWC.

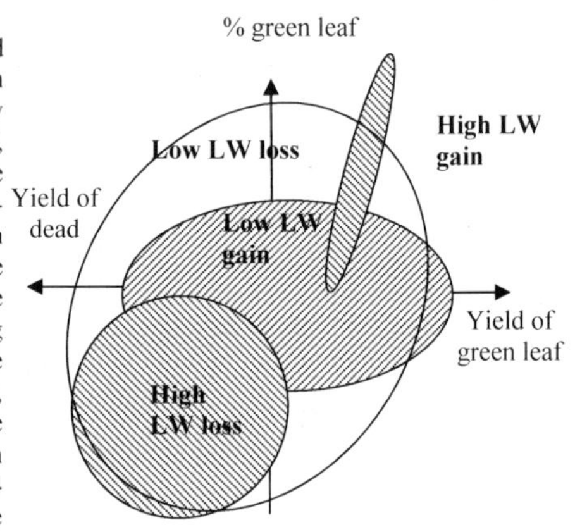

Figure 1 Schematic representation of relationship between LWC and pasture attributes

Conclusion The study has demonstrated the difficulties in using simple pasture or climate-based empirical relationships to predict LWC in the sub-tropics. Future efforts at improving our predictive capability of LWC in tropical and sub-tropical pasture systems should be based on more mechanistic approaches that take better account of year-to-year variation in pasture quality and seasonal animal diet selection. This will require measurement of field attributes more appropriate to modelling animal production.

References

Hirata, M., C.K. McDonald & R.M. Jones (1993) Prediction of liveweight gain of cattle grazing pastures on brigalow soil. *Proc. XVIIth International Grassland Congress, Palmerston North, New Zealand,* 1902-1903.

McKeon, G.M., A.J. Ash, W. Hall & D.M. Stafford-Smith (2000) Simulation of grazing strategies for beef production in north-east Queensland. In: G.L.Hammer *et al.* (eds.) The Australian Experience. Kluwer Academic Publishers, Netherlands, 227-252.

Effect of urea-treated *Pennisetum pedicellatum* and supplementation of concentrates with urea on milk production of "Mossi" ewes

V.M.C Bougouma-Yameogo and A.J. Nianogo
Institut of Rural Development (IDR) Polytechnic University of Bobo-Dioulasso, 01 BP 1091 Bobo-Dioulasso Burkina Faso, Email:Bouval2000@yahoo.fr

Keywords: *Pennisetum pedicellatum*, urea treatment, urea supplementation, milk production, milk composition, "Mossi" ewes

Introduction The "Mossi" sheep is a near parent of "Djallonke" sheep that live in sudano-sahelian area of Burkina Faso. However, there are few available results on dairy production from this breed. The treatment of straw with urea is a technique used in several developing countries to improve the nutritional value of gramineous forages (Sourabié *et al.*, 1995). The aim of this study was to test the influence on the performance of "Mossi" ewes and on milk composition of treatment of *Pennisetum pedicellatum* (Pp) with urea in comparison with addition of urea to the concentrate feed.

Materials and methods Twenty-three "Mossi" ewes in early lactation were used. Three dietary rations were tested: (1) untreated Pp + 22% concentrate (treatment NoU); (2) untreated Pp + 2.8% urea + 20% concentrate (UCo); (3) Pp treated with 6% urea + 22% concentrate (UPp) (Table 1). The Pp was harvested at the straw stage at the beginning of the dry season. The composition of the concentrate was: 25% whole cottonseed, 25% cottonseed cake and 50% of ground corn grain. The diet dry matter was offered at 4.4-6 % of bodyweight (BW). Measurements were made of milk yield, by the oxytocin method, milk composition (AOAC, 1984), fat (Babcock) and body condition (Russel *et al*.,. 1969).

Results The effects on milk yield, body weight changes, feed intake and milk composition are given in Table 2. The ADY was significantly higher for UPp than for UCo, but not higher than for the NoU treatment, which had no added urea. The differences in average daily milk yield were reflected in differences between treatments in the yields of milk solids, milk protein and milk fat. There were no significant differences in milk protein %. Body weight (BW), however, showed a clear advantage for UPp over the other two treatments.

Table 1 Chemical composition of straw and experimental diets

	Pp-U	Pp+U	UPp	NoU	UCo
CP	3.7	13.9	8.0	15.9	15.9
EFUL	0.58	0.66	0.69	0.68	0.75

EFUL=French Energy Feed Unit for lactation/DM; Pp-U, *P. pedicellatum* without urea; Pp+U, *P. pedicellatum* treated with urea

Table 2 Effect of urea treatment on milk yield, body weight and condition changes, and on feed intake and on milk composition

	NoU	UCo	UPp	Prob
ADY (g)	257^{ab}	197^{b}	316^{a}	0.0001
BW	-3.3	-1.3	+0.2	-
TFI/MW	93	88	96	NS
Milk composition				
ES(g/d)	44^{b}	35^{b}	56.7^{a}	0.0001
Ash(g/d)	2.8^{ab}	2.3^{b}	3.1^{a}	0.01
CP(g/d)	9.5^{b}	7.5^{ab}	12.9^{a}	0.0001
Fat(g/d)	17.2	13.8	21.5	NS
CP(%)	3.8^{b}	4.0^{a}	4.1^{a}	0.05

a,b: means in a row with the same superscript are not significantly different ($p>0.05$). ADY, average daily milk yield; BW, bodywieight; TFI/MW, total feed intake g/kgBW$^{0.75}$; ES, milk solids

Conclusion Whilst milk production in local ewes cannot be sustained by poor quality straws alone, natural grasses such as Pp may be significantly improved by treatment with urea. This treatment gave better results than addition of urea at feeding.

References

AOAC (1993). Official Methods of Analysis, Association of Official Analytical Chemists. Washington DC, 114 pp.

Russel, A.J.F., J.M. Doney & R.G. Gunn (1969). Subjective assessment of body fat in live sheep. *Journal of Agricultural Science, Cambridge*, 72, 451-454.

Sourabié, K.M., C. Kayouli & C. Dalibard (1995). Le traitement des fourrages grossiers à l'urée : une technique très promoteuse au Niger. *World. Animal Review*, 82, 3-13.

Accumulation of polyphenols and major bioactive compounds in *Plantago lanceolata* L. as a medicinal plant for animal health and production

Y. Tamura[1] and K. Yamaki[2]
[1]*National Agricultural Research Centre for Tohoku Region, Shimokuriyagawa Morioka, Iwate 020-0198, Japan, Email: fumi@affrc.go.jp, and* [2]*Japan International Research Centre for Agriculture Sciences, 1-2, Ohwashi, Tsukuba, Ibaraki, 305-8686, Japan*

Keywords: animal health, bioactive compounds, medicinal plants, *Plantago lanceolata* L., polyphenols

Introduction Producing animals without the use of feed-grade antibiotic growth promoters and chemical medicines is sought. Scientific studies with this aim have focused on medicinal plants to identify and quantify any beneficial effects that they might have on animal production. *Plantago lanceolata* L. has been used in herbal medicines and is currently being evaluated as a potential pasture species because of its beneficial effects on animal health. In the present study, the accumulation of polyphenols in *P. Lanceolata* is compared to that in principal pasture species, and genetic variation and environmental changes in the major bioactive compounds in *P. Lanceplata* are investigated.

Materials and methods The total polyphenol content of plantain leaves was determined using the Folin-Denis method, and was compared to the values for the following principal pasture species: *Dactylis glomerata* L., *Phleum pratense* L., *Lolium perenne* L., *Phalaris arundinacea* L. and *Trifolium repens* L.. The catalpol, aucubin and acteoside contents of two varieties and one ecotype of *P. lanceolata* were quantitatively determined using high-performance liquid chromatography (HPLC) of plant samples obtained throughout the growing season (Tamura & Nishibe, 2002), and under different environmental and cultivation conditions.

Results The highest total polyphenol content was observed in *P. lanceolata*, and was up to twice that of the average value for the pasture species tested (Figure 1). The aucubin and acteoside contents were relatively high, and significant genetic variation was detected within the cultivars and in comparison with the ecotype (Figure 2). By contrast, the catalpol content was low and this compound was absent from one variety. Plants grown under a high light intensity with a low nitrogen fertiliser application accumulated greater amounts of these compounds.

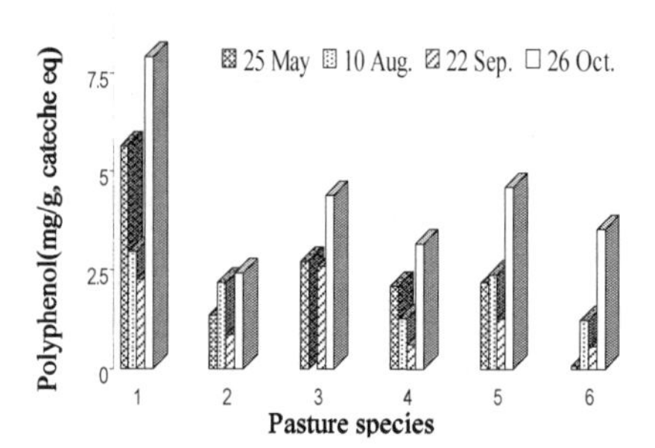

Figure 1 Polyphenol content of *P. lanceolata* and Pastures
Note: 1; *Plantago lanceolata* L., 2; Dactylus glomerata L., 3; Phleum pratense L., 4; Lolium perenne L., 5; Phalaris arundinacia L., 6; Trifolium repens L.

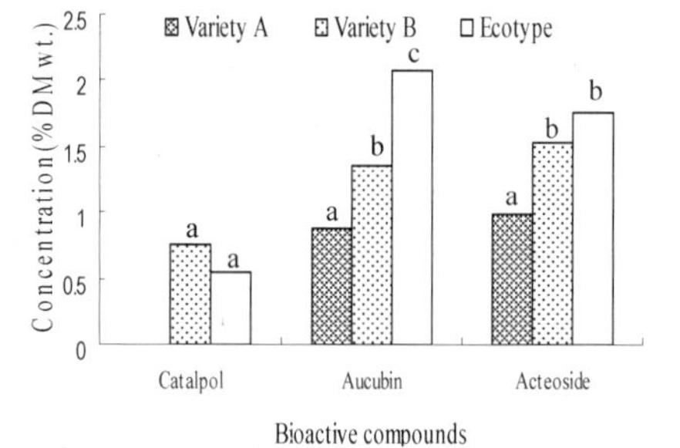

Figure 2 Bioactive compounds in leaves of varieties and ecotype of *Plantago lanceolata* L.
Note: Different letters indicate statistically significant differences between groups at the 5% level

Conclusions *P. lanceolata* accumulated greater amounts of polyphenols than the principal pasture species, and significant genetic variation was observed in the levels of bioactive compounds. Several previous studies have examined the effect of *P. lanceolata* on animal health and meat quality, and have reported positive effects, such as a decreased n-6/n-3 fatty-acid ratio in chickens, and decreased blood glucose levels and a higher meat grade in pigs. Further studies will be necessary to clarify the precise effects of bioactive compounds on animals.

References

Tamura, Y. & N. Sansei (2002). Changes in the contents of bioactive compounds in plantain leaves. *Journal of Agricultural and Food Chemistry,* 50, 2514-2518.

Effect of stocking rate on a *Stipa breviflora* Desert Steppe community of Inner Mongolia

G. Han[1], W.D. Willms[2], M. Zhao[1], A. Gao[1], S. Jiao[1] and D. Kemp[3]
[1]*College of Ecol. and Env. Sci., Inner Mongolia Agric. Univ., Hohhot, Inner Mongolia 010018 P.R. of China, Email: nmghand@hotmail.com,* [2]*Agric. and AgriFood Canada, PO Box 3000, Lethbridge, Alberta, Canada T1J 4B1,* [3]*University of Sydney, PO Box 883, ORANGE, New South Wales 2800, Australia*

Keywords: plant cover, above ground annual production

Introduction Stocking rate is an important factor in grazing management. The stocking rate defines utilization and ultimately grazing pressure, which in turn affects grassland sustainability. Grassland sustainability is partly defined by its species composition and ultimately by its productivity. These attributes are unique for specific plant communities and the effect of stocking rate must be established for each in order to understand the community response to grazing and to determine its carrying capacity. While some information exists on the effects of stocking rate on livestock production in the *Stipa breviflora* Griseb. Desert Steppe (Wei *et al.*, 2000), the effects on the plant community are not understood well. This study aimed to determine the effects of stocking rate on the species composition and productivity of that community.

Materials and methods The study was conducted on the Inner Mongolian Plateau (41° 47' N, 111° 54' E, average annual precipitation, 280 mm: elevation, 1450 m asl: soil, Light Chestnut) in May, 2004. The dominant species were *S. breviflora, Artemisia frigida* Willd. and *Cleistogenes songorica* (Roshev.) Ohwi. The site had been moderately degraded. A 4 (stocking rates) x 3 (replicates) study commenced in 2004. Each paddock was 4.4 ha and stocked with 0 (control), 0.46 (light), 0.91 (moderate) or 1.35 (heavy) sheep units/ha per yr in a randomized complete block design. Standing crop was estimated monthly from May to October using 10 moveable cages (1.5 x 1.5 m) that were randomly located within each paddock. Total above ground annual production (ANPP), and ANPP of individual species, were estimated by the cumulative difference method of paired grazed and ungrazed plots (1 x 1 m). The cages were moved to a new location after each harvest.

Results Total ANPP was greater ($P<0.05$) with the control and light stocking rate, than with the moderate and heavy stocking rate (Table 1). The ground cover of the plant community decreased linearly with increasing stocking rate. Similar trends were observed for individual species, with the greatest difference occurring between the control and light grazing (Table 1).

Table 1 Effect of stocking rate on the ANPP (kg/ha per yr) and ground cover (%) of the whole plant community and of selected species in the Desert Steppe of Inner Mongolia

Stocking rate	Plant community	*Stipa breviflora*	*Artemisia frigida*	*Cleistogenes songorica*	*Convolvulus ammannii*	*Heteropappus altaicus*
	-------------------------Above ground net primary production (Mean±SD)----------------------					
Control	108.8±5.3a	12.9±2.0a	71.9±5.3a	5.0±0.8a	6.7±1.1a	1.5±0.3a
Light	81.4±4.8b	6.2±2.0b	53.2±5.7b	3.5±0.8ab	1.3±0.9b	0.2±0.1b
Moderate	62.3±6.8c	6.3±1.5b	56.0±3.6b	3.1±1.2ab	0.1±0.0b	0.1±0.1b
Heavy	51.1±3.7c	6.9±2.0b	34.9±4.8c	1.0±0.6b	0.8±0.5b	0.1±0.1b
	--Ground cover (Mean±SD)--					
Control	23.3±0.9a	3.8±0.8a	15.0±1.1b	2.6±4.2a	1.4±0.2a	0.6±0.1a
Light	22.2±0.9a	3.3±0.6ab	18.2±1.3a	1.8±2.4ab	0.9±0.2ab	0.6±0.1a
Moderate	16.6±0.8b	2.6±0.3ab	11.8±0.7c	2.3±2.0ab	1.1±0.2ab	0.2±0.1b
Heavy	10.6±0.4c	2.3±0.1b	7.8±0.3d	1.3±0.8b	0.8±0.1b	0.1±0.0b

Means within a subset of a column having a common letter are not different ($P > 0.05$)

Conclusion A sustainable stocking rate on the *Stipa breviflora* Desert Steppe community appears to be <0.91 (moderate) sheep/ha per yr. This stocking rate is below 1.1 sheep/ha per yr, which was recommended by Han *et al.* (2000), based on liveweight gain. These results are not unexpected because the optimal stocking rate for livestock often exceeds the optimal to maintain a desirable plant community.

Acknowledgements This work was funded by the National Natural Science Foundation of China (30060056,30360022)

References

Wei, Z., G. Han, J. Yang & Xiong Lu (2000). The response of *S. beviflora* community to stocking rate. *Grassland of China,* 6: 1-5.

Han, G., B. Li, Z. Wei & H. Li (2000). Liveweight change of sheep under 5 stocking rates in *Stipa breviflora* Desert Steppe. *Grassland of China*, 6: 4-6.

Using the n-alkane technique to estimate the herbage intake and diet composition of cattle grazing a *Miscanthus sinensis* grassland

Y. Zhang[1], Y. Togamura[2] and K. Otsuki[2]
[1]*Animal Science and Technology College, China Agricultural University, West Road 2 Yuan Ming Yuan, Beijing, P.R. China 100094, Email: zhangyj@cau.edu.cn,* [2]*National Institute of Livestock and Grassland Science, Senbenmatsu, Nishinasuno, Tochigi, Japan 329-2793*

Keywords: n-alkane technique, intake, diet composition, *Miscanthus sinensis*

Introduction Plant wax alkanes are now widely used as marker substances (Dove & Mayes 1991) for the estimation of forage intake and diet composition of grazing herbivores. The objective of this study was to evaluate this method with cattle grazing a *M. sinensis* grassland in Japan.

Materials and methods Four cattle were continuously stocked on the grassland from 28 July to 24 August. The sward consisted of 85% *M. sinensis*, 5% *Pleiablastus chino* and 10% *Aralia elata* and *Lespedeza bicolour*. Animals were dosed with a controlled release device capsule (Captec™, New Zealand).. Faecal samples of individual animals were taken once daily by hand, immediately after faecal excretion on the ground. Meanwhile, herbage samples of each species were hand plucked. Alkane concentrations were determined in the samples as described by Zhang *et al.* (2002). The proportion of each species consumed were calculated using the software package 'Eatwhat' (Dove & Moore, 1995). The alkane concentrations in the diet were corrected using the proportions of the pasture species and the alkane concentrations in each species. Herbage intake was calculated on the basis of the C_{33}/C_{32} ratio using the equation described by Dove & Mayes (1991).

Results The animals had consumed 10 - 40% *P. chino* and 3 - 12% *L. bicolour* each day. Average DM intakes ranged from 1.6 - 2.3% of live weight (MBW) and were significantly. ($P < 0.05$) related to the change in body weight (Table 1).

Table 1 Daily herbage intake on *M. sinensis* grassland

Animal	MBW (kg)	Daily herbage intake (kg) Mean ±SD	DMI as % of body weight	Change in body weight (kg/day)
1	388.5	6.17±1.02	1.59	0.04
2	523.5	11.74±0.55	2.24	0.78
3	385	8.69±0.71	2.26	0.22
4	418.5	6.98±1.04	1.67	-0.19

MBW: mean body weight over a 27 days period; SD: standard deviation.

Conclusions The relative proportions of the different species in the diet was successfully estimated using the alkane technique. Using the diet composition correction, herbage intake of *M. sinensis* grassland can be successfully determined.

References

Dove, H. & R.W. Mayes (1991). The use of plant wax alkanes as marker substances in studies of the nutrition of herbivores: a review. *Australian Journal of Agricultural Research* 42, 913-925.

Dove, H. & A.D. Moore (1995). Using a least-squares optimization procedure to estimate diet composition based on the alkanes of plant cuticular wax. *Australian Journal of Agricultural Research* 46, 1535-1544.

Zhang, Y. J., Y. Togamura & K. Otsuki (2002). Differences in the n-alkane concentration of four wild plants species in Japan. *Grassland Science* 48, 50-52.

Section 3

Multifunctional pastoral systems: biodiversity, landscape and social issues

An ecosystem modelling approach to rehabilitating semi-desert rangelands of North Horr, Kenya

G.A. Olukoye[1], W.N. Wamicha[1] and J.I. Kinyamario[2]
[1] *Department of Environmental Sciences, Kenyatta University, P.O. Box 43844, 00100 GPO, Nairobi, Kenya, E-Mail: golukoye@hotmail.com and* [2] *Department of Botany, University of Nairobi, P.O Box 30197, Kenya*

Keywords: ecosytem, modelling, rangelands, sand dune, *Suaeda monoica*

Introduction Decreased rainfall, recurrent droughts and increased anthropogenic activities have led to a dramatic increase in wind erosion on pastoral lands of North Horr resulting in the reactivation of the once-stable sand dunes. This has degraded the vegetation and impoverished the local community. Mobile sand has a severe impact on dry season grazing areas (Omar & Abdal, 1994) and, therefore, affects pastoral livestock production. In North Horr, *Suaeda monoica* is important in camel production and for stabilising sand dunes but it has been over-utilized over the years. The objective of this study was to use ecosystem modelling approaches to examine the issue of land rehabilitation in North Horr taking cognisance of emerging perspectives on interactions among climate, plants and herbivory in such rangelands.

Materials and methods The study site is situated in North Horr Division on the north-western edge of the Chalbi Desert in northern Kenya. It falls under agro-climatic zone VI and receives an annual mean rainfall of 157 mm (Schwartz *et al*., 1991). A Range Utilization Model (RUM) that relates potential forage production to herbivory energy demand was used to derive the herbivory pressure index for the camels in utilizing *S. monoica* in the years 1999 and 2000. The PRY (Prying livestock productivity) model (Baptist, 1990) was used to model the *S. monoica* subsystem in terms of economic and ecological functioning, using 56 of the 70 households that own camels.

Results Monthly camel herbivory pressure status indices for the years 1999 and 2000 were all less than 10%. In 1999 and 2000, camel herbivory pressure and corresponding vegetation degradation in the *S. monoica* complex was highest in February (1%) and April (3%) respectively. The difference between years in vegetative degradation may not have been attributable to an imbalance between energy demand and *S.monoica* forage production associated with a high camel population since the 1999 and 2000 monthly average camel numbers of 265±79 and 292±21 were not statistically different ($P>0.05$). Modelling showed that within the traditional management system (no culling and female age limit at 300 months) camel productivity was least efficient with total output value per animal per year of US$ 58 and a dry matter intake of 989 kg per animal per year resulting in an efficiency level of 72%. This was 33% less efficient compared to the optimal culling practice given by setting a breeding-female age limit at 274 months. In terms of pastoral management, the traditional camel management translated to a productivity index of US$ 59 per ton of dry matter intake. This value was less than that which could be obtained at an optimal culling practice of US$ 86 per ton of dry matter intake.

Conclusions The PRY model confirmed the results of the RUM model in that the observed imbalance between camel herbivory demand and supply could be related more to climate-induced vegetation degradation than herbivory demand. It may be concluded that *S. monoica* can perform the dual ecological and economic functions of dune stabilization and camel browsing respectively, if flexible camel stocking rates are applied within the *S. monoica* vegetation complex. This would require either an adjustment in the camel management system with respect to culling strategies to increase offtake or alternatively increase in the *S.monoica* cover. *S. monoica* cover could be increased through natural regeneration, which should take advantage of the symbiotic relationship between camels and this shrub. It is therefore, important to integrate both the biophysical and socio-economic aspects in sustainable *nebkha* –dune stabilization and hence land rehabilitation in North Horr.

Acknowledgement The financial support of the German Ministry of Economic Cooperation and Development through Deutsche Gesellschaft für Technische Zusammenarbeit is appreciated.

References

Baptist, R. (1990) Simulated livestock dynamics-effects of pastoral offtake practice and drift on cattle wealth. Proceedings of the 8th Scientific Workshop of the Small Ruminant Collaborative Research Support Program, Nairobi, Kenya, pp. 207-215.

Omar, S.A.S. & M.Abdal (1994). Sustainable agricultural development and halting desertification in Kuwait. *Desertification Control Bulletin,* 24, 23-26.

Schwartz, H.J., S.Shaabani & D.Walther (eds) (1991). Range Management. Handbook of Kenya. Vol.II, 1. Republic of Kenya, Ministry of Livestock Development, Range Management Division, Nairobi.

Riparian management in intensive grazing systems for improved biodiversity and environmental quality: productive grazing, healthy rivers

S.R. Aarons[1], M. Jones-Lennon[1], P. Papas, N. Ainsworth, F. Ede and J. Davies
[1]*Ellinbank Research Institute, PIRVic Ellinbank, RMB 2460 Hazeldean Rd, Ellinbank, Victoria, 3821 Australia, Email: sharon.aarons@dpi.vic.gov.au & www.dpi.vic.gov.au/vro/biodiversity/riparian*

Keywords: biodiversity, weeds, water quality, woody debris

Introduction Within high rainfall intensive grazing systems of southern Victoria, riparian zones are often degraded due to vegetation clearing, stock access and inappropriate farm management. Streams in these landscapes often have poor water quality and reduced biodiversity due to degraded terrestrial and aquatic ecosystems. Improved management of riparian zones depends on developing tools and practices for integration into productive grazing systems. This paper describes the approaches used and the tools developed in the '*Productive Grazing, Healthy Rivers: Improving riparian and in-stream biodiversity*' project.

Materials and methods The project sites are on intensive beef and dairy properties that occur within the 4 high rainfall (>750mm) bioregions (NRE 1997) of southern Victoria (Figure 1). During the project development phase (September 2001-April 2002) an extensive literature and data review was undertaken followed by consultation with key stakeholders involved in riparian zone management in the grazing industries. Research gaps identified in this process were used to develop the 6 activities currently underway in the research and development phase (June 2002-July 2005) of the project.

Results The review of riparian biodiversity assets (NRE 2002) identified the need to survey riparian biodiversity actively on grazing properties in the study area. Current riparian management actions often differed from recommended guidelines. Eight key riparian management issues were identified, which were used to develop 6 research and development modules. These are: biodiversity surveys of 40 paired, fenced and unfenced, riparian sites on dairy and beef properties in southern Victoria; investigation of the impact of introducing small woody debris on biodiversity of fenced replanted riparian sites; surveys of 36 sites to identify factors affecting native tree recruitment in fenced riparian sites; development of a weed decision support tool to assist landholders manage riparian weeds; assessment of riparian condition of 107 dairy farm sites; water and soil monitoring along the riparian zone of 2 commercial dairy farms to identify farm management impacts and to monitor changes after fencing and revegetation (Aarons *et al*., 2005).

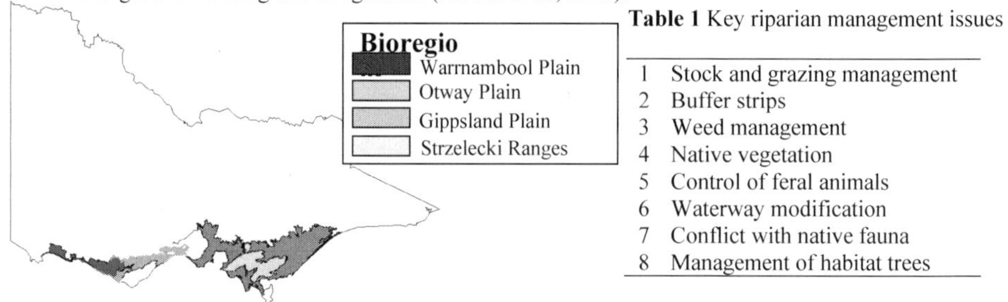

Table 1 Key riparian management issues

1	Stock and grazing management
2	Buffer strips
3	Weed management
4	Native vegetation
5	Control of feral animals
6	Waterway modification
7	Conflict with native fauna
8	Management of habitat trees

Figure 1 Project location showing the 4 study bioregions

Conclusions Key riparian management issues identified by the team and stakeholders were the basis of the research activities developed. Biodiversity survey data indicates that fenced sites on beef and dairy farms have greater native biodiversity. Unfenced riparian areas on dairy farms in SE Victoria are generally in poor condition, with on-farm activities contributing to degraded riparian zones. On the other hand, weed species often are associated with fenced riparian areas. A riparian weed decision support tool was developed to assist farmers to manage fenced riparian land. Information from this project is disseminated regularly to landholders and natural resource managers in the project study area.

References

Aarons, S.R., A. Melland & C.J.P. Gourley (2005). Grazing management impacts on the riparian zone and on water quality. *Proceedings of the XX International Grasslands Congress*.

NRE (1997). Victoria's biodiversity: Directions in management. Department of Natural Resources and Environment, East Melbourne, Victoria.

NRE (2002). Biodiversity conservation in intensive grazing systems: Riparian and in-stream management. Department of Natural Resources and Environment, Ellinbank, Victoria.

Contributions of the United States Department of Agriculture Natural Resources Conservation Service to conserving grasslands on private lands in the United States

L.P. Heard
USDA/NRCS, Wildlife Habitat Management Institute, 100 Webster Circle, Suite 2, Madison, Mississippi 39110 USA, Email pete.heard@ms.usda.gov

Keywords: grassland, private lands, biodiversity

Introduction The future of biodiversity in the USA is tied inseparably to activities taking place on private lands. Agriculture is by far the most important user of these lands, with about 50% or 900M acres managed as private cropland, grassland or rangeland. Decisions made by America's farmers and ranchers directly affect grasslands and their impact on food supply, biodiversity, soil protection and water quality. Agricultural programs and policies in the USA have had a large influence on the choices available to farmers and ranchers in land management. Since the 1930s, USDA's Natural Resource Conservation Service (NRCS) has been working with farmers, ranchers, and other land managers to promote conservation of natural resource through the nation's 3000 soil and water conservation districts. The Conservation Title of the 1985 Farm Bill, amended in 1996, raised the importance of biodiversity /wildlife in the delivery of conservation programs to the nation's privately owned lands. NRCS is charged with developing and delivering the proper grassland establishment techniques to landowners and evaluating the results. Recognising the opportunities and challenges related to conserving and enhancing fish and wildlife habitat, NRCS created the Wildlife Habitat Management Institute (WHMI) in 1997 as part of the NRCS National Science and Technology Consortium. WHMI was to interact with academic institutions, partner agencies, non-government organisations and others to develop and disseminate scientifically based technical materials to NRCS field staffs and others to enhance delivery of sound habitat management principles and practices, including grasslands to America's land users.

Materials and methods NRCS/WHMI works with scientists from various institutions to develop tools useable in the field and to evaluate the response of the utilised tools to develop native grasslands. Work on tool development was started first, recognising results could be 5-6 years away. As tool development began, an evaluation of wildlife response to previous Farm Bill programs was undertaken. The effort was intended to get a starting point to guide future planning for WHMI. Planning by WHMI staff is critical, as only 7 staff guide the effort to develop technical tools, such as jobsheets and technical notes used in management recommendations on grassland and other private land habitats across the USA.

Results A Comprehensive Review of Farm Bill Contributions to Wildlife Conservation (Heard *et al.*, 2000) included reviews of impacts on grasslands. The Conservation Reserve Program (CRP) added 18M acres of grassland, mostly native, in the Midwest,. These grasslands were beneficial to grassland birds and contributed to a 30% (10.5M) increase in ducks from the northern Great Plains. The new grasslands did not produce an immediate increase in population of many grassland birds but decreased the precipitous decline that was being experienced. Conversion to native grassland in the southeast has been limited by the use of exotic forage grasses, rapid succession, and mowing. Most of the southeast habitat improvement from native grasses has come from the development of agricultural field edges. Although the above report reviewed wildlife responses, the response of native grass establishment on soil protection, water quality and biodiversity has been positive also.

In addition to the early report above, NRCS/WHMI continues to work with partners to develop techniques and to evaluate the impacts of native grassland establishment on wildlife. The following are current projects designed to move NRCS forward in its ability assist agricultural producers to conserve and increase native grasslands within their operations: (1) Bird response to grassland management in the northeast, (2) Lesser prairie chicken habitat in the southwest plains, (3) Evaluating wetland restorations in the Gueydan Prairie in coastal Louisiana. WHMI will continue to work to enhance the habitat of species of concern such as northern bobwhite and sage grouse.

References

Heard, L. P., A. W. Allen, L. B. Best, S. J. Brady, W. Burger, A. J. Esser, E. Hackett, D. H. Johnson, R. L. Pederson, R. E. Reynolds, C. Rewa, M. R. Ryan, R. T. Molleur, P. Buck (2000). A comprehensive review of Farm Bill contributions to wildlife conservation, 1985-2000. In W. L. Hohman and D. J. Halloum (eds). U.S. Department of Agriculture, Natural Resources Conservation Service, Wildlife Habitat Management Institute, Technical Report, USDA/NRCS/WHMI-2000.

Protection of agrobiodiversity: model calculations in Rhineland-Palatia: costs and implications for farmers

H. Bergmann
Department of Agricultural Economics, University of Göttingen, Platz der Göttinger Sieben 5, D-37073 Göttingen, Email: hbergma1@gwdg.de

Keywords: agri-environmental schemes, cost efficiency, grassland extensification, agrobiodiversity

Introduction Biological conservation and production use the same areas of land in less favoured areas. Grassland in these areas makes an important contribution to the protection of agro-biodiversity. However, under existing market conditions and production needs, the use of low yielding grasslands is not economically efficient. The objective of this study was to analyze the economic consequences of different mowing strategies in a small region in Rhineland-Palatia (Germany) that served the protection of two butterfly species.

Materials and methods The impacts of extensification measures for nature protection on herbage quality and yield have been described from a literature analysis. The effects of different cutting dates on quality and yield were calculated, based on a function by Opitz von Boberfeld (1994). Before these calculations, the calculated yields/ha were qualified in relation to nutritional requirements of cattle. The method of standard gross margin calculations was used to determine the compensation payments for the profit foregone by farmers that adopt alternative mowing regimes. It was assumed that farmers purchased concentrates as an additional fodder to compensate for the loss of energy yields and the compensation payments were calculated accordingly. Because most farmers in the region used meadows for silage, the calculations were based on silage production with a base energy yield of 52 GJ NEL/ha.

Results The literature on cattle and horse nutrition suggests that fodder with an energy content >6 MJ NEL/kg DM is usable in intensive cattle production, while fodder of 4-6 MJ NEL/kg DM is mostly usable in horse and heifer nutrition. Due to its low quality, silage harvested with a first cut in August is not recommended for use in horse nutrition or for cattle (except for yearlings between 12-18 months old; Figure 1). Therefore, for mowing regimes with a first cut after the beginning of August, farmers must be compensated for the complete loss of use of the meadow. The curve of calculated compensations (Figure 2) shows three stages, (1) until 15 June, with compensation costs of about 200 €/ha, (2) from 15 June to 1 August, with compensation costs increasing to 1000 €/ha and (3) after 1 August, when a total loss was presumed, with compensation costs of 1156 €/ha.

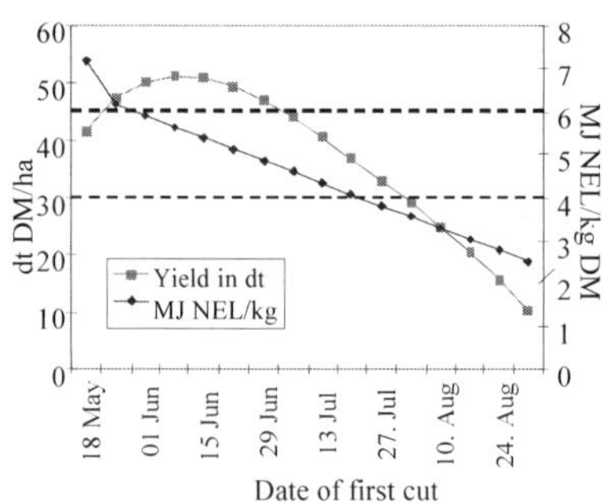

Figure 2 Effect of different mowing dates on MJ NEL/kg, DM/ha and their usability
Source: Own calculations based on Opitz von Boberfeld (1994)

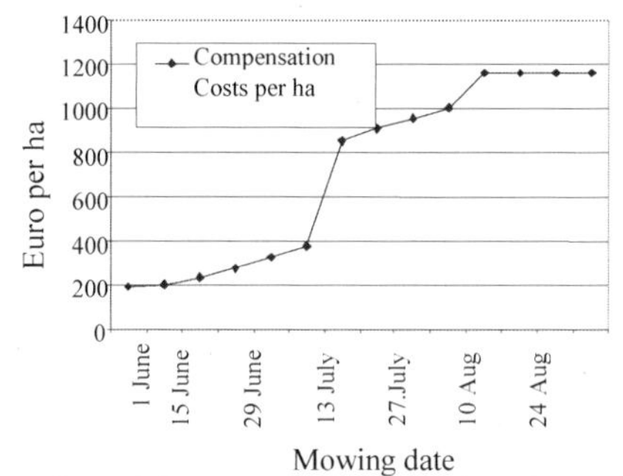

Figure 3 Compensation Costs in dependence of mowing date (1 cut a year)
Source: Own calculations

Conclusions Linearity in compensation calculations is not a realistic way to reimburse farmers for extensification requirements. Justifiable compensation amounts can be calculated only in combination with specific grassland science, cattle nutrition and ecological knowledge.

References

Opitz von Boberfeld (1994). Grünlandlehre, UTB 1770, Stuttgart.

Stocking rate theory and profit drivers in north Australian rangeland grazing enterprises

N.D. MacLeod, A.J. Ash and J.G. McIvor
CSIRO Sustainable Ecosystems, 306 Carmody Rd, St Lucia Q 4067, Australia, Email: neil.macleod@csiro.au

Keywords: stocking rate, economics, rangelands

Introduction Setting correct stock numbers is a key decision for successful pastoralism. In marginal environments, typified by northern Australia, this involves careful cattle herd management across landscapes and seasons characterised by heterogeneous land condition and extreme climatic uncertainty. Stocking rate theory which links animal production to stocking rates concentrates only on liveweight gain of sale animals and ignores complex herd (e.g. reproduction, mortality) and pasture dynamics (e.g. land condition) and costs of maintaining stock numbers (e.g. supplementary feeding). Related economic models are generally naïve and incomplete, being based on liveweight gain, meat prices and variable husbandry costs (e.g. Workman, 1986). Modelling approaches, which simulate whole herds (including breeding animals) with dynamic links between animal numbers, pasture availability, management effort and profits are more realistic. This paper explores the economic implications of changing stocking rates, animal production and management effort in the form of supplementary feeding.

Methods Modelling is used to examine the relationship between stocking rates, land condition, animal production and net profit for a 28,000 ha farm in Queensland. A pasture simulator was linked to an economic herd model (MacLeod *et al.*, 2004) for 3 levels of stocking rate, 10.0, 6.7, and 5.0 ha/adult equivalent (AE) and 2 land conditions (State 1 = Perennial grasses dominated by digestible tussock spp.; State 2: Perennial-annual grasses dominated by less digestible perennial grasses, annual grasses and forbs) for 100 simulations. If annual liveweight gain of animals fell below a threshold (<50kg/animal), supplementary feeding (8% urea:molasses) is imposed, consistent with local practice.

Results While average number of stock carried is similar between States 1 and 2, highest profit per animal is at the lowest stocking rate for the State 2 pasture (Table 1). Profit per hectare is highest at the highest stocking rate for State 1 pasture and almost the same as for the lowest stocking rate for State 2. Although mean branding rates are similar for between land classes, and decline with stocking rate, the ability of the herd to self-replace breeding cows declines, and is more difficult for State 2 land. Feeding costs per hectare are much higher for State 2 pasture for a given stocking rate, contributing to the rapid decline in profitability as stocking rate increases.

Table 1 Mean values of stock carried, net profit and feeding costs from 100-year simulation

	State 1			State 2		
Stocking rate (ha/AE)	10.0	6.7	5.0	10.0	6.7	5.0
Total stock carried (AE)	2948	4429	5922	3014	4425	5922
Net profit/AE ($AU)	35	34	26	50	21	-15
Net profit/ha ($AU)	3.7	5.3	5.5	5.4	3.3	-3.3
Branding rate (calves branded/100 breeders mated)	68	64	60	72	66	59
Proportion of years that heifer calves are insufficient to maintain current herd size (%)	10	22	26	14	28	39
Feeding costs/ha ($AU)	0.4	1.7	3.5	1.7	5.1	12.5
Proportion of years that net profit is negative (%)	22	23	21	19	29	40

Conclusions Profitability is co-determined by stocking rate and land condition. While total stock carried is similar between land classes, moderately degraded pastures can yield profits similar to good condition pastures, but only at lower stocking rates. Profitability is the interplay of herd management (especially reproduction efficiency and feeding costs), climatic patterns and land condition. Stocking rates based on a less than complete understanding of the complex relationship between herd dynamics, climate and pasture response can be misleading.

References

MacLeod, N.D., A.A. Ash & J.G. McIvor (2004). An economic assessment of the impact of grazing land condition on livestock performance in tropical woodlands. *Rangeland Journal,* 26, 49-71.

Workman, J.P. (1986). *Range Economics*, Macmillan Publishing, New York.

An ecological and economic risk avoidance drought management decision support system

R.K. Heitschmidt and L.T. Vermeire
USDA Agricultural Research Service, Fort Keogh Livestock & Range Research Laboratory, 243 Fort Keogh Road, Miles City, MT 59301 USA, Email: *rod@larrl.ars.usda.gov*

Keywords: net primary production, precipitation, rangeland, temporal dynamics

Introduction Ecologists have long recognized the fundamental impacts of drought on rangeland structure and function. Simulation models have been developed to increase our understanding of these impacts as they relate to forage production, particularly for predictive purposes. Although the capacity of these models to accurately predict quantity and quality of forage produced under varying climatic conditions is often quite good, their ability to serve as an effective and proactive drought management decision support system is often limited. This is in large part because their complexity impedes their use by on-the-ground managers. The objective of this research was to develop a very simple and user-friendly drought management decision support system for rangeland managers in the Northern Great Plains of North America.

Materials and methods The proportion of herbage produced on a monthly basis was estimated using aboveground net primary production (ANPP) estimates from 14 native rangeland data sets (Heitschmidt *et al.* 1995; 1999; 2004) collected over 11 years at the Fort Keogh Livestock and Range Research Laboratory located near Miles City, MT. Correlations coefficients were calculated between monthly precipitation and annual ANPP estimates. Published (http://hprcc.unl.edu/index.html) site-specific, monthly, rainfall probabilities were also incorporated into the system to enhance predictive capacity.

Results Both total and perennial grass production were significantly ($P<0.10$) correlated with precipitation in January and April whereas cool-season perennial grass production, the over-whelming dominant functional group in this herbage complex, was significantly correlated with precipitation in January and May. In these analyses, the correlation in January is viewed with suspicion (i.e., a spurious correlation) as precipitation is snow and only averages 12 mm which is < 4% of the 338 mm of the annual total. The correlation with spring precipitation is biologically sound in that the sum total of precipitation in April and May averages 84 mm. In a similar study at this location, Kruse (2004) also found significant relationships between total herbage production and precipitation in April and May.

Analyses of the proportion of ANPP of perennial grasses produced on a monthly basis showed 35%, 69% and 91% was produced by 1 May, 1 June, and 1 July, respectively. The 95% confidence limit for the estimate on 1 July was 24% which indicates that at least 67% of annual production of perennial grasses would be completed by 1 July in 19 out of 20 years. Thus, in this region grazing managers can estimate appropriate end-of-growing season stocking rates by early July with a relatively high level of confidence. Moreover, confidence in these decisions can be further bolstered by examining long-term, post-1 July probabilities of rainfall. For example, the probability of receiving 25.4 mm of precipitation in July and August at Miles City, MT is 59% and 42%, respectively, whereas probability estimates for receiving 50.8 mm in July and August are 22% and 17%, respectively.

Conclusions The results provide insight into temporal dynamics of forage production in the Northern Great Plains of USA. Results demonstrate that forage production is largely a function of precipitation received prior to 1 July. Thus, proactive stocking rate adjustments can be made by early July, with considerable certainty, thereby reducing ecological and economic risks that arise from late-season forage demand/availability imbalances.

References

Heitschmidt, R. K., E. E. Grings, M. R., Haferkamp & M. G. Karl (1995). Herbage dynamics on two Northern Great Plains range sites. *Journal of Range Management,* 48, 211-217.
Heitschmidt, R. K., M. R. Haferkamp, M. G. Karl & A. L. Hild (1999). Drought and grazing: I. Effects on quantity of forage produced. *Journal of Range Management,* 52, 440-446.
Heitschmidt, R.K., K. D. Klement & M. R. Haferkamp (2004). Interaction effects of drought and grazing on Northern Great Plains rangelands. *Journal of Range Management* (in press)
Kruse, R. E. (2004). Beef cattle management decisions relating to drought in the Northern Great Plains. M.S. thesis, Department Animal & Range Sciences, Montana State University, Bozeman, MT, 138 pp.

Predicting the effects of management on upland birds, economy and employment

S.M. Gardner[1], G.M. Buchanan[2], J.W. Pearce-Higgins[2], M.C. Grant[2] and A. Waterhouse[3]
[1]*ADAS Preston, 15 Eastway Business Village, Oliver's Place, Fulwood, Preston PR2 4WT, U.K, Email: sarah.gardner@adas.co.uk,* [2]*RSPB Scotland, Dunedin House, 25 Ravelston Terrace, Edinburgh EH4 3TP, UK and* [3]*SAC, Sustainable Livestock Systems, West Mains Road, Edinburgh, EH9 3JG, UK*

Keywords: livestock systems, bird populations, vegetation dynamics, upland farming

Introduction Livestock farming systems play a significant role in the economy and conservation of the UK uplands and rely heavily upon public financial support. Changes in that support could have far-reaching impacts on the wildlife interest and socio-economics of upland areas. Predicting the impacts of such changes is difficult, since they arise from responses to new economic circumstances. The effect of management change is also influenced by natural variation, such as the mosaic of plant communities, already present in the upland landscape. This paper sets out an approach that integrates theoretical models with field studies to investigate the effects of management change on birds, economics and employment in the UK uplands.

Methods The approach combines outputs from separate models of bio-economics, vegetation dynamics and bird abundance. The bio-economic model uses livestock energy demands and thresholds together with standard costs and assumptions, to trigger economic costs and output changes. The vegetation model uses a grid-based modelling approach in which vegetation change is driven by plant competition, spatial distribution, growth and management. Field data characterising the plant communities present on a site, their distribution, composition, growth phases and management, are used as input data. The bird models were derived from field data collected from 85 2-km^2 plots in southern Scotland. Generalised Linear Models were used to identify variables that significantly affect the abundance of bird species, and incorporated both site and management effects, and variables describing vegetation composition and structure. To determine the effect of management change on bird abundance and farm economics, different grazing regimes were simulated within the vegetation dynamics model, the outputs of which were used in the bio-economic and appropriate bird (red grouse and meadow pipit) abundance models. The scenario analysed was a 200 ha wet heath mosaic of *Calluna, Molinia, Nardus,* sedges and rushes. The regimes were: all-year high sheep grazing (4.5 ewes/ha), zero grazing and all-year mixed sheep grazing (0.66 ewes/ha) with 0.75 cattle/ha during June-August. Vegetation simulations were run for 10 years.

Results and discussion Economic modelling predicted a financial turnover of £71,576, £53,549 and zero, for the high, mixed and zero grazing regimes respectively. The first two regimes included CAP area payments of £7878 whilst farm labour units employed were 0.80, 0.54 and zero respectively. Calluna cover was predicted to rise under each regime and drove the predicted increases in red grouse abundance (Figure 1). The effect was greatest under the mixed and no grazing regimes. Fine-leaved grasses and sedges were predicted to decline under each grazing regime, with the smallest decline occurring under mixed grazing. Meadow pipit numbers were predicted to decline, but faired best under the mixed-grazing regime, potentially reflecting changes in the balance of Calluna to fine-leaved grasses and the declines in sedge cover compared to the starting year. By linking field data to theoretical models, changes in management practice, arising from external economic or policy decisions, can be analysed in relation to their direct and indirect effects on economy, employment and biodiversity. These models can indicate the magnitude and direction of expected change and thus inform decisions about sustainable land management.

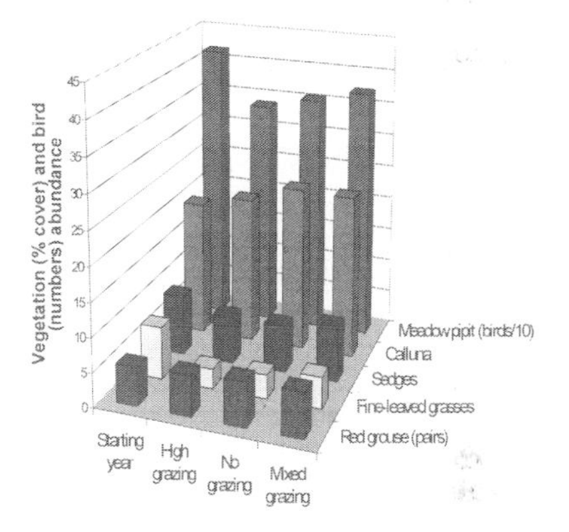

Figure 1 Predicted effect of grazing regime on specific plant and bird species (after 10 yrs)

Acknowledgement This work was funded by Defra, English Nature and the Countryside Council for Wales.

Managing resources by grazing in grasslands dominated by dominant shrub species

D. Magda[2], C. Agreil[1], M. Meuret[1], E. Chambon-Dubreuil[2] and P.-L. Osty[2]
[1]INRA-Sad Ecodéveloppement. Site Agroparc. F-84 914 Avignon Cedex 9, France, Email: agreil@avignon.inra.fr and [2]INRA-Sad Orphée, BP 27, Chemin de Borde-rouge,. F-31 326 Castanet-Tolosan Cedex, France

Keywords: dominant species, plant part, small ruminant, grazing, paddock

Introduction The European natural grasslands are attracting new attention because of their environmental value as habitats for threatened fauna and flora species and their contribution to the diversity of landscapes. Those responsible for the implementation of the European agri-environmental policy are hence encouraging livestock farmers to adopt grazing practices that contribute to the conservation of grassland biodiversity especially by limiting encroachment by dominant shrubs. However, current scientific knowledge and technical information are often insufficient to connect flock feeding and the impact of grazing on shrub dynamics and livestock farmers are not very enthusiastic about restoring or conserving "plant mosaics" including shrubs that support biodiversity in their fields. This paper presents results of an interdisciplinary study on interactions between small ruminant feeding strategy and population dynamics of dominant shrub species with the objective of managing by grazing the structure of plant community and thus to provide the renewal of resources on a multi-year scale.

Methods Concerning the ruminants' feeding strategy, experiments with dry ewes were carried out on farm, by recording their foraging behaviour and adjustments in response to varied and variable feeds on offer. The direct observation method of bites was recently improved for highly diversified environments, which allowed the recording of intake rate changes during bouts and the estimation of daily intake. In order to analyse the demographic strategy of the dominant plant population, recording of the different demographic parameters of the dynamics of broom populations (seed germination and dormant rate, adult fecundity and survival rates) were carried out on an ungrazed shrubland. A grazing experiment was conducted at different phenological stages of the broom population to identify the edible plant parts and quantify their consumption rate. The demography was modelled using Leslie matrices

Results During a paddock-grazing sequence, the dry ewes progressively expanded the range of their daily bite masses which contributed to maintaining the stability of daily intake. This adjustment was associated with a regular temporal pattern: the alternating between bouts of high and low intake rate with a pseudo-period of approx. 15 min. The development of approaches that considers the "functional feeds" is advocated. These cannot only be described by grass height and nutritive value alone. Concerning the dominant shrub population, several plant parts are consumed (flowers, young pods, young shoots and mature stems) offering multiple potential demographic interactions. A major response of repeated grazing was the process of adult bushes becoming vegetative and then creating a new demographic category within the population structure. Simulations of potential grazing impact at different seasons pointed out the importance of the role of rejuvenation of adults and mortality of juveniles. These results led to the building of a model capable of linking two biological processes that are generally treated separately, i.e. the feeding strategy of small ruminants and the population dynamics of species with a strong dominance capacity. The model was designed as an account of a real grazing situation in which a livestock farmer pursues a two-fold aim: to feed a flock of small ruminants and, simultaneously, to maintain the species diversity in a plant community by controlling the dynamics of a dominant species (in this case broom) at a paddock scale. By distinguishing four time scales, we argue in favour of interlinking the two processes at the level of the plant part. Plant parts can be classified according to the role that they play in the organization of the feeding strategy and to the effects of their removal on the dominance dynamics. The model is useful for identifying "target plant parts" to be grazed. From the management point of view, paddock adjustment practices and the season of utilisation (in relation to the functional feeds on offer) should be tailored to fit the animals' motivation to consume these target plant parts. From a scientific point of view, the model encourages an in-depth study of the spatial dimension of processes and related practices.

A decision support system for rangeland management in degrading environments

R.G. Bennett and F.J. Mitchell
Department of Agriculture and Environmental Affairs, Private Bag X9059, Pietermaritzburg, 3200, South Africa, Email: bennettr@cedara.kzntl.gov.za

Keywords: remote sensing, sustainability, desired state, production

Introduction The continued viability and productivity of commercial and emerging agriculture in KwaZulu-Natal Province, South Africa, depends on the accurate assessment and sustainable utilization of available natural resources. Sustainability implies that growth and development must take place, and be maintained over time, within the limits set by natural ecosystems. Utilizing an extensive GIS database, field surveys and remote sensing technology, a land assessment decision support system (LADSS) has been developed in an attempt to define these limits for the Province. This system has been developed to assess the appropriate use of existing resources as well as the suitability of current land use practices. LADSS includes a predictive tool which allows the impact of a proposed change in land use to be forecast within 590 agro-ecological zones of the Province.

Materials and Methods Two case study areas were selected which represented widely varying agricultural potential. The northern area, suited to extensive farming practices such as beef and game, is restricted by shallow erodible soils and low rainfall. The southern area, with a higher production potential including intensive cultivation, has deeper fertile soils and higher rainfall. Both areas encompass a wide variety of range management policies from multi-species game or conservation areas, highly managed commercial beef and dairy enterprises to areas which are heavily stocked, continuously grazed and communally managed (Table 1). An existing natural resource classification system, unique to the Province, combined with remote sensing data is utilized to provide an accurate assessment of (i) the extent of cultivation including fodder crops, (ii) extent and severity of degradation, (iii) areas of high biodiversity, (iv) loss of production potential and (v) areas requiring rehabilitation. Soil, climate and crop model information is used to determine optimum land use options and productivity of rangeland and cultivated land.

Results Resource plans, using LADDS, were developed to assist planners in determining appropriate land uses and to develop an agricultural strategy for each agro-ecological zone, based on GIS, land assessments and remote sensing data.

Table 1 Comparison of Land Use Patterns between two study areas

Land Cover	Northern Area (%)	Southern Area (%)
Indigenous forest	0	3.2
Bushland thicket	31.5	23.5
Natural grassland	21.5	20.5
Exotic Plantations	5.6	16.4
Wetlands/waterbodies	0.2	3.2
Degraded land /eroded areas	30.2	9.4
Cultivated land and pastures	10.0	17.0
Urban areas/communal settlements	1.0	6.8

Mentoring and training in natural resource management were undertaken to limit further degradation and to encourage community-driven sustainable strategies to be formulated and adopted. The gap between potential production and current land uses enabled the identification of areas which offer an opportunity to improve livelihoods, reduce risk and apply integrated intervention strategies which might ensure sustainable managed systems.

Conclusions The "best practice" approach was identified, intervention strategies were implemented and assistance with emerging farmer settlement schemes was provided. LADSS was found to assist land users and planners with information which is fundamental to the requirements of sustainable agricultural practices and allows for monitoring future changes in land use. Strategies, which were identified from the derived data, included community-driven approaches, participatory approaches to prioritize community needs, and the identification of drivers to ensure adoption and successful implementation of the strategy.

Modelling the encroachment of farmhouse culture on private village pastures and its environmental fall- out in Northern Western Ghats, India

S.B. Nalavade, K.R. Sahasrabuddhe and A.A. Patwardhan
RANWA, C-26/1, Ketan Heights, Kothrud, Pune-411038. Maharashtra, India, Email: sbnalavade@vsnl.net

Keywords: private village pastures, urbanization, farmhouses, rural economy, access corridors

Introduction Tropical India harbours numerous pasturelands across small landholdings ranging up to few hectares which are covered with grass that is suitable as fodder. These grazing lands are commonly known as '*Gairan*' in urbanised northern Western Ghats mountain tract in Western India). Such grasslands comprise about 20% of the total area of a village (Jodha, 1986), support livestock and supplement the agro-economy of the village. These pasturelands are being replaced by fenced 'farmhouses' of the urban elite, resulting in land use changes that caused drastic qualitative and quantitative changes in terms of area, fodder species composition and livestock they support (Patwardhan *et al.,* 2003). The study area has faced large changes in the last few decades with increases in the area under settlement by 240%s as well as a decrease in the area of agriculture land and grasslands-scrub vegetation by 31 % and 39 % respectively (Nalavade, 2003). The present paper documents socio-cultural, economic and environmental changes in private village pastures across the Mumbai-Pune urban belt.

Materials and methods A dozen villages from the study area were selected. Revenue maps were used to map the land use. Present land use was mapped by conducting field visits and by using government records. Sem-i structured interviews with villagers revealed the past land use. A concentric circle model was developed, and found to be effective tool for comparing changes in land use and to generate response options (Nalavade, 2003).

Results The traditional set-up of a village shows a typical land use pattern with agriculture, private pastures, public pastures and forest land surrounding the settlement. Current land use involves traditional pasturelands being replaced by fenced 'farmhouses' of the urban elite, restricting movement of wildlife besides blocking the approach of local grazers and fuel wood collectors towards the public pastures and state forest lands. A decrease in the grazing area has both changed the composition of livestock by forcing farmers with small landholdings to keep fewer cattle than previously and through a shift to hybrid cattle. Shifting pressure from private pastures to public pastures and forest lands has led to degradation of these habitats. This has also resulted in change in employment options leaving landowners to work as caretakers, gardeners or security persons in farmhouses.

Conclusions To restrain the situation, there is need for access corridors between settlement and public pastures and forest land. Local specific communication, education and public awareness strategies and action plans need to be developed to overcome the situation. Otherwise in the future, present pastoral ecology may be gradually replaced by urban culture thereby hampering village ecosystem.

References

Jodha, N.S. (1986). CPR and rural poor in dry regions of India. *Economic and Political Weekly*, 21 (27).
Patwardhan, A., Kanade, R., Sahasrabuddhe, K., Nalavade, S., Ghotge, N., and U. Ghate. (2003). Social review of pastoral ecosystem services in Western India. *African Journal of Range and Forage Science,* 20, 194-197.

Rangeland as a common property resource: contrasting insights from communal areas of central Eastern Cape Province, South Africa

J.E. Bennett and H.R. Barrett
Geography and Environmental Science Subject Group, School of Science and the Environment, Coventry University, Coventry, CV1 5FB UK, E-mail: j.bennett@coventry.ac.uk

Keywords: South Africa, rangeland, livestock, common property regimes, grazing management

Introduction In communal areas of South Africa, grazing systems are held under a variety of different common property regimes. However, the social and ecological realities of these communal grazing systems remain poorly understood, particularly with regard to the use of land allocated for crop production. Little is known about how these arable areas are utilised as a common grazing resource but the wide array of tenure arrangements under which they are held suggests that they facilitate some interesting departures from recognised common property systems. A clearer understanding of how common property regimes function at an integrated level in South Africa will be fundamental in developing an empirical foundation for effective institutional capacity building at both the local and national level as well as other policy recommendations. This research outlines the diversity of grazing management regimes operating in communal areas, relating it to key social and ecological factors and emphasising the critical role played by the arable land allocations.

Method Research was undertaken at two socially- and ecologically-contrasting villages, Guquka and Koloni, in central Eastern Cape Province. Guquka lies in an area of relatively poor grazing, better suited to rain-fed crop production, and with a turbulent history of forced resettlement under apartheid. Koloni lies in an area of higher quality grazing, with a history of relative social stability. Information was collected through a semi-structured interview schedule administered to key informants at both villages and supplemented by participant observation.

Results The research indicated that three different types of grazing management regime are in operation at the two villages (Figure 1). Type 1, identified at Koloni, is the egalitarian governance of grazing resources as part of a true communal property regime, made possible primarily through the relatively low pressure on resources at the village. The remaining two were identified in the environment of high environmental pressure experienced at Guquka. Type 2 is the generally recognised scenario of 'open-access' grazing involving little or no community control over the grazing system. The other, unexpected, typology (Type 3) is devolution of control over grazing management to the individual level through the use of fenced arable fields.

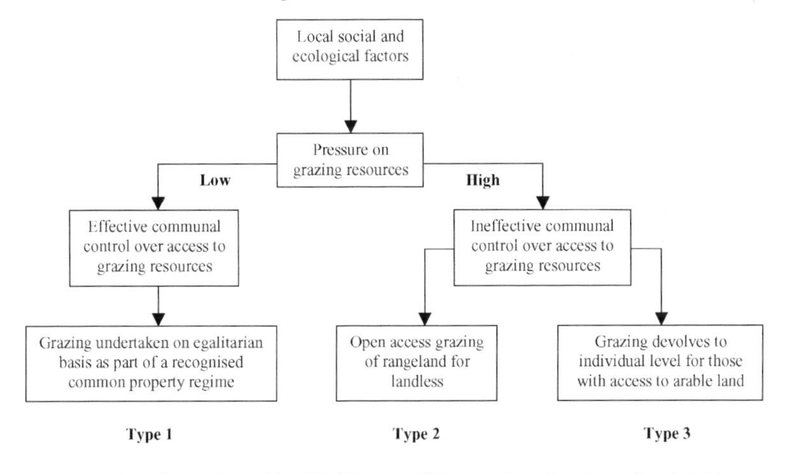

Figure 1 Typology of grazing regimes identified in central Eastern Cape Province, South Africa

Conclusions Although open-access grazing is believed to be representative of the general scenario in central Eastern Cape, the identification of two additional property regimes operating in the region has ramifications for policy-makers. The use of the arable land allocations for private grazing is of particular note as it exemplifies the enclosure of common land in the face of resource pressure as has happened elsewhere in Africa. If land reform policies in the communal areas are to be effective and inclusive, they must give adequate recognition to the existence of these different management regimes and the enormous social and ecological heterogeneity at the local level that has created them.

Andean pastures in the fourth region of Chile: marginal lands and vital spaces for a transhumance system

T.S. Koné[1], R. Osorio[2] and J.-M. Fotsing[1]

[1]*Laboratoire ERMES, Institut de Recherche pour le Développement (IRD), 5 rue du Carbone 45072 Orléans Cedex 2, France, E-mail: Tchansia.Kone@orleans.ird.fr, and* [2]*Departamento de Biología, Facultad de Ciencias, Universidad de la Serena, La Serena, Chile*

Keywords: pastoralism, aridity, andean grasslands, transhumance, forage availability

Introduction In the fourth region of Chile, the high Andean pastures between Chile and Argentina are the summer destination for transhumant shepherds and sustain a part of the regional livestock. Since 2000, Chile has prohibited the passage of livestock to Argentina for animal health reasons in spite of official registers indicating that 60 to 75% of the summer transhumance livestock had an Argentine destination. Under those conditions it is questionable whether the Andean Chilean grasslands can absorb the increased pastoral demand without suffering damage. The objective is to provide elements of an answer to this question regarding the distribution and availability of the Andean forage resource and its modalities of explotation in the local transhumance system.

Materials and methods Two approaches have been used for this study. The first one, concerning the forage resource, used remote sensing tools, GIS and land occupation cartography methods (Etienne & Prado, 1982) to describe the types and distribution of the Andean pastures in two cordilleras of the Limari province, Fourth Region of Chile, during the summer of 2003-2004. For the forage availability of grasslands, floristic registers and quantitative methods (point quadrat, lineal transects, biomass harvest) were used. The second approach was based on direct observation of the pastoral practices and interviews with the users of the Andean forage resource and the different actors of the transhumance (shepherds, pastures' landowners and the regional livestock service).

Results Table 1 shows the extent of the different types of pastures in the area mapped and their animal use. The mapped area is dominated by strongly sloped land with bare soils and low vegetation cover (<25%). Denser shrub pastures, that constitute 25-30% of the mapped area, are of occasional use due to their low forage quality according to observations and shepherds' interviews. These types of pastures are grazed by goats mainly. The dense grasslands (>50% cover) account for less than 6% of the mapped area and its productivity is presented in Table 2. Access to the Andean pastures is determined by seniority and relationships between landowners and shepherds. In this social model, those who used Chilean Andean pastures before the close of the frontier have now access to the best pastures while those who used Argentine pastures have to accept poorer pastures. In this latter case the shepherds must consider pasture heterogeneity, forage availability and the possibility of access to dense grasslands according to the type of their livestock. Nevertheless, at the end of the season, the dense grasslands are of vital importance as they still remain grazable when others are not.

Table 1 Type and cover of Andean pastures in the Tascadero and El Maitén cordilleras in the Fourth Region.

Cordillera	Total area (ha)	Mapped area (ha)	Pasture types in mapped area (%) grasslands<25% (goats)	grasslands>50% (goats, sheep, cattle)	shrubs<25% (goats)	25%<shrubs<50% (goats)
Tascadero	23127	10292	37.0	3.6	33.3	25.4
El Maitén	3844	3844	41.2	5.7	22.5	30.6

Table 2 Dense grasslands (>50%) productivity (t/ha) in the Tascadero and El Maitén cordilleras.

Cordillera	Dense grasslands types productivity (t/ha) Non-irrigated (goats, sheep)	irrigated (sheep, cattle)	forage production *(Medicago sativa)* (cattle)
Tascadero	3.7 – 5.6	11.1 – 13.8	9.3 – 14.2
El Maitén	5.5 – 7.6	2.6 – 9.1	-

Conclusions The results of this study show that particular attention should be paid to dense herbaceous pastures even though they constitute a small area of the sampled cordilleras. The low-quality shrub pastures do not show a significant increase in their usage while dense grassland appear to be more exploited, especially in critical periods. The transhumance system has to resolve itself because eventually there will be a lack of alternative pastures to relieve the pastoral pressure in the Chilean Andes after the close of the frontier.

Reference

Etienne, M. & C. Prado (1982). Descripción de la Vegetación mediante la Cartografía de Ocupación de Tierras. Ciencias Agrícolas N° 10. Facultad de Ciencias Agrarias, Veterinarias y Forestales, Universidad de Chile. 120 pp.

Australian pasture systems: the perennial compromise

L.W. Bell and M.A. Ewing
CRC for Plant-based Management of Dryland Salinity & School of Plant Biology, The University of Western Australia, 35 Stirling Highway, Crawley WA 6009, Australia, Email: lbell@agric.uwa.edu.au

Keywords: integration, annual pastures, perennial pastures, sustainability, agricultural systems

Introduction Dryland salinity, soil acidification and weed herbicide resistance challenge traditional agricultural production systems in south Australia. The pasture component of such systems rely on annuals like *Trifolium subterraneum* and *Medicago* spp. Replacing annual with perennial pastures allows some redress of the sustainability challenges, but few well-adapted species are available (Ewing & Dolling 2003). A range of perennial species are under evaluation to supplement current options. Some of these new perennial pastures may need modified production systems that allow full expression of their productive potential, especially when integrated with annual crops including cereals, pulses and oil seeds. Integrated systems rely on spatial or temporal segregation of pastures from crops. The necessary characteristics of plants for likely systems are discussed.

Systems Very important factors to design systems into which new species might be embedded are: (1) There is a trade-off between persistence and productivity. In very arid, low productivity environments, new pasture species need to have specific advantages to warrant adoption. (2) New species may vary in plant form (i.e. woody shrubs or trees to herbaceous species). Woody species, though less productive, have deep roots, maintain green leaf area and differ in growth pattern. Their purpose may be only to boost water use and fill gaps in feed supply. (3) Production systems must be flexible enough to respond to temporal changes in profitability between component enterprises. (4) The system must reflect feasible investment and input needs and have enough low risk associated with the technology to encourage wide-scale and rapid uptake.

Phase farming rotates pasture with crops. Pasture phases are flexible and allow farmers to change to crop production when desired. Farmers can integrate perennials into the pasture phase. However, the need to re-establish pastures at the start of each phase means that establishment cost must be low. Species well suited to this system must have a cheap source of seed, high early vigour, compete with weeds and withstand grazing in the year of sowing. Lucerne (*Medicago sativa*) is successful. It can be used to manage dryland salinity by establishing a dry soil buffer below the root depth of annual species to reduce recharge of groundwater tables (Latta *et al*., 2001). It also aids weed management and fixes nitrogen for subsequent crops.

Alley systems are used for woody species when spatial separation of plants is more suitable. This enables annual pastures or crops to be grown between rows of fodder shrubs or trees. Plants used in this system generally are slow to establish but are long-lived. Tagasaste (Chamaecytisus proliferus), Leucaena (Leucaena leucocephala) and Saltbush (Atriplex spp.) are successful. These have deep roots and can access the water table to increase water use of agricultural systems (Lefroy *et al*., 2001). They also improve the continuity of feed supply.

Intercropping (or companion cropping) is where another species is grown amongst a crop. This has been explored little with perennial pasture species. Suitable species would be leguminous, have low winter activity to reduce the competition on the accompanying crop, prostrate habit to avoid contamination of crop products and responsiveness to opportunities outside the crop-growing season. This system may enable sustainability objectives such as high water use to be achieved whilst continuing to produce crops.

Conclusion Integration of perennial pastures can improve sustainability of agricultural systems. Development of new species should address essential considerations. Production systems that integrate perennial pastures may need to evolve as new species are developed.

References

Ewing M.A. & P.J. Dolling (2003). Herbaceous perennial pasture legumes: their role and development in southern Australian farming systems to enhance system stability and profitability. In S. J Bennett. (ed.) New perennial legumes for sustainable agriculture, Univ of Western Australia Press: Perth, 3-14.

Latta R.A., L.J. Blacklow & P.S. Cocks (2001). Comparative soil water, pasture production, and crop yields in phase farming systems with lucerne and annual pasture in Western Australia. *Australian Journal of Agricultural Research,* 52, 295-303.

Lefroy E.C., R.J. Stirzaker & J.S. Pate (2001) The influence of Tagasaste (Chamaecytisus proliferus Link.) trees on the water balance of an alley cropping system on deep sand in south-western Australia. *Australian Journal of Agricultural Research,* 52, 235-246.

The effect of alternative soil amendments on the botanical composition, basal cover, dry matter production and chemical properties of re-vegetated mine land

W.F. Truter and N.F.G. Rethman
The Department of Plant Production and Soil Science, University of Pretoria, Pretoria South Africa 0002.
Email: wayne.truter@up.ac.za

Keywords: acidic soils, infertile soils, fly ash, sewage sludge, soil amendment

Introduction Coal mining impacts large grassland areas of the Mpumalanga Province of South Africa. To mitigate such impacts, it is imperative to restore the once productive soils to the best possible condition. The re-vegetation of mine land presents a particular challenge. Soils being rehabilitated are often acidic and nutrient-deficient, which are major limiting factors in re-vegetation programmes. Conventional methods of liming and inorganic fertilisation have been used to improve the productivity of impacted soils. In the past few years the use of a coal combustion by-product, class F fly ash, and an organic material, such as sewage sludge, have demonstrated the feasibility of using such materials to amend acidic and infertile substrates (Truter, 2002; Norton *et al.*, 1998). The objective of this research was to determine if alternative amendments can create a more sustainable system where botanical composition, basal cover, dry matter production and soil chemical properties can be improved.

Materials and methods A field experiment was established in January 2000 at an opencast coal mine in the Mpumalanga Province. These soils were amended with class F fly ash, a mixture of fly ash and sewage sludge, dolomitic lime and compared to the standard mine treatment (conventional lime and inorganic fertilisers) and a control (no treatment). Soils were re-vegetated with a mixture of Teff (*Eragrostis tef*), Rhodesgrass (*Chloris gayana),* Bermuda grass (*Cynodon dactylon)*, Smutsfinger grass (*Digitaria eriantha)* and lucerne (*Medicago sativa).* Botanical composition, basal cover, dry matter production and soil chemical properties (pH, P, K, Ca and Mg) were monitored seasonally.

Results The percentage basal cover and botanical composition in 2004 is given in Figures 1 and 2. It is evident, from the observations made four years after establishment, that soils receiving a mixture of fly ash and sewage sludge (S) had a higher percentage of Rhodesgrass, and a higher production, whereas the control (no treatment) had a higher plant diversity.

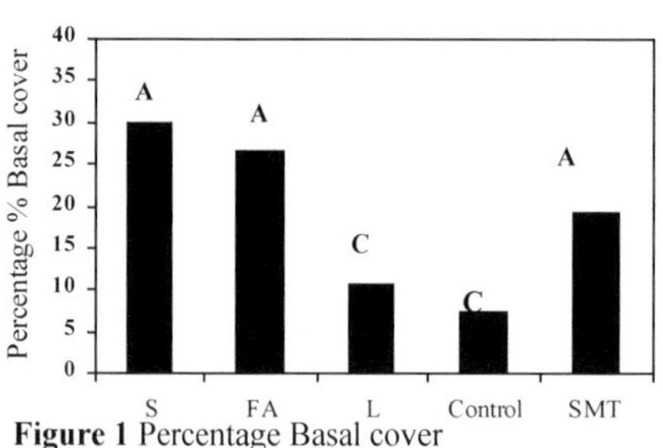

Figure 1 Percentage Basal cover

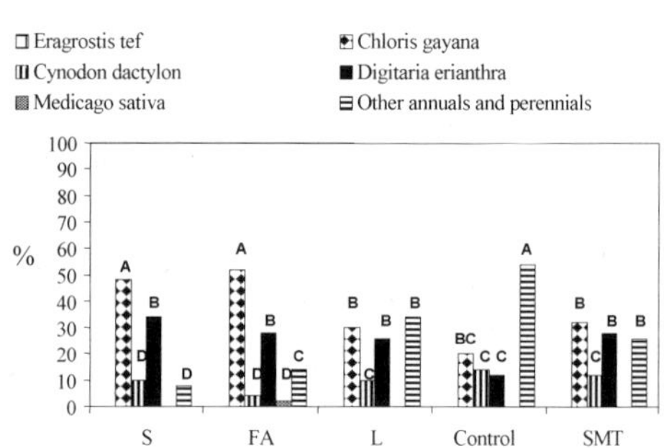

Figure 2 Botanical composition on soils receiving different amendments

*Means with same letter are not significantly different (P> 0.05) Tukey's Studentised Test.

Conclusions Results indicate that alternative ameliorants (fly ash and organic materials) can have a marked beneficial effect, which is still evident in the fifth season, despite no fertiliser having been applied since the first season. This would appear to indicate that such ameliorants produce a more sustainable vegetation than the current practice.

References

Norton,L.D., R. Altiefri & C. Johnston (1998). Co-utilization of by-products for creation of synthetic soil. In: S. Brown, J.S. Angle and L. Jacobs (eds.) Beneficial co-utilization of agricultural, municipal and industrial by-products. Kluwer Academic Publishers, Netherlands, 163-174.

Truter, W. F. (2002). Use of waste products to enhance plant productivity on acidic and infertile substrates. MSc (Agric) Thesis, University of Pretoria, South Africa.

Optimization of the pasture resource in boundary environments as a basis for regional nature management

M.V. Rogova
Institute of Geography SB RAS, Ulan-Batorskaya St House 1, Office 410, Irkutsk 66403, Russia, E-mail: traveller-irk@yandex.ru

Keywords: grazing capacity, optimal nature management, boundary landscapes

Introduction In spite of the globalization processes encompassing all spheres of human life and activity, land remains the main resource and provides the feeding source for population and the fodder base for livestock rearing. On the other hand, the activity of local communities can have important global consequences. The study area that includes Lake Baikal's western shore (East Siberia) and the lake's largest island exemplifies the traditional type of nature management, namely, grazing management which was originated by an indigenous population within the context of suitable natural climatic conditions. This investigation furnished an opportunity to make an assessment of the status of this sector and of the district's ecological situation, as well as to propose an optimal nature management scheme.

Materials and methods The work reported here was done using field investigations, documentary and archival materials from the local district administration, photographic and cartographic data of long-term ecological monitoring, as well as the data from case interviews with specialists and representatives of local residents. A method of comparing the above-mentioned materials was used to analyze the district's nature management process and to predict pasture management changes. Based on the traditional regional land use as well as on a careful study of the natural-climatic component, recommendations were formulated with regards to the scheme of eco-friendly nature management practices.
Contact zones of several types of landscapes make for sustainability of the territory's nature management by providing resources of different types. The study district combines steppe, forest and transitional communities (Ryabtsev, 2003). Geographical location determines the main, historically established, types of nature management, fishery, grazing animal husbandry, hunting and forest utilization. Thus the combination of these resources constitutes the district's ethos as discussed by Ragulina (2004). Not only did the nomads and cattle-breeders determine the culture and traditional types of economic relation but they were also responsible for the state of environment..

Results Investigations revealed that the years of human presence in the Prebaikalia have seen an intensification of the processes of steppe formation. However, the highest index of anthropogenic stress took place in the 20th century when in the 15,900 km^2 of the district, the number of livestock reaches 60,000 or more (Kuznetsov *et al.*, 2003). The intervening time period has shown that the neglect of the natural climatic factor has led to regression of pastoral lands and to a disturbance of plant communities and soil cover integrity.

Conclusions The above-stated challenges, together with the economic crisis that emerged in the 1990s, have dictated the need to search for concepts of optimal nature management for the territory of Baikal's western shore area. The above-mentioned geographical, historical and ecological factors of influence on boundary landscapes should be taken into account when devising a relevant concept. A key objective of this study was to carry out a detailed calculation of the grazing capacity for this territory and, as the result of the investigation, withdrawing the particular pastoral areas from exploitation for 10-15 years. Such a time-span is required for the re-establishment of soil and vegetation cover under Southern Siberia conditions.

References

Ryabtsev, V.V. (2003). Vascular Plants of the Pribaikalsky National Park. Collection of Scientific Papers, Irkutsk: Izd-vo «Oblmashinform», 112 pp. (in Russian)

Ragulina M.V. (2004). Cultural geography: theory, methods and regional synthesis. IG RAS SB Publishers, Irkutsk. 172 pp. (in Russian)

Kuznetsov M.A., A .P. Suhodolov, N.M. Sysoeva and D.U. Fedotov (2003). The scheme of development and location of the Irkutsk region's productive forces into the year 2005, Irkutsk,, pp. 178-186. (in Russian)

Grazing, biodiversity and pastoral vegetation in the South Sudanien area of Burkina Faso

E. Botoni-Liehoun[1] and P. Daget[2]
[1]*INERA Farako-Bâ, BP 910 Bobo Dioulasso (Burkina Faso), Email: edwigebot@hotmail.com, Cirad-emvt, TA 30/E, campus international de Baillarguet 34398 Montpellier cedex5, (France)*

Keywords: biodiversity, grazing pressure, *Isoberlinia doka* forest

Introduction Grazing impact on plant diversity is dominated by two contradictory views. In some studies, it has been found to lead to an increase in diversity and in other studies to a decrease associated with dominance of a few species (Nösberger *et al.*, 1998, Hiernaux, 1998). In an *Isoberlinia doka* forest ecosystem, considered as the climax vegetation in the South Sudanien area of Burkina Faso, a study was carried out to assess the impact of grazing on the diversity of herbaceous species. The *Isoberlinia doka* forest is one type of South Sudanien savanna. The woody stratum is open and allowed development of a continuous stratum of graminae dominated by Andropogonea such *Andropogon ascinodis and Hyparrhenia spp.* .

Materials and methods Seven sites (4 m X4 m) had been protected in three areas which had been submitted to three levels of grazing pressure according to the duration and the season of grazing:
Level 0: No grazing pressure. Two sites had been surveyed in a protected forest
Level 1: Low grazing pressure. Two sites had been also surveyed in a pastoral area. This unoccupied area had been managed only for pastoral use since 2001. It received cattle from May to February.
Level 2: High grazing pressure. Three sites had been surveyed in the village of Torokoro which is submitted to silvopastoral pressure
Individual animals pressure cannot be identified because of common use of the pastureland. Measurements of floristic richness, forage production and forage quality (pastoral value) were made according to Daget & Poisssonet (1972).

Results The higher the grazing pressure the greater was the floristic richness of herbaceous plants. However, the added species were unpalatable (e.g. *Spermacoce* spp. and *Indifofera* spp.). Species diversity, measured by the Shannon index, was higher in grazed than in the ungrazed vegetation. Grazing allowed other species to alter the balance of the native grasses such as *Andropogon ascinodis*, *Andropogon shirensis*, *Schyzachyrium sanguineum* and *Hyparrhenia* spp. Forage production and its quality was lower when plant biodiversity (floristic richness, specific diversity) increased.

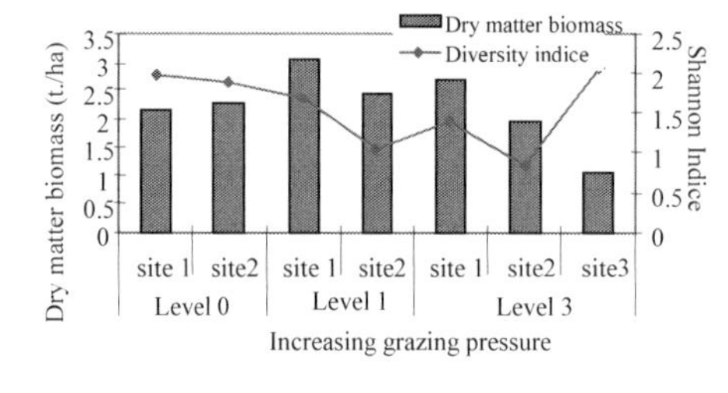

Figure 4 Forage production evolution according to grazing pressure

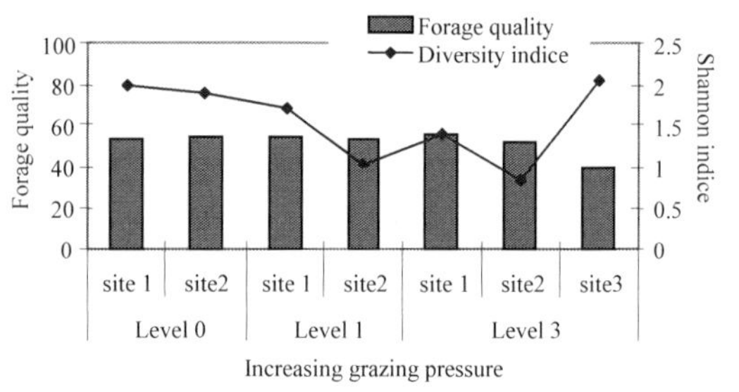

Figure 5 Forage quality according to grazing pressure

Conclusion In the South Sudanien savannah of Burkina Faso, grazing pressure led to increased plant diversity. But this is not favourable to livestock sustainability because of a reduction in forage productivity and its quality. These results show that a high biodiversity is not a good indicator for high productivity of pastoral vegetation.

References

Hiernaux P. (1998). Effects of grazing on plant species composition and spatial distribution in rangelands of the sahel. *Plant Ecology*, 33, 387-399.
Nösberger, J., M. Messerli & C. Carlen (1998). Biodiversity in grassland. *Annales de Zootechnie*, 47, 383-393.
Daget P. & Poissonet J., (1972). Un procédé d'estimation de la valeur pastorale des fourrages. *Fourrages*, 46, 31-39

Effects of landscape structure on plants species richness in small grassland remnants in two different landscapes

S.A.O. Cousins and O. Eriksson
Department of Botany, Stockholm University, 106 91 Stockholm, Sweden, Email: sara.cousins@botan.su.se

Keywords: area, connectivity, GIS, history, land use

Introduction There is an increasing interest in using the landscape as the operational scale in many ecological studies. Current species richness in the landscape may be explained by past land use, and habitats may harbour species favoured by an environment that no longer exists. In this study we have included both a landscape scale and a temporal scale. The objective was to explain species pattern and the effect of isolation, habitat size and surrounding land use, and past land use change, on small grassland remnants in rural landscapes.

Materials and methods Two different landscapes were analysed in south-eastern Sweden, a modern open agricultural landscape (Selaön), with little (5%) semi-natural grassland left and a forested traditional rural landscape (Nynäs), with more (11%) semi-natural grassland left. Both landscapes have a long rural management history. Maps, 100-years-old, and aerial photos from 1950 and 1990 were used in a GIS. Plant species presence and abundance were recorded in 4 m^2-plots, 40 plots along road-verges and in all midfield islets in each landscape. Size of midfield islet and connectivity to other habitats, and roadside vegetation were also analysed.

Results A hundred years ago both landscapes had 60% managed grassland; 50 years ago Selaön had less than 11% grassland left whilst Nynäs had 19%. Table 1 shows the number of species found. Species richness was 54% higher in road verges (R.V.) and 20% higher in midfield islets (M.I.) at Nynäs compared to Selaön. More species were found at Selaön, although with a lower abundance. Landscape structures were more important for species richness and numbers at Selaön, as 60% of the species in midfield islets and 48% in road verges was influenced by area, connectivity, or surrounding vegetation compared to 20 and 22% at Nynäs respectively. Nynäs not having reached a fragmentation threshold or a shorter time-range since the threshold was passed could explain the differences, thus local populations are still remnant. A conceptual model based on the results (Figure 1) illustrates that small grassland remnants are sensitive to extinction and fragmentation effects can be detected in landscapes that have been fragmented for >50 years. Past land use history was not particularly important for present species richness or incidence in small grassland remnants, contrary to other studies in the region (Cousins & Eriksson 2002, Lindborg & Eriksson 2004). If the focal habitat has an internal heterogeneity, which may be the case in these studies, there are local processes that uphold species richness within the habitat compared to small relatively homogenous areas. However, it is important to stress that difference in species incidence and landscape structures between the two landscapes is still a result of land use history.

Table 1 Total number of species and mean species richness (sd)/plot

	Selaön	Nynäs
Total number		
road verges	146	135
midfield islets	237	160
	(173 islets)	(53 islets)
Mean species richness		
road verges	18.8 (±5.8)	29.0 (±6.4)
midfield islets	13.5 (±5.5)	16.2 (±6.4)

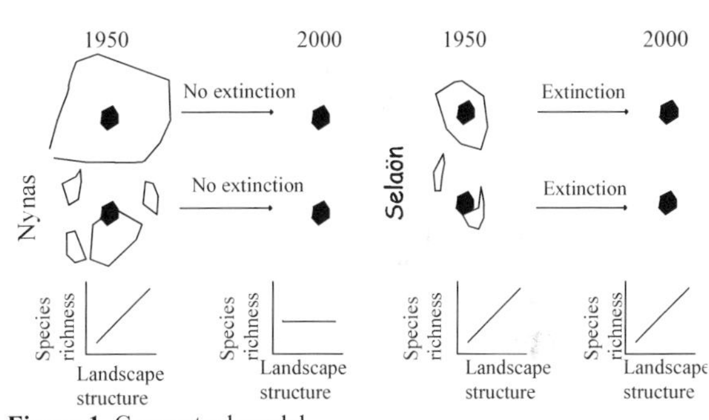

Figure 1 Conceptual model

Conclusions The study demonstrates the importance of conducting studies in different landscapes and not only using grassland "hotspots". The legacy of surrounding landscape remains in local species pools for at least 50 years. Small grassland remnants are more sensitive to fragmentation effects compared to larger grasslands and can thus be useful in finding fragmentation thresholds for plants. They also encompass a substantial part of the grassland species pool and can be valuable to include in reconstructions of grassland management.

References

Cousins S.A.O. & O. Eriksson. (2002). The influence of management history and habitat on plant species richness in a rural hemiboreal landscape, Sweden. *Landscape Ecology,* 17, 517-529.

Lindborg, R. & O. Eriksson (2004). Historical landscape connectivity affects present plant species diversity. *Ecology,* 85, 1840-1845.

Is biodiversity declining in the traditional haymeadows of Skye and Lochalsh, Scotland?

G.E.D. Tiley[1] and D.G.L. Jones[2]
[1]*Scottish Agricultural College (SAC), Auchincruive, Ayr, KA6 5HW, UK, Email: Karen.Crighton@sac.co.uk,*
[2]*SAC Advisory Office, Somerled Square, Portree, Isle of Skye, IV51 9EH, UK*

Keywords: biodiversity, haymeadows, management

Introduction Species-rich haymeadows have developed on crofts in the Isle of Skye and Lochalsh Districts of north-west Scotland as a result of a century or more of traditional land use. This has involved long rotations of late cutting for hay with aftermath grazing by cattle and short breaks for cropping. The traditional haymeadows are increasingly coming under threat from changes taking place in the countryside. A survey of the main haymeadows still remaining in Skye and Lochalsh was carried out during 2003 to assess the current botanical composition, management and conservation value, and to compare with earlier surveys.

Materials and methods Grassland sites (31 in Skye; 18 in Lochalsh) were recorded, including several surveyed earlier (Orange, 1987; Hutcheon, 1997). The presence and estimated abundance of plant species were recorded as Dafor values for communities and Domin scores in 2m x 2m quadrats. Croft owners or managers provided data on present, past and intended future managements. Based on floristic composition, uniformity, stability and area of the constituent communities, each site was assigned a subjective conservation value.

Results Conservation values of 5/13 sites on Skye, recorded in the Orange (1987) survey, have deteriorated radically due to neglect, invasion by *Juncus effusus* and heavy grazing by sheep. The other 8/13 sites showed a sharp reduction in the occurrence of orchid species and of globeflower (*Trollius europaeus* L.). As compared with the survey by Hutcheon (1997), deterioration of conservation values was similar on sites in Lochalsh. Table 1 shows the frequencies of current managements observed during the 2003 survey.

Table 1 Haymeadow management observed

Management	No of sites (%)
Not cut or grazed	18
Cut only	8
Cut and grazed	34
Grazed only	40
Of which % grazed	
By cattle	47
By sheep	28
By cattle and sheep	25

Discussion and conclusions The widespread traditional croft management that led to the development of haymeadows has only a tenuous presence on a few scattered crofts in Skye and Lochalsh. Older crofters still make hay but big bale silage is the norm now, mainly because it allows for flexibile management to cope with the prevailing weather. Other threats to the maintenance of biodiversity are replacement of cattle grazing by intensive sheep, invasion of rushes and changes in social structure leading to decrofting. Croft managers may require incentives to encourage them to continue to use traditional management systems.

Acknowledgement In conjunction with Scottish Natural Heritage, the Highland Regional Council Biodiversity Project commissioned the 2003 survey of haymeadows in Skye and Lochalsh.

References

Hutcheon,K.(1997). Vegetation survey of Drumbuie, Kyle of Lochalsh. National Trust for Scotland.
Orange,A.(1987). A survey of haymeadows and associated grasslands in Skye, Ardnamuchan, Sunart and Lochaber. Nature Conservancy Council, Edinburgh.

Forage Development in the Nepal mid-hills: new perspectives

A.D. Robertson
"Oaky Creek" Wilson's Downfall, MS 1983, Stanthorpe 4380, Australia, Email: halfmoon@halenet.com.au

Keywords: groups, participatory, quick start, Nepal

Introduction Nepali hill farming communities are typically poor and remote, and are currently severely affected by conflict. The challenge is to define simple approaches which can generate results within this context. Livestock are central to livelihoods and to the sustainability of farming, with rain-fed agriculture dependent on inputs of manure-based compost. Stall feeding has increased dramatically with the adoption of community forestry and general preclusion of grazing. A broad landscape approach to forage development is increasingly being adopted, with concurrent on-farm interventions, such as intercropping and back-yard forage, and off-farm interventions, such as landslide stabilisation with forages, development of forest understory, and reinforcement of degraded grazing areas with forage. Considerable work has been undertaken in the mid-hills (below 1800m ASL) with very limited higher altitude programs to 4,000m ASL. Until the late 1990s forage development was restricted to use of a very narrow array of genetic material. To accommodate the agro-ecological diversity, broad mixtures are now commonly used, encompassing species with known potential locally, and some peripheral commercial and pre-release material for testing to refine recommendations. Productive erect cut-and-carry grasses including Mott Napier (*Pennisetum purpureum* cv Mott) are popular. A suite of legumes including *Stylosanthes.guianensis*, *Chamaecrista rotundifolia* cv Wynn, *Aeschynomene americana* cv Glenn, *Aeschynomene villosa*, *Neonotonia wightii*, forage arachis (*Arachis pintoi)*, and *Leucaena leucocephala* have been successful in various niches.

Scale and benefits Since the late 1990s, it is estimated that at least 40,000 households have been involved in the adoption of the newly introduced genetic material. There are now more than 20 new species actively promoted. Additionally, more than 35,000 school students became involved in only two districts during 2004. Adoption rates of the forage packages are as high as 80%, with preliminary "social mobilization" unnecessary. Farmer-farmer exchanges have been encouraging. Most forage from on-farm and cultivated communal plantings is used in cut-and-carry systems for feeding goats and dairy and draft bovines, with increasing use also for poultry and swine. On-farm forage development has dramatically reduced labour requirements for forage collection. Farmers report benefits from forage legume introduction to crop areas, in terms of stabilizing crop production. Landslide stabilisation from direct seeding has been successful on many sites.

Major reasons for success

Farmer attitudes and farmer groups Farmers perceive the lack of good quality forage to be a major constraint on livestock productivity. Women commonly spend more than four hours per day on fodder collection; they welcome any intervention to reduce this burden. Traditional involvement of the majority of households in milk production for home consumption or sales is considered to be a major factor in achieving higher adoption rates than could be expected, for example, in South-east Asia or most of Africa. Nepali hill farmers are commonly coordinated into focus groups, such as livestock groups, community forest user groups (of which there are now more than 10,000) and various women's empowerment groups; this presents exceptional opportunities for intervention and efficient delivery of technology, with high rates of farmer-farmer adoption locally.

Farmers and participatory research The Nepali Government capacity for conventional forage research is negligible in the context of the vast agro-ecological diversity. Hence large numbers of widely scattered farmers and farmer groups are now directly involved in screening new development strategies and genetic material. Experience has shown the necessity of including simple and reliable strategies (such as back-yard or terrace-riser forage) and some conspicuous and reliable genetic material including Mott napier and forage arachis. Such species have had a high rate of spontaneous lateral adoption locally, although technology transfer over larger distances has typically been slow; this reinforces the importance of initiating work at many widely scattered sites.

The future Remoteness of communities and the current conflict preclude regular visitation. Simple and flexible technical packages and technology delivery mechanisms, with the capacity to provide quick and conspicuous results for poor and remote communities, are central to success; recent programmes have demonstrated the potential for reaching large numbers of poor hill farmers. It is now necessary to maintain access to improved genetic material, to improve the supply of good quality seed, to involve more development agencies and community based organisations in delivery of the technology, to trial technologies in new environments, and to streamline adoption by facilitating exchange visits for farmers from new areas.

Forage Arachis in Nepal: a simple success

A.D. Robertson
"Oaky Creek" Wilson's Downfall, MS 1983, Stanthorpe 4380, Australia, Email: halfmoon@halenet.com.au

Keywords: adoption, participatory, farming systems, Nepal

Introduction Nepali farming systems are remarkably diverse. Livestock play a central role in livelihoods and sustainable farming on most farms. There is a need for productive forage legumes that can fit existing farming patterns and that can be multiplied easily. A wide array of genetic material has been introduced recently into the cropping, cut-and-carry, grazing, and forestry systems, mainly in the Terai (Ganges Plain) and in the "mid-hills" to about 2km ASL. In 1999/2000, 8 lines of *Arachis pintoi* were introduced from CIAT, and additional *A. pintoi* and *A. glabrata* lines from Queensland. The introduced arachis was established on a small number of permanent sites to enable close observation, crossing and continual selection, and a reliable long-term supply of planting material. Concurrently, small samples were provided to a large number of smallholder farmers (>1000 in the first season alone) over a very diverse agro-ecological range, for evaluation, local demonstration, and the supply of planting material within the community. The programme has been based entirely on vegetative material since 2000. Most arachis establishment has been in intensive smallholder systems, involving cover cropping, mixed planting with productive grasses in back-yard areas, and establishment on terrace risers. There also have been trial plantings in ley systems and on communal land, including land-slips/slides and roadside cuttings.

Scale and benefits Forage arachis has become popularised in many communities; it is likely that 15-20K households are already involved. Farmers continue to expand on-farm areas and refine systems of management and utilisation. Some households have bulked up from a few slips to 1000-2000m^2 within 3 years. The arachis is used commonly in cut-and-carry systems for supplementary feeding of milking buffalo and cattle, goats, pigs, and poultry. Due to positive feeding responses, rapid expansion in plantings is occurring within individual farms. Establishment on communal areas, such as roadside cuttings, has been on a smaller scale and is much less significant.

Major reasons for success

Farming systems and bulking-up Most Nepali mid-hill farming systems are very intensive and offer many niches for forage arachis. Small farm sizes, and the small unit areas, initially targeted for on-farm forage development, are well suited to vegetative propagation. Nepali farmers generally prefer this to the use of seed, partly because of the quick and conspicuous results. Some bulking-up has been undertaken on a contract basis.

Farmer attitudes New forage interventions in Nepal typically have very high uptake rates. This may be attributed partly to the long tradition of back-yard dairying that most households practise. Participating farmers commonly undertake their own screening, feeding new material to all classes of livestock and poultry. Forage arachis has stimulated more interest than most other forage introductions.

Groups and networks Nepali farming communities are characterised by a high degree of organisation into enterprise groups, including livestock, milk marketing, and forest user groups. These groups facilitate rapid farmer-to-farmer exchange of planting material. Providing the forage arachis to selected farmers, who act as "farmer resource centres" for screening, demonstration, bulking up and local supply of cuttings to other group members, has been successful. This approach is important in the context of reaching remote communities, particularly in a conflict situation that precludes regular follow-up.

Participatory approaches The immediate and direct involvement of smallholder farmers have streamlined screening, demonstration and multiplication. Development agencies have coordinated initial inter- and intra-district exposure visits and distribution of planting material to newly participating areas.

Mobilisation of field staff Field development workers from diverse disciplines, including forestry, soil conservation, livestock, and women's empowerment, have been mobilised to assist in the delivery of the programme. This multi-disciplinary approach has enabled quicker and more widespread adoption.

The future Further establishment of scattered nucleus sites in new areas in the Terai and lower mid hills could result in secondary adoption of forage arachis by vast numbers of farmers at very low cost. The current range of genetic material should be expanded, particularly in terms of introducing more cold tolerance and more erect lines for cut-and-carry.

Grazing prohibition programme and sustainable development of grassland in China

X.Y. Hou and L. Yang
Department of Research Management, Chinese Academy of Agricultural Sciences. Zhongguancun Nandajie 12, Beijing, 100081,China, Email: houxy16@caas.net.cn

Keywords: grazing prohibition programme, grassland sustainable development

Introduction Prohibition of grazing is now the main grassland management measure in China. From 1999, prohibition of grazing has been implemented on a trial basis in some areas. From 2001, the grazing prohibition programme (GPP) has been carried out in five provinces (Shaanxi, Gansu, Hebei, Jilin and Yunnan) and two autonomous regions (Inner Mongolia and Ningxia), with the objective of protecting and restoring grassland by seasonal or yearly banning of grazing with subsidiary assistances. The area within which grazing was prohibited of 2.93×10^7 ha in 2001 was increased to over 3.33×10^7 ha in 2004. With a view to improving the GPP and ecological reconstruction, we conducted a survey in some counties to review the relationship between GPP and the sustainable development of grassland.

Methods Six sampled counties (Chinba'erhuzuo county, Ewenke county, Kerqin youyizhong county, Hangjin county, Etuoke county, Wulatehou county) were in the Inner Mongolia autonomous region, two sampled counties (Gangcha county and Haiyan county) were in Qinghai province, and one sampled county (Songpan county) was in Sichuan province. One hundred and sixty-three households, 42 officers in the rural and pasture area and 11 officers in counties were interviewed individually with a survey questionnaire used during the interviews.

Results From 1998, the methods of feeding livestock from grassland in China began to change. The ratio of households which herded livestock dropped from 73.8% in 1998 to 65.9% in 2002, and the percentage of households which fed livestock indoors rose from 4.3% in 1998 to 15.2% in 2002. Some 62.4% of surveyed herders agreed with GPP, while only 22.4% disagreed (Table 1). Economic concerns were the main barrier in the course of promoting GPP. The survey showed that just 8.5% of herders had benefited from GPP, while 27.9% saw losses. Meanwhile, 72.5% of interviewed officers believed GPP will decrease the benefit to herders in the long term and 94.0% of interviewed herders believed that they should receive subsidies because of GPP. More than half of the surveyed herders proposed that subsidies of 200 Yuan/ ha per year would be acceptable. The survey of herders showed that several factors presented difficulties in promoting GPP. These included shortage of starting capital (85.7%), lack of forage (66.7%), absence of technical guidance (42.9%) and traditional concepts (42.9%).

Table 1 Some survey data relating to the grazing prohibition programme

Attitude toward GPP	Percent	Critical factors in GPP	Ranking	Subsidies required (Yuan/ ha per yr)	Percent
Agree	62.4%	Overpopulation	1	50	31.0%
Disagree	22.4%	Risk resist intention	2	100	18.0%
Neutral	9.7%	Lack of forage	3	200	51.0%

Conclusions An increase in costs, a decrease of benefits and a shortage of starting capital should affect implementation of GPP. Traditional grazing systems, which lack modern scientific approaches and techniques cannot support sustainable development of grassland in China any more. So it is suggested that: firstly, government should provide practical financial aids for herders to solve the current difficulties in this special transitional period; secondly, technical and scientific assistances should be provided in order to transform the traditional grazing system; for example, construction of basic facilities, subsidised loans, scientific guidance on grazing and suitable livestock species for indoor feeding

Hedgerow systems and livestock in Philippine grasslands: GHG emissions

D.B. Magcale-Macandog, E. Abucay, R.G. Visco, R.N. Miole, E.L. Abas, G.M. Comajig and A.D. Calub
Institute of Biological Science, University of the Philippines Los Baños, College, Laguna, Philippines, Email: macandog@pacific.net.ph

Keywords: hedgerow system, livestock, N_2O emission, methane emission, GHG emissions

Introduction Hedgerow systems are widely adopted in the smallholder farms in the sloping grassland areas of Claveria, Mindanao, Philippines. The system is effective in addressing soil erosion problems and in conserving the topsoil. *Gmelina arborea* and *Eucalyptus deglupta* are two fast-growing timber species that are planted in hedgerow systems while maize is planted in the alley areas in between the hedgerows. Livestock holdings are widespread in Claveria, with 74% of the households having livestock. Cattle and carabao are the most common livestock in smallholder farms providing draught power for land preparation and transportation. In hedgerow systems, fodder tree leaves and crop residues are fed to livestock, while animal manure is added to the soil. Thus, these systems may serve as both a source and sink of methane and nitrogen oxides, depending on the management practices and component trees and crops of the system. This study aims to estimate methane emissions from livestock holdings and nitrogen oxide emissions through fertilization, tree litterfall and decomposition, maize residue incorporation and livestock manure from *G. arborea* and *E. deglupta* hedgerow systems.

Materials and methods Experimental plots were established in 1 and 7-yr old *E. deglupta*- and *G. arborea*-hedgerow systems with maize planted in the alley areas. The treatments are different combinations of tree species, tree age, and tree spacing. Inorganic N and P fertilizer, and maize crop residues were applied in the maize crop. Maize biomass, grain yield, tree litterfall and leaf litter decomposition were measured. A survey of 300 households in Claveria was conducted to gather information on livestock holding and management.

Results The major sources of N inputs in the different hedgerow systems are the maize crop residues (FCR) and synthetic nitrogen fertilizer (FSN) (Fig.1). Other sources include animal manure (FAW) and tree leaf litter. Since the average animal holding is quite small, nitrogen input from animal waste is small. Direct soil N_2O emissions from the plots range from 2.11 to 5.17 kg N $ha^{-1}yr^{-1}$. Direct soil N_2O emissions from 1-year old hedgerow systems are significantly higher than emissions from 7-year old hedgerow system. Local values for N excretion from cattle and carabao were 12.3 kg and 14.2 kg, respectively; much lower than the default values of 40 kg for both non-dairy cattle and carabao given by IPCC (1997). Enteric fermentation of cattle and carabao (11,352 kg and 3,410 kg, respectively) and swine manure management (2,786 kg) were the main sources of CH_4 emissions from livestock holdings in 300 Claveria households (Table 1).

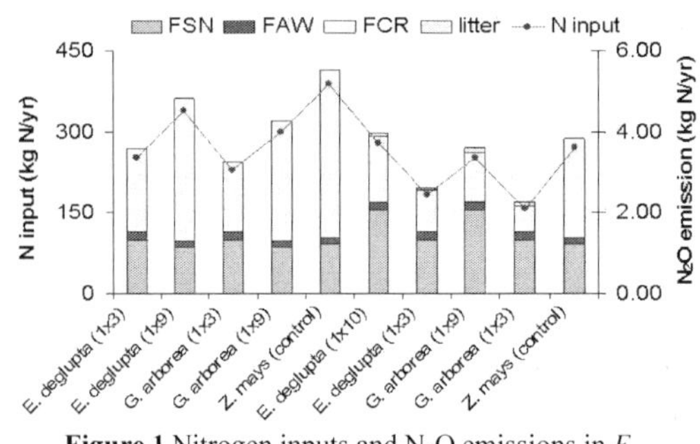

Figure 1 Nitrogen inputs and N_2O emissions in *E. deglupta* and *G. arborea* hedgerow systems

Table 1 Methane (CH4) emissions from enteric fermentation and manure management per animal type

Livestock Type	Number of animals	CH_4 emission (kg)			
		EF	Enteric Fermentation	EF	Manure Management
Non-dairy cattle	258	44	11,352	2	516
Carabao	62	55	3,410	3	186
Goat	46	5	230	0.22	10.12
Swine	398	1.5	597	7	2,786
Poultry	1252	-	-	0.023	28.8
Total			15,589		3,526.92

Conclusions In tree-based hedgerow systems, crop residue incorporation and fertilizer application are the major sources of nitrogen inputs. Direct soil N_2O emissions from the plots range from 2.11 to 5.17 (kg N yr^{-1}) with significant N_2O emissions in 1-year old hedgerow systems than 7-year old hedgerow system. Use of local values for N excretion factors will reduce uncertainties in the estimates of N excretion from animal manure. Enteric fermentation of cattle and carabao and swine manure management were the main sources of CH_4 emissions from livestock holdings in 300 Claveria households.

Reference

IPCC. (1997). The Revised 1996 Guidelines for National Greenhouse Gas Inventories (Workbook and Reference Manuals). Intergovernmental Panel on climate Change. OECD. Paris, France.

Agroforestry systems in Cuba: some aspects of animal production

J.M. Iglesias, L. Simón, L. Lamela, I. Hernández, M. Milera and T. Sánchez
Estación Experimental de Pastos y Forrajes "Indio Hatuey ", CP 44280, Matanzas, Cuba Email: iglesias@indio.atenas.inf.cu

Keywords: agroforestry systems, animal production

Introduction The silvopastoral systems, that nowadays constitute scientific achievements of the Grasses and Forages Research Station "Indio Hatuey ", have been developed from the results of investigations that were carried out since the 1980s, to improve the productivity of natural pastures through the introduction of valuable herbaceous species and tree legumes. Those investigations also determined the essential elements of pasture management such as the optimal stocking rates for low input systems and suitable methods of grazing to obtain sustainability of grasslands.

Material and Methods Among the diverse types of Silvopastoral systems under study, the protein banks and multiple associations of legumes and grasses have contributed much to the development of sustainable dairy and meat production, and could be considered as systems that can be extended and to the farmers and that integrate well with the production objectives of Cuban cattle production.

Results *Leucaena leucocephala* has been the most frequently used tree in Cuban silvopastoral systems and it has also contributed much to experimental data that demonstrate the real advantages of agroforestry (Table 1). However, it is not the only species used. Others such as *Albizia lebbeck, Erythrina berteroana, E. poeppigiana, Gliricidia sepium, Bauhinia purpurea* and *Morus alba*, have been tested with success and appear to be important elements of diversification of plant communities in silvopastoral systems in Cuba.

Table1 Effect of different silvopastoral systems with low external inputs on performance of young fattening bulls

Production System	Genotype	Accumulated gain (g/day)	Live weight at slaughter (kg)	Age (months)
Association of P. maximum with L. leucocephala	Zebu	621,8	413,7	24
Association of P. maximum with L. leucocephala	1/2 H x 1/2 C	525,6	376,3	26
Association of P. maximum with L. leucocephala	5/8 H x 3/8 C	491,6	357,1	28
Protein Bank of L. leucocephala (25 % of the total area) + Natural pasture	Zebu	394,0	355,0	24
Protein Bank of L. leucocephala (25 %) + P. maximum (80 kg of N)	Zebu	555,0	372,5	25
Protein Bank of L. leucocephala (25 % of the total area) + A. gayanus	Zebu	487,0	449,0	29
Association of P. maximum with Albizia lebbeck	Zebu	729,0	409.2	24
Association of P. maximum with L. leucocephala	Zebu	788,0	424,0	24

Conclusion The main results obtained on the use of agroforestry for animal production in Cuba are 1. Daily live weight gains of between 500 and 600 g in young bulls for fattening, with an average production of around 800 kg of meat per ha annually. 2. Daily milk production of 7-10 kg/cow (14-25 kg/ha), without supplements. 3. Daily live weight gains of between 400-525 g in growing replacement heifers, which allows a live weight for reproduction of 290-300 kg at 20-27 months of age. 4. Minimal use of external inputs to the system

Optimising forage production on degraded lands in the dry tropics through silvopastoral systems

P.S. Pathak
Indian Grassland and Fodder Research Institute, Jhansi 284003, India, Email: pathak36@yahoo.com

Keywords: silvopastoral system, degraded land, livestock production, ecological restoration

Introduction In India, 187 M ha out of a total area of 328 M ha face the problem of land degradation, mostly due to water and wind erosion. The problems are aggravated by poor land cover and increasing pressure of human and livestock populations. There is over-exploitation of the scarce resources of forage and firewood. Several techniques, including watershed based silvopastoral land use have been proposed (Patil & Pathak, 1977). Tree, grass and legume based systems have been tried after land treatment to reduce runoff and soil loss while meeting the forage needs of the livestock and firewood for cooking in many studies (Debroy & Pathak, 1983). Results of an operational research project on silvopastoral systems are reported in this paper.

Materials and methods Watershed-based land treatments such as staggered contour trenches, bunds, diversion drains, gabion structures were coupled with an appropriate silvopastoral model to rehabilitate four different types of degraded lands: degraded revenue lands and forests, salt-affected ravines and undulating terrains around Jhansi (central India). The experiment involved the tree species *Acacia tortilis, Albizia amara, A. lebbeck, Hardwickia binata* and *Leucaena leucocephala* planted at three spacings in associations with *Cenchrus ciliaris, Chrysopogon fulvus, C. setigerus, Dichanthium annulatum* and *Stylosanthes hamata, Macroptilium atropurpureum* on the degraded lands. Control plots were not given any soil conservation treatments, except protection from grazing. Initial application of fertiliser was 40 kg N, 40 kg P and 20 kg K/ha followed by 20 kg N/ha every year. Use of leguminous nurse crops seeded on the mounds on the side of trenches to assist better tree growth and edaphic enrichments, controlled grazing by livestock for testing the value of biomass produced and bio-economic modelling were attempted to evaluate the fitness of different interventions (Pathak *et al.*, 1995).

Results The treatment with 600 contour trenches (3x0.4x0.5 m) per ha followed by seeding of grasses and legumes and tree planting (4x4 m) reduced soil loss from 17.8 to 1.26 t/ha/yr in 4 years. It also increased species richness and biodiversity of the natural vegetation. There were also improvements in the physical, chemical and biological properties of the soil. The production of herbaceous and woody biomass of less than 1 t/ha/yr was increased to 10 t/ha/yr with a 10-year rotation. There was a 7-fold improvement in crude protein. The system was evaluated by a mixed herd of cattle, sheep and goats grazing year-round in a deferred rotational system at the equivalent of 1 adult cattle unit per ha stocking rate after 7 years of establishment of the system. During the monsoon period the heifers grew 500 g/head/day without any supplemental feeding. During the remaining part of the year the animals gained 250-350 g/head/day after supplementation of concentrates and tree leaves. The models were able to predict the harvest cycle along with the yield to assist land managers. The benefit/cost ratio and the internal rate of return (IRR) from this project was 1.42 and 18 % respectively, ensuring the possibility of getting support from banks for the rehabilitation of dry degraded lands. Based on these studies, the species have been identified for scaling up forage and firewood production from different types of degraded lands in India (Pathak *et al.*, 1995). Currently the Government is allocating >20 M ha degraded forest to the landless under Joint Forest Management, where this technology has a great scope to assure environmental conservation together with forage and firewood supply. It is assumed that if 50 % of the degraded lands are allocated to this technology, it will be possible to meet the deficits of fodder in the country (Pathak & Roy, 1994).

Conclusions Silvopastoral systems of degraded land management assured conservation of natural resources along with supply of fodder, grazing and firewood in addition to the environmental amelioration. Grazing livestock within the carrying capacity produced high levels of individual livestock production. Such projects have been found economically viable and environmentally sound. It has a great applicability as a technology.

References

Deb Roy , R. & P.S. Pathak. (1983). Silvipastoral research and development in India. *Indian Review of Life Sciences*. 3, 247-264.

Pathak, P.S. & M.M. Roy (1994). Silvipastoral system of production, *Bulletin*, IGFRI, Jhansi. pp. 55.

Pathak, P.S., S.K. Gupta & P. Singh (1995). IGFRI approaches: Rehabilitation of degraded lands : *Bulletin*, IDRC - IGFRI, Jhansi. pp. 23.

Patil,B.D. & P.S.Pathak (1977) Energy plantation and silvipastoral systems for rural areas. *Invention Intelligence*, 12(1-2), 79-87.

How to simplify tools for natural grassland characterisation based on biological measures without losing too much information?

P. Ansquer, P. Cruz, J.P. Theau, E. Lecloux and M. Duru
UMR 1248 ARCHE, INRA -ENSAT, Chemin de Borde rouge, BP 27, 31326 Castanet-Tolosan Cedex, France, Email: ansquer@toulouse.inra.fr

Keywords: natural grasslands, biological traits, tools, management

Introduction In marginal areas, such as the Pyrenees, natural grasslands are the only available resource for livestock feeding. Despite this, there is a lack of simple and efficient tools for advisers to aid the management of the complex vegetation of these grasslands. Therefore, we tested an approach derived from functional ecology, to construct such tools: using biological traits to inform on the agronomic characteristics and the way farmers' practices act on them (Ansquer *et al.*, 2004). Nevertheless, the required protocol of measurement is still time-consuming and difficult. In this paper, we test different ways of simplifying this protocol by reducing the number of species measured and not considering specific abundances.

Materials and methods The experimental design, located in the French Pyrenees (Ariège), consisted in 6 "treatments" derived from two factors corresponding to the farmers' practices: the fertility level (N and P) and the intensity of mowing and/or grazing. Three biological traits were measured: height of the plant, specific leaf area (SLA) and leaf dry matter content (LDMC). Measurements were made at the vegetative stage, as described by Cornelissen *et al.* (2003), on each species contributing at least 80% of the grassland standing biomass. The first level of simplification tested corresponds to comparisons between traits values calculated at 3 different levels of the grassland community: all the measured species, the grass community and the 2 dominant grass species only. At a second level, comparisons were made between traits values weighted (w) or not (nw) by the specific abundances of the species considered. Spearman rank correlation tests and one-way Anova were used respectively to compare the different levels of simplification and the capacity of traits to separate the treatments.

Results Table 1 shows that the weighted trait values obtained at the whole community level are well correlated with those calculated using the grass community and both dominant grass species. The order of SLA values is the best conserved. The weighted and not weighted values are also well correlated at the 2 "grass levels" (Table 2); no correlation between w and nw LDMC was obtained for the whole community. Significant differences among treatments for each trait were found, weighting or not by abundances, at the "grass levels" ($p<0.05$) but not for LDMC calculated on all species. These results can be explained by the different ranges of LDMC values between growth forms so mixing different growth forms as for the "all species" value is not relevant.

Table 1 Rank correlations (r and level of significance[#]) between weighted traits values calculated at the grassland level (All species) and on the grass community (Grasses) and between All species and the 2 dominant grass species together (2 dom. Grasses).

Trait	SLA	Height	LDM
All sp. *vs* grasses	1.00***	0.94 **	0.83 *
All sp. *vs* 2 dom. grasses	0.94 *	0.83 *	0.83 *

* $P<0.05$; ** $P<0.01$; *** $P<0.001$; ns, not significant

Table 2 Rank correlations (r and level of significance[#]) between w and nw traits values calculated at the grassland level (All species), on the grass community (Grasses) and on the 2 dominant grass species (2 dom. Grasses).

Trait	SLA	Height	LDMC
All species	1.00 ***	0.94 **	ns
Grasses	0.94 **	1.00 ***	0.94 **
2 dom. grasses	1.00 ***	1.00 ***	0.94 **

Conclusions Our results emphasize the possibility of measuring functional traits of the grass community or even only of the 2 dominant grasses without taking into account the specific abundances. Thus, tools for characterising grasslands, using a simple protocol of trait measurement, can be constructed.

References

Ansquer, P., J.P. Theau, P. Cruz, R. Al Haj Khaled & M. Duru (2004). Caractérisation de la diversité fonctionnelle des prairies à flore complexe: vers la construction d'outils de gestion . *Fourrages,* 179, 353-368.

Cornelissen, J.H.C., S. Lavorel, E. Garnier, S. Diaz, N. Buchmann, D.E. Gurvich, P.B. Reich, H. ter Steege, H.D. Morgan, M.G.A. van der Heijden, J.G. Pausas & H. Poorter (2003). A handbook of protocols for standardised and easy measurement of plant functional traits worldwide. *Australian Journal of Botany*, 51, 335-380.

Cow-calf production on perennial pastures in the central semi-arid region of Argentina

C.A. Frasinelli[1], K. Frigerio[1], J. Martínez Ferrer[2] and J.H. Veneciano[1]
[1]*INTA San Luis, CC N° 17 (5730) Villa Mercedes, San Luis, Argentina, Email: cfrasinelli@sanluis.inta.gov.ar, and* [2]*INTA Manfredi, Córdoba, Argentina*

Keywords: cow-calf production, *Digitaria eriantha*

Introduction In cow-calf production systems in San Luis, the annual crops are heavily used (eastern region) or the natural pastures are over-used (western region). Both cases constitute unsustainable systems. In the present study, the possibility of structuring an efficient and stable stockbreeding agroecosystem based exclusively on perennial summer grasses was tested. The objectives were: 1) to test a cow-calf production system based on perennial pastures: Digitaria (*Digitaria eriantha* Steudel subsp. *eriantha* cv Irene), without protein supplement during winter, and 2) to determine the physical and economical efficiency of the production system and sustainability of such indicators in the short-term (6 years of performance).

Material and methods The experiment was conducted in Villa Mercedes, San Luis, Argentina (33° 39' S, 65° 22' W) and at an altitude of 515 m. This region has a dry continental climate. The soil is classified as Ustic Torripsament. The system was tested on an experimental area of 90 ha of only one perennial summer grass, Digitaria, using information about its production, quality and management from the region. The pasture was divided into 2 plots to allow rotational grazing. The cow-calf herd was composed of the Hereford breed. In both cows and heifers, live weight (LW) and body condition score (BCS) were recorded. Birth date and LW of calves at birth, 60 days and at weaning were recorded.

Results Table 1 shows the cows' LW and BCS throughout the year. By the end of the summer they had the highest LW and BCS. It is necessary to achieve this state of body reserves (BCS=3.4) to survive the winter period without intake of a protein supplement, as the nutritional value of the pasture is quite low (crude protein and neutral-detergent fibre concentrations of 40 and 750 g/kg DM respectively: in vitro dry matter digestibility of 0.34). During this period, the cows lost 4.1% of LW and 11.7% of BCS. Although this response did not affect the animals' reproductive performance (Table 2), it is slightly below the expected values according to those obtained from animals using deferred Digitaria in winter (Stritzler *et al.*, 1986). Although this production system is characterised by simplicity of management, it is necessary to perform periodic checks of the cows' BCS and pasture availability to avoid intake restrictions. The present system is compatible with: a) high reproductive rates (93% weaning), b) stability in mean calving date, c) acceptable weaning LW (174 kg at 150 days), d) high productivity per unit area (106kg LW ha^{-1} $year^{-1}$), and e) good economic indices (production cost: 0.14U$S kg^{-1} LW; gross margin: 45U$S ha^{-1}).

Table 1 Production system using Digitaria during the whole year. Live weight (kg) and body condition score (BCS, scale 1-5) of cows at different physiological states. Mean of 6 years' data (n= 225 cows)

State	Month	LW (kg)	BCS
Pre-calving	Sept.	468 ± 3.2^{a}	3.0 ± 0.03^{a}
Pre-mating	Dec.	474 ± 3.6^{a}	3.3 ± 0.04^{b}
Weaning	March	488 ± 4.0^{b}	3.4 ± 0.05^{c}
Pregnancy	June	492 ± 3.4^{b}	3.1 ± 0.03^{b}

Table 2 Productivity of cow-calf system. Mean of 6 years' data

Records	Min.	Mean	Max.
Stocking rate (CE $ha^{-1}year^{-1}$)	0.57	0.69	0.85
Weaning rate (%)	90	93	95
Calving date (day/month)	09/10	17/10	26/10
Weaning live weight (kg)	161	174	198
Age at weaning (days)	128	150	168
Daily liveweight gain (g $calf^{-1}$ day^{-1})	844	904	956
Liveweight gain per unit area (kg $ha^{-1}year^{-1}$)	95	106	125

Conclusions The reproductive performance of cows, the live weight of calves at weaning, the live-weight production per hectare with low costs and the pasture stability indicate that the implementation of this system

References

Stritzler, N.P., C. M. Rabotnikof, H. Lorda & A. Pordomingo (1986). Evaluación de especies forrajeras estivales en la región pampeana semiárida III. Digestibilidad y consumo de Digitaria eriantha y Bothriochloa intermedia bajo condiciones de diferimiento. Revista Argentina de Producción Animal, *6, 67-72* with the stocking rates used in this study is completely feasible.

Growth performance of crossbred steers on unfertilised mountain pastures at low stocking rates

A. Chassot and J. Troxler
Agroscope Liebefeld-Posieux (ALP), Swiss Federal Research Station for Animal Production and Dairy Products, 1725 Posieux, Switzerland, Email: andre.chassot@alp.admin.ch

Keywords: extensification, beef, stocking rate, mountain pasture

Introduction As a consequence of increasing economic pressure on Swiss agriculture, marginal areas are threatened by abandonment, especially in the mountainous regions. Using these areas for extensive beef production might preserve an open landscape and favour biodiversity. A grazing experiment was conducted with steers on an unfertilised mountain pasture to study the effects of a reduction of stocking rate on the growth of the animals and on changes in the vegetation.

Materials and methods The grazing experiment took place on a mountain pasture in the Swiss Jura that had not been fertilized since 1987 (Les Verrières, 1126 m asl, mean total precipitation from May to September: 675 mm, mean July temperature: 14.6 °C, vegetation dominated by Festuca rubra L. and Agrostis capillaris L.). No supplementary food was offered with the exception of minerals. Crossbred steers (Limousin x Red Pied) that weighed approximately 400 kg were used. The steers had no access to housing. The experimental area was divided into 3 sections grazed at fixed stocking rates (SR) of 1.8, 1.2 and 0.6 AU/ha (1 AU = 600 kg live weight).Each section was subdivided into 3 paddocks which were grazed in rotation. The steers were blocked by live weight at turnout and randomly assigned to the SR treatments.

Results The average cumulative liveweight gains (LWG) per animal and the corresponding average daily gains (ADG) per rotation are presented in Figure 1. The initial ADG were generally higher than 1.0 kg/day for all SR. This indicates adequate quality and quantity of feed. Nevertheless, growth rates decreased drastically in summer. With decreasing SR from 1.8 to 0.6 AU/ha, the grazing period could be extended by 43 % and 69 % in 2001 and 2002, respectively. As a consequence, and combined with higher ADG, the cumulative LWG per animal realised on the unfertilised mountain pasture was more than doubled in both years at 0.6 compared to 1.8 AU/ha. The LWG per ha was not affected by a reduction of SR from 1.8 to 1.2 AU/ha. At the lowest SR, higher ADG combined with longer grazing period could not compensate for the low stocking rate, leading thus to the lowest LWG per hectare.

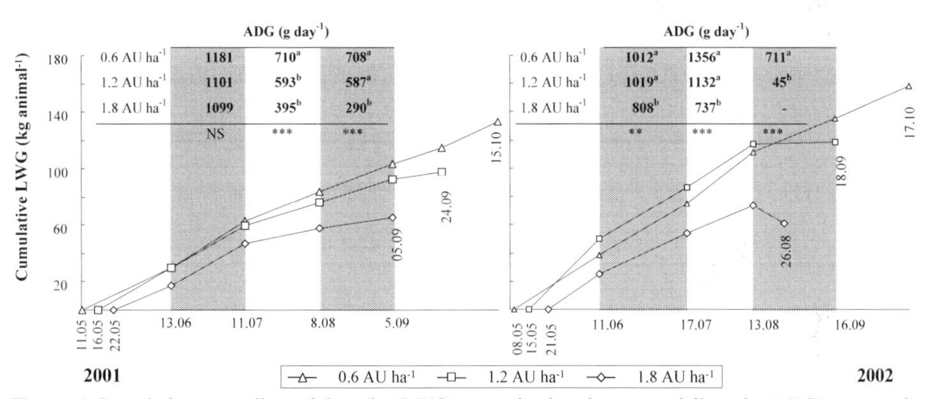

Figure 1 Cumulative mean liveweight gain (LWG) per animal and average daily gain (ADG) per rotation at three stocking rates during the grazing period in Les Verrières (1126 m a.s.l.). Means within one rotation with a same letter are not statistically different (P= 0.05).

Conclusion This study shows that a reduction of the SR on mountain pastures in Switzerland and their use to fatten crossbred steers can be an alternative to abandoning these areas for grazing.

The milk yield by Cinisara cows in different management systems: 1. Effect of season of calving

C. Giosuè, M. Alabiso, M.L. Alicata and G. Parrino
School of Agriculture, University of Palermo, Viale delle Scienze. Palermo, Sicily 90100, Italy, Email: crigio_24@yahoo.it

Keywords: dairy cows, management, quality milk

Introduction The Cinisara cow is Sicilian autochthonous breed; the milk has very interesting qualitative characteristics (chemical, physical and technologic parameters, principally due to its high part β of K casein) and is processed to make Caciocavallo cheese. Characteristics of milk yield are influenced by exogenous factors, such as management system, lactation number and season of calving. Pastures provide the basic feed but grazing is not continuous through the year. The aim of this research was to optimize the distribution of production over the year through the study of the effect of season of calving on qualitative characteristics of milk from Cinisara cows on three farms located at different altitudes (P=plain, C=hill and M=mountain) near Trapani.

Materials and methods The study was carried out over 16 months, from August 2001 to November 2002, on three different farms and on 60 cows, 52 pluriparous (P) and 8 primiparous (p), that completed their lactation over the period of study. The farms and the pastures were located: P, at 280 m a.s.l.; C, at 750 m a.s.l.; and M, at 1000 m a.s.l. The manual milking was made with the calf that ingested approximately 0.25 of the milk yield.

Table 1 Description of system of milk production

Farm	Cows	Grassland	Feeding daily supplement (kg/cow)
P	13 P 1p	Mainly sulla	Lucerne hay (12-14); feed (6-8)
C	28 P 2p	Mainly clover; at the end of the winter and beginning of the spring, sulla and vetch	Sulla hay (10); feed (4-5)
M	11P 5p	Non-homogeneous pasture	Sulla hay (10-12); feed (6-7)

Measurements were made of daily bulk milk yield and quality through the analysis of individual and bulk milk samples collected every month. Experimental data were analysed using ANOVA, a factorial model that considered the effect of season of calving, farm and lactation number.

Results The cows mainly calved in spring (38.3%) and autumn (35.0%) with less in winter (15.0%) and in summer (11.7%). The results are reported in Table 2. Spring calving was supported by a good availability of forage. Autumn as the season of calving produced the highest total milk yields. The length of lactation was significantly shorter in summer than in other seasons. The cows that calved in the autumn and in winter could use pasture which was of high quality over a longer time. Grazing and lactation yields for summer-calving cows were limited by high temperatures. The protein content of the milk was higher in autumn- and summer-calving cows than winter and spring-calving cows. Farm M recorded longer lactations than Farm C.

Table 2 Effect of season of calving, farm and lactation number on production variables

Factor	Treatment	Total milk yield (kg)	Length of lactation (d)	Milk (kg/d)	Fat (%)	Protein (%)	Lactose (%)	SCC (x 1000/ml)
Season of calving	Autumn	2241^{A}	214^{Aa}	10.2^{a}	3.7	3.7^{Aa}	5.0	549^{a}
	Winter	2126^{A}	225^{Aa}	9.4^{ab}	3.8	3.4^{ABbc}	5.0	417^{a}
	Spring	1964^{A}	196^{ABa}	10.2^{a}	3.6	3.4^{Bb}	5.2	395^{a}
	Summer	1116^{B}	155^{Bb}	7.2^{b}	3.8	3.7^{ABac}	4.9	940^{b}
Farm	P	1922	197^{AB}	9.6	3.6^{ABa}	3.6	5.1	471
	C	1748	177^{A}	9.8	3.5^{Aa}	3.5	5.1	532
	M	1916	219^{B}	8.3	4.0^{Bb}	3.6	4.9	723
Lactation number	1st	1595^{a}	174^{A}	9.0	3.9	3.6	5.1	339^{A}
	2nd and over	2129^{b}	221^{B}	9.5	3.6	3.5	4.9	812^{B}

Within columns values with different capital and small letters are significant at $P \leq 0.01$ and $P \leq 0.05$.

Conclusions The Cinisara breed of cow has considerable potential, through optimising the distribution of births during the year, to produce milk, cheese and ricotta in different seasons of the year.

The milk yield by Cinisara cows in different management systems: 2. Effect of season of production

M. Alabiso, C. Giosuè, M.L. Alicata and G. Parrino

School of Agriculture, University of Palermo, Viale delle Scienze, Palermo, Sicily 90100, Italy, Email: malabiso@unipa.it

Keywords: cows, milk, management, season of production

Introduction Cinisara is a Sicilian autochthonous breed of dairy cow. Qualitative characteristics of milk yield are influenced by composition of pasture and its changes over the seasons (Di Grigoli *et al.*, 2000). The grazing resource is characterized by notable variability, even in adjacent areas, because of different environmental conditions, such as aspect, the nature of soil and altitude. The aim of this research is to study the effect of season of production on qualitative characteristics of Cinisara cow milk in farms located at different altitudes.

Materials and methods The study was carried out from August 2001 to November 2002, in three farms, near Trapani on 116 cows, 98 pluriparous (P) and 18 primiparous (p). The pastures were located at three different altitudes: plain (P), at 280, hill (C), at 750, and mountain (M), at 1000 m a.s.l. Manual milking was performed with the calf present that ingested approximately 0.25 of the milk yield. A description of the feeding system is given by Giosuè *et al.* (2005). Measurements were made of daily bulk milk yield and its quality through the analysis of individual and bulked milk samples collected every month. Experimental data were analysed using a factorial model that considered the effect of season of production, farm, lactation number, stage of lactation and the interaction between season of production and farm.

Results The pattern of calving of cows was autumn 55.2%, winter 13.8%, spring 19.0%, and summer 12.1%. The winter and the spring were the seasons most favourable for milk production. On farm P the milk daily yield in summer was higher than on the other farms, as a result of the management system. The quality of milk was very variable, differing among farms, and over the seasons on the same farm. On farm P the milk protein content was higher probably because lucerne hay was given. In winter the content of urea in milk was high in all farms probably leading to an excess of protein in the diet. On the P and M farms the casein index (CI) was high in winter (Table 1).

Table 1 Daily mean milk yield, chemical, physical and other measurements on mass milk samples

Season	Farm	Milk yield (kg/d)	Fat (%)	Protein (%)	Lactose (%)	SCC	Urea	CI (%)	CBTUFC*‰ml	R	K_{20}	A_{30}
	P	7.9	3.82^{ABa}	3.78	4.95	654^{B}	29.3	75.3	344^{a}	17.0	2.1	40.7
Autumn	C	7.1	3.52^{Aa}	3.66	4.93	1606^{A}	20.5	74.0	174^{b}	20.6	2.8	32.5
	M	7.1	4.15^{Bb}	3.70	4.87	244^{B}	14.6	77.3	130^{b}	15.4	2.7	37.6
	P	11.5^{ab}	3.75^{Aa}	3.77^{b}	5.13	420^{B}	30.5	779	362	12.4	3.1	55.1
Winter	C	11.7^{a}	3.32^{Ab}	3.52^{a}	4.97	1124^{A}	35,5	77.1	286	14.0	2.0	49.3
	M	10.2^{b}	4.31^{Bc}	3.72^{b}	5.13	549^{B}	31.4	77.9	143	12,8	2.4	53.5
	P	11.3^{b}	3.97	3.62	5.07	427B	25.7	76.4	150	15.1	2.3	50.5
Spring	C	9.5^{a}	3.68	3.48	4.98	1616^{A}	28.0	77.4	152	14.8	2.1	47.0
	M	10.9^{b}	3.71	3.60	5.07	892^{B}	24.9	76.8	210	15.6	2.5	49.7
	P	10.3^{a}	3.46	3.48^{A}	4.95^{a}	793^{B}	27.7	74.4	289^{a}	12.0	2.7	51.1^{a}
Summer	C	8.1^{b}	3.40	3.30^{AB}	5.23^{b}	1625^{A}	15.1	75.2	109^{b}	12.2	2.1	45.2^{ab}
	M	8.6^{b}	3.28	3.20^{B}	5.05^{ab}	507^{B}	12.2	77.5	--	9.8	1.9	36.6^{b}

Within a column A and B refers to values significantly different at $P \leq 0.01$ and a, b and c at $P \leq 0.05$.

Conclusions The management system and altitude are the factors of immense variability. The present system of production leads to great heterogeneity over the year. The standardization of production is difficult even as niche product.

References

Di Grigoli A., A. Bonanno, D. Giambalvo, M. L. Alicata, M. Alabiso & A.S. Frenda (2000). Influenza del pascolo e dell'integrazione con concentrato sulla produzione di latte e Caciocavallo Palermitano di bovine Cinisare. Atti del 35° Simposio Internazionale di Zootecnia, Ragusa, pp. 207-216.

Giosuè C., M. Alabiso, M. L. Alicata & G. Parrino (2005). The milk yield by Cinisara cows in different management systems: 1. Effect of season of calving. *(This volume)*

Eating biodiversity: investigating the links between grassland biodiversity and quality food production

A. Hopkins[1], H. Buller[2], C. Morris[2] and J.D. Wood[3]

[1]Institute of Grassland and Environmental Research, North Wyke, Okehampton, Devon EX20 2SB, UK, Email: *alan.hopkins@bbsrc.ac.uk*, [2]*Centre for Rural Research, University of Exeter, EX4 6TL, UK and* [3]*Division of Food Animal Science, University of Bristol, BS40 5DU, UK*

Keywords: biodiversity, meat, cheese

Introduction Modern food production systems are generally detrimental to biodiversity, and the widespread loss of species-diverse grassland as a consequence of intensive farming methods is well documented. Since the 1980s, a range of policy measures and financial incentives for farmers have been introduced in Europe to halt (and in some cases, reverse) this trend, primarily to meet environmental objectives of species and habitat conservation and landscape protection. Biodiversity, where associated with agricultural production, has largely been regarded as a positive 'externality' to the process of food production; a 'product' which benefits wider society without necessarily conferring an agricultural benefit to the producer. However, with increasing emphasis on food quality, and the marketing of food products by geographical origin, method of production, gastronomic value and nutritional and health properties, there is potential to improve financial returns for farmers and the wider rural economy. Production in which grassland biodiversity is an 'input' to the livestock production food chain are embedded in some speciality systems, notably in mountain areas of Europe (Peeters and Frame, 2002). In the context of conserving grassland biodiversity there is a need to improve our understanding of the links between food products and animal diets, including pasture composition. This paper outlines a 3-year project funded by the UK RELU programme (RELU, 2005) which commenced in 2005. Results are not yet available so this summary focuses on the strategy being followed and the wider implications of linking enhanced food-product value to biodiversity.

Methodology This project is investigating the extent to which environmental distinctiveness (specifically grassland biodiversity) in UK food production sites can be actively valorised through the food product chain. Thus, it addresses the links between product value and pasture composition, and realising potential enhanced values of grassland biodiversity not just in terms of conservation objectives but to deliver socio-economic benefits for producers and rural communities. The approach combines the resources of agro-ecologists, food scientists and social scientists, through:

1. Examining examples of food products and production practices in UK where biodiversity or local distinctiveness in forage resources is an important input in food production, and in doing so create data sets on botanical assessments of grasslands, mineral composition, analyses of components of animal diet including pasture, hay/ silage and bought-in feed;
2. Providing biochemical data relating to final food products (e.g. meat composition / quality testing);
3. Determining the nature and the perception of the quality of food products derived specifically from animals fed on inputs from high biodiversity sources, compared with a control sample of products.

Interpretation and Outcomes Through the involvement of social scientists and rural economists the project will assess the actual and potential role of naturally embedded food products in rural development; draw up effective management prescriptions and identify examples of good practice for the integration of biodiversity as an element of product value; and explore varied regulatory, contractual and other instruments for delivering naturally embedded food products. In developing these outcomes evaluation will be made of existing knowledge in both the natural and social sciences concerning the links between biodiversity and food products and processes, including the impact on rural development.

Discussion The focus on identifying potential for increased value of dairy and meat products associated with biodiversity in forage resources extends societal valuation of species-rich grassland. Maintenance of farmland biodiversity has become increasingly dependent on payments to farmers for land management in ways consistent with delivering environmental goods, recognising that low-input, biodiverse systems have low levels of output. The research challenge is to identify the potential for additional product value linked to grassland biodiversity and thereby unlock benefits in terms of both conservation and rural development objectives.

References

Peeters, A. and Frame, J. (2002) Quality and Promotion of Animal Products in Mountains. FAO/CIHEAM Inter-Regional Cooperative Research and Development Network for Pastures and Fodder Crops, REU Technical Series 66. FAO, Rome. 147pp

RELU (2005) Rural Economy and Land Use Programme. www.relu.ac.uk.

GLM+ delivers improved natural resource management and production outcomes to extensive grazing properties in the savannas of semi-arid north Queensland, Australia

J. Rolfe and K. Shaw

Department of Primary Industries and Fisheries, Kairi, Queensland 4872 Australia, Email: joe.rolfe@dpi.qld.gov.au

Keywords: cattle grazing, savannas, land management

Introduction Native pastures are the main feed resource on extensive cattle grazing properties (each usually >25,000 ha) in the savannas of semi-arid north Australia and it is widely accepted that condition of many important land types is declining. A wealth of resource information is publicly available but it is usually complex in nature, diffuse and not presented in terms readily understood by land managers. Extension agencies have also moved towards information delivery using group processes that are not readily accepted by remote land managers. The GLM+ program uses concepts and tools from the Grazing Land Management (GLM) workshop (Chilcott *et al.*, 2003) and also incorporates and builds on producer experience. It is delivered on-property to individual management teams who identify their own resources, the condition of those resources, and opportunities to manage for improved land condition. Its use is described in this paper.

Methods A 3-stage programme (GLM+) is delivered over 12 months (preferably mid- to late-pasture growing season, at the end of growing season and at the end of the dry season). At stage 1, delivered over 1.5 days, the management team identifies, describes and maps natural resources and infrastructure available to the business in local terms. Reasonable estimates for the extent of land type areas within paddocks rather than complex measurements are used although some accuracy is sought on paddock areas. Basic pasture ecology and an introduction to the ABCD land condition framework is then given. This respectively describes the amalgam of current soil surface condition, pasture composition, exotic weed invasion and extent of woodland thickening to assess whether current sustainable carrying capacity is at 100%, 75%, 45% or 20% of original carrying capacity. Land condition assessments are practised in the field and the condition of land types in at least one paddock is completed and the implications of this considered in economic and environmental contexts. In stage 2 (1 day), after managers have completed a condition assessment for all land types in all paddocks, the implications of current land condition on the overall management plan for the business are considered. Grazing management options are assessed by the management team in terms of feasibility, affordability and profitability. Stage 3 (1 day) examines whether changed management decisions achieved the identified goals or whether further modification is needed.

Essential materials include a satellite image at an appropriate scale (c. 1:50,000), mapping kit, paddock sheets and a toolkit containing regionally sourced land type photo-standards for land condition and pasture yield. Valuable back-up resources are local climate information and relevant land resource information.

Results and Conclusions Grazing businesses completing the GLM+ program have a full and contemporary understanding of type, extent and condition of available resources under their care. The ABCD land condition framework is easily understood and used by land managers to objectively describe and document land condition at a paddock level where any management modifications need to be made. Locally derived photo-standards of pasture yields for major land types allows an accurate assessment of available feed which, when linked with climate information, easily translates into a stocking rate. Because assessments are recorded, it is a simple procedure to verify success or otherwise of changed management decisions before the beginning of the next wet season.

Alone, or when added to an overall property plan, GLM+ achieves multiple objectives. It can be used to address land condition issues, identify and build a case for targeted property development to increase carrying capacity and to provide an objective argument for support from financial institutions. As well, it can be used to demonstrate duty of care to support applications for lease renewal and answer other environmental queries. It is also useful to support a market price if the property is being sold.

Reference

Chilcott, C.R., B.S. McCallum, M.F Quirk and C.J Paton (2003). Grazing Land Management Education Package Workshop Notes – Burdekin, Meat and Livestock Australia Limited, Sydney, Australia.

Profitable and sustainable grazing systems for livestock producers with saline land in southern Australia

N.J. Edwards[1], D. Masters[2], E. Barrett-Lennard[3], M. Hebart[1], M. McCaskill[4], W. King[5] and W. Mason[6]
[1]*South Australian Research and Development Institute, Struan Agricultural Centre, PO Box 618, Naracoorte, SA, 5271, Australia, Email: edwards.nick@saugov.sa.gov.au,* [2]*CSIRO Livestock Industries, Private Bag No. 5, Wembley, WA, 6913, Australia,* [3]*Department of Agriculture Western Australia, Locked Bag No. 4, Bentley Delivery Centre, WA, 6983, Australia,* [4]*Department of Primary Industries, Private Bag 105, Hamilton, Victoria, 3300, Australia,* [5]*NSW Dept of Primary Industries, Orange Agricultural Institute, Forest Road, Orange, NSW, 2800, Australia,* [6]*SGSL Coordinator, PO Box 2157 Orange, NSW, 2800, Australia*

Keywords: dryland salinity, saltbush, puccinellia, tall wheat grass

Introduction Dryland salinity affects over 2.5 M ha in Australia, mostly in southern states and is expanding at 3-5% per year (NLWRA, 2001). The prognosis is for considerable expansion of the area affected by salinity and waterlogging (12–17 M ha at equilibrium), because groundwater levels continue to rise and only small-scale land management programmes have been implemented. In addition, many waterways are increasingly saline, especially in the Murray Darling Basin and in Western Australia (WA). Sustainable Grazing on Saline Land (SGSL) addresses the need to make productive use of saline land and water resources. Its research component operates at 12 sites across WA, South Australia (SA), Victoria and New South Wales (NSW) and consists of coordinated activities that have regional relevance and contribute nationally. The programme seeks to develop and demonstrate profitable and sustainable grazing systems on saline land that have positive environmental and social impacts. Whilst there are different priority research issues at each site, data collection is governed by common measurement protocols for salt and water movement, biodiversity, and pasture and animal performance in order to make comparisons and data sharing across sites practical.

The research programme In WA research is spread across seven sites, representing about 4.3M ha of salt affected land. These include two large (about 50 ha) sites (near Tammin and Yealering) that allow comparisons between unimproved land and land improved to current best practice, using a saltbush (*Atriplex* spp.)-based system with and without improved understorey species. Other sites have been established at Yealering, Lake Grace, Wubin, Meckering and Grong Grong (NSW) to examine factors affecting the composition, growth, grazing management, utilisation and value of saltbush-based pastures to sheep. Saline areas in the upper south east of SA are subject to both severe waterlogging and inundation in winter, when rising groundwater brings salt to the root zone and soil surface, inhibiting plant growth, seed set and survival in spring and early summer. Research here is focused on a puccinellia (*Puccinellia ciliata*)-based pasture where the impacts of fertiliser and addition of balansa clover (*Trifolium michelianum*) into existing puccinellia stands are being assessed under continuous and strategic grazing. Maintaining the persistence of balansa clover is a key challenge. In Victoria the targeted areas are characterised by shallow water tables, which are often saline but where winter waterlogging and inundation are an added challenge. The research here is focusing on use of tall wheat grass (*Lothopyron ponticum*) and annual legumes to provide quality out of season grazing compared to unimproved pastures. The targeted areas in NSW, the Lachlan and Macquarie catchments in the central west of the state, are characterised by high and rising salt load and electrical conductivity levels and generally small discharge areas close to waterways. Research is assessing the impacts of a salt-tolerant, perennial grass-based pasture (tall wheat grass dominant), compared to volunteer/naturalised pasture, on pasture and animal production, and water, soil and salt movement off-site. All projects will be assessed for their impact on whole farm economics.

Conclusions This ambitious project is testing current best-bet options for animal production from saline land. Outputs will include clarifying the environmental impacts and quantification of the production and economic benefits of grazing saline land. Extension products to assist farmers to make better decisions about managing these land types will boost their confidence to incorporate more saline land into their whole-farm management plans for environmental, economic and social outcomes. A significant component of this national network of projects and sites is its links with, and the participation of, farmers through the research being located on commercial farms, the involvement of local advisory groups and formal and informal links with a national network of over 125 farmer initiated small-scale projects testing locally relevant options for managing saline land.

Acknowledgements SGSL is an initiative of Australian Wool Innovation Pty Ltd, with Land and Water Australia and support from the Cooperative Research Centre for Plant-based Management of Dryland Salinity and Meat and Livestock Australia.

References

NLWRA. (2001). Australian dryland salinity assessment 2000: extent, impacts, processes, monitoring and management options. National Land & Water Resources Audit, Canberra.

Alternative land use options for Philippine grasslands: a bioeconomic modeling approach using the WaNuLCAS model

D.B. Magcale-Macandog, E. Abucay and P.A.B. Ani
Institute of Biological Science, University of the Philippines Los Baños, College, Laguna, Philippines 4031, Email: macandog@pacific.net.ph

Keywords: *Imperata cylindrica, Eucalyptus deglupta,* hedgerow, WaNuLCAS model, land-use change, bioeconomic modelling

Introduction In the Philippines, pure grasslands occupy 1.8 million ha and another 10.8 million ha (33% of the country's total land area) is under extensive cultivation mixed with grasslands and scrub. Most of these grasslands are under-utilised and dominated by *Imperata cylindrica. Imperata* grasslands generally represent areas of degraded soils that are acidic, low in organic matter and susceptible to erosion. However, conversion of these grassland areas into upland farms planted to annual crops and perennial trees is proliferating at a fast rate. This is triggered by the interacting factors of rapidly increasing population, the system of landholding, scarcity of jobs and the declining arable area in the lowlands.

Materials and methods The biophysical and economic consequences of land-use change from *Imperata* grasslands to continuous maize and agroforestry (*Eucalyptus deglupta* + maize hedgerow) systems were assessed using bioeconomic modeling. The Water, Nutrient and Light Capture in Agroforestry Systems (WaNuLCAS) model (van Noordwijk, Lusiana & Khasanah, 2004) was used to examine tree and crop growth and productivity, soil fertility changes, soil erosion and water balance. The different land-uses were modeled in the sloping upland areas of Southern Philippines characterised by rugged topography, clayey soils and annual rainfall of about 2500 mm.

Results Simulation showed that the dynamics of nutrients (N and P) in the systems differ. More than half of the total nitrogen in the three systems is tied up in the soil organic matter (SOM). Leaching and lateral flow are the main avenues of nitrogen losses in the three systems. Much of the P (90%) is tied up in SOM and immobilised in the *Imperata* grasslands.

Results of modeling the water balance of the three systems showed that *Eucalyptus*-maize hedgerow system had the highest subsurface flow and surface run-off (Table 1) compared with the other two systems. Maize cropping and *Imperata* grassland had significantly more drainage compared with the agroforestry system.

Simulation results also showed significant competition for light between trees and crops under the *Eucalyptus*-maize hedgerow system. Maize yield was initially higher in the continuous annual cropping system (2.4 t/ha) than under the *Eucalyptus*-maize hedgerow system (1.8 t/ha).

The benefits obtained from the maize cropping system is the grain yield, from the *Eucalyptus*-maize hedgerow system the benefits are maize grain yield and *Eucalyptus* timber, while biomass from the *Imperata* grassland is the harvested and sold as roofing material. Cost benefit analysis showed that the *Eucalyptus*-maize hedgerow system had the highest NPV after 9 years of simulation (P 304,323), compared with the *Imperata* grassland (P 10,722) and continuous maize (P 20,872).

Table 2 Water balance (li /m^2) in the three land use systems

Component	Agroforestry	Continuous crop	*Imperata* grassland
Surface	18,311	18,284	18,243
Subsurface flow	214,206	204,186	205,121
Drainage	4,151	153,844	156,255
Soil	274	9,555	7,331
Canopy	1,962	342	342
Crop	4,753	21,514	21,514
Tree	9,237		
Total	249,150	407,720	408,800

Conclusion This study has shown that land-use change from *Imperata* grasslands or continuous maize cropping system to *Eucalyptus*-maize hedgerow systems provide significant improvements to a range of biophysical and economic measures of productivity and sustainability.

Reference

Van Noordwijk, M., B. Lusiana & N. Khasanah (2004). WaNuLCAS version 3.1, Background on a model of water nutrient and light capture in agroforestry systems. International Centre for Research in Agroforestry (ICRAF), Bogor, Indonesia.

Sustainable semi-arid grazing management based on indigenous Shona practices prior to introduction of western ideas in Zimbabwe

O. Mugweni and R. Mugweni
Njeremoto Biodiversity Institute, P.O. Box 135 Masvingo, Zimbabwe, E-mail:muweni@zol.co.zw

Keywords: semi-arid areas, grazing management, Southern Africa, grazing systems

Introduction In the Shona culture the land, i.e. the plants, animals, soil, water, air and others, evolved with herding animals. Hence, the absence of one results in the destruction of the other. It is argued that the conventional grazing management belief that too many animals cause overgrazing is a misconception of the semi-arid savanna environments of Southern Africa where these environments evolved with thousands of herding grazers and mega-faunas such as elephants, wildebeests and buffalo. The objective of the research is to establish that grazing with an adequate recovery period for grazed plants, as a result of domesticated animals being managed effectively rather than staying on the same piece of land too long (continuous grazing) or returning too soon to the grazed area (rapid rotational grazing systems), can reverse the process of land degradation and the low water table of semi-arid rangelands, and can improve biodiversity by engaging in communal herding of livestock.

Materials and Methods The research is a four-year study which started in October 2002 at Njeremoto Biodiversity Institute near Chatsworth ICA in Zimbabwe. The results of the first two years of the study are reported. The systems study involves intervention strategies which are being implemented in two areas, which are grazing and arable areas. The grazing area, which is 200 hectares in extent, is divided into three zones, A, B and C, and is grazed as shown in Table 1 below:

Table 1 Period of controlled grazing

Period	A	B	C	Arable
Early summer grazing (Nov. to Jan.)	X			
Late summer grazing (Feb. to May)		X		
Full summer recovery (Nov. to May)			X	
Dry season grazing in arable area (June to Oct.)				X

Communal herding of livestock is practiced in summer once in each zone. Monitoring of vegetation is done through transect and fixed-point methods. As well as this data, information is also recorded by means of fixed-point photography, video recordings and field work notes.

Intervention strategies in the arable area, which is 50 ha in extent, include organic farming, permaculture, water-harvesting, fodder production, and planting multi-purpose and fruit trees. The area is grazed in the dry season (June to October) after harvest.

Results The following are the results of the study obtained to date. Controlled grazing with high animal impact causes many plants to grow with tight plant spacing and increased vigour. The recovery times used has produced a multi-species pasture of healthy, tight plant communities with a good age distribution. There is increased grass cover, as well as reduced mature capping, bare ground, gully formation and sheet erosion. There is an increase in productivity of stover, feed, fruits and fodder on the arable area. The social structure through the community herding of cattle is changed and is beneficial to the young children who will be the environmental managers of the future.

Conclusion The research findings to date reveal that indigenous knowledge systems complemented with modern methods of investigation may result in effective and productive, and potentially sustainable land management practices for semi-arid rangelands.

Herders and wetland degradation in northern Cameroon

E.T. Pamo, F. Tendonkeng and J.R. Kana
University of Dschang, FASA, Dept. of Animal Sci. P.B. Box 222 Dschang, Cameroon, Email: pamo_te@yahoo.fr

Keywords: Cameroon, wetland, degradation, pastoralists, adaptation

Introduction Livestock rearing in Northern Cameroon is carried out under two majors systems: the nomadic and the transhumance production systems (Pamo & Pamo, 1991). Nomadism is the practice of wandering from place to place, while transhumance involves seasonal displacement of flocks from one area to another by herders. These production systems involved large grazing areas, which may encompass different ecosystems. The Yaére, the only wetland of the northern Cameroon, is the major dry season grazing lands for livestock and wildlife. The main characteristic of this wetland is that the whole area is excluded from grazing during the growing season as a result of large scale flooding. Thus the major forage species (i.e. *Echinochloa pyramidalis, Oryza longistaminata, Hyparrhenia rufa, Echinochloa stagnina*) can set seed thereby ensuring their continued dispersal, establishment, and survival during the subsequent rainy season. In 1979, an upstream dam of 28 km with an additional 20 km embankment along the Logone river was build to store water for a rice irrigation project. This suppressed flooding over some 60 000 ha, and seriously affected the hydrological regime over another 200 000 ha. Major perennial forage species were gradually replaced by less palatable annual species such as *Sorghum arundinaceum*. This paper investigates how herders coped with the induced degradation of this dry season grazing land.

Wetland use before the dam construction The Yaére was a multifunctional human use area. Exploitation varied by site and by season, corresponding to the dynamic character of the flood plain and the cultural background of the floodplain users. During the dry season the area played a fundamental role in sustaining the rural economy on a regional scale. Fish and muskwari were exported from the area and herds from many parts of the North Cameroon, Chad, and Nigeria were provided with fresh pasture and water. During this period, the crude protein content of rain-fed pasture declines and the surrounding savanna pasture becomes desiccated. The accessibility of the flood plain and its primary products are of vital importance to livestock and wildlife. Pastoralists with cattle, sheep and goats moved into the area as the dry season progressed.

Adaptation to wetland degradation Pastoralists used flexible strategies to mitigate the effect of a high-risk environment. This required a formal or mainly informal set of rules allowing herders access to different ecological areas of the region to use different resources at different seasons of the year. Their survival is attributable to a wide spectrum of adaptive strategies. Some were ecologically based, while others depended on socio-economic and cultural mechanisms. The ecologically, as well as economically, based strategies rely on herd maximization which is achieved by herd diversification. The use of different livestock species has ecological and economic implications. Wetland degradation creates forage scarcity and leads to the poor spatial herd distribution. Different species then fill different ecological niches which may be more efficient as each species prefers to graze certain plant species. Increased mobility was also widely used and involved resource exploitation mobility (Oba & Lusigi, 1987), carried out in response to unpredictable forage and water availability, and escape mobility involving long distance migration to escape the combined effects of range degradation and decreased rainfall. Resource exploitation mobility allowed utilization of a widely dispersed forage resource at the times when it was rare. The distance covered, the routes followed, the length of stay in an area and the degree of flexibility built into the system varied from year to year. The number of displacements during the dry season depended on the state of available resources and of livestock. Such patterns of land use allowed for a high degree of fluidity and variation in the pastoralist system and provided an opportunity to individual herd owners to respond independently to seasonal fluctuations. Escape mobility, involving long-distance migration, was implemented to escape the combined effects of range degradation and reduced rainfall.

Conclusion Degradation of rangeland creates long-term economic and ecological disasters and diverts scarce resources to relief programmes. These observations from Northern Cameroon demonstrate that policy options for economic development require the assessment of economic, social and environmental function of any ecosystem before deciding to initiate and implement any project. In fragile environments, such as North Cameroon, measures should be taken to reinforce the ability of pastoralists to move between different ecological areas. Inevitably such a process requires limits to be placed on the numbers allowed to use a particular area in any season. Social and cultural background of the local population suggests that this will never be an easy strategy to implement in the short term.

References

Oba, G. & W..J. Lusigi (1987). An overview of drought strategies and land use in African Pastoral system. Pastoral Network Paper 23a, ODI, London, 1987.

Pamo, T.E. & C.T. Pamo (1991). An evaluation of the problems of open range use system in northern Cameroon. *Tropicultura,* 9.3, 125-128. Belgique.

Inner Mongolian herders move toward sustainability and elevate their incomes from Cashmere goat production by reducing grazing pressure on fragile grasslands

B.P. Fritz[1] and M. Zhao[2]

[1]*CIDA China-Canada Sustainable Agriculture Project, Email: bfritz@sasktel.net and* [2]*Department of Ecology and Environmental Science, Inner Mongolia Agricultural University, Hohhot, Inner Mongolia Autonomous Region, PRC 010018, China*

Keywords: cashmere goats, sustainability, biodiversity

Introduction Overgrazing, mainly caused by Cashmere goats, is contributing to the desertification of West Erdos fragile grasslands resulting in the threat of extinction of several endangered wild plant species. This transition area between desert and grassland includes some 400,000 ha and some 72 unique, relic and endangered plant species. The area is home to 5,000 inhabitants, mainly subsistent goat farming families and coal mining activity. Industrial land use in the reserve adds additional economic pressure to herders operating on a shrinking land base. This phenomenon has elicited the entrenched, traditional response of producing more livestock thus jeopardizing current levels of production and risking inevitable total desertification of this fragile rangeland. The objectives of this study are to return the land to full and sustainable biodiversity levels and to increase the incomes of the traditional West Erdos herders. Lessons learned may be transferable to other areas of Inner Mongolia where desertification is active.

Materials and Methods An integrated approach to planning a range/livestock management programme was planned and initiated with a definition of goals involving the herders. Data collection followed as deemed necessary to achieve thegoals. A rangeland inventory and evaluation was completed, which included an inventory of the endangered plant species and a baseline study of the area. A small group of leading herders was approached to become involved in a programme demonstrating an integrated approach to ranch and livestock management. Benchmark levels of production and income levels were established with the group, and feed and wool samples were collected to access possible nutritional deficiencies. A confinement and range feeding programme was designed which complements a planned breed improvement programme aimed at doubling cashmere production. An integrated grass/range management programme operates in conjunction with the livestock production programme and with the main objective being to balance the forage supply and demand. The processes of resource inventory and evaluation are the pillars of the range management plan.

Results The baseline survey suggested that herders are typically environmentally friendly, have an innate desire to achieve sustainability and preserve the grasslands for future generations. It is notable that herders consider the loss of nomadic herding, now being replaced by fixed grazing, along with global warming as being key reasons responsible for grassland desertification. Herders point to many friendly practices of today, like the use of cattle dung for heating/cooking, preventing the use of grasslands as burial sites, stopping the collection of medicinal herbs/plants and not planting trees as examples of being environmentally friendly to the grassland.

The maintaining of simple records is the key to measuring improvement. This is a new concept, and one that is very difficult to entrench and sustain. Feed test results revealed nutrient shortages/imbalances of five minerals, phosphorous, cobalt, zinc, copper and selenium. These deficiencies are being correlated with cashmere quantity and quality, with the aid of a designer mineral formulated to supplement existing feed supplies. A management system was designed by co-operators to promote a healthy grass/livestock balance. This system combines the practical application of herders with quantitative techniques in monitoring, range evaluation and setting stocking rates, as introduced by the Inner Mongolian Agricultural University and CIDA's Sustainable Agriculture Development project.

Conclusions The results suggest that many of the practices of Inner Mongolian herders are very traditional in nature and thus are very resistant to change. The practice of co-operating with selected demonstration herders promotes neighbouring herders to consider new management methods. Participatory research, planning and management are new concepts, but are affecting change; to date progress in increasing production and reducing stocking rates has been made. The overall objective of rejuvenating/ preserving the grassland and raising incomes of herders is taking place and winning ideas here will be transferred throughout Inner Mongolia, helping to halt desertification and the resulting sand storms which are having global impacts.

Keyword index

Author index